W0036935

Genetics: Analysis of Genes and Genomes

Genetics: Analysis of Genes and Genomes

Contributors

Jianliang Ni, Shuangfei Hu et al.

AURIS
Reference

www.aurisreference.com

Genetics: Analysis of Genes and Genomes

Contributors: Jianliang Ni, Shuangfei Hu et al.

Published by Auris Reference Limited

www.aurisreference.com

United Kingdom

Copyright 2016
Printed in 2017 for Sale in the Indian Subcontinent

The information in this book has been obtained from highly regarded resources. The copyrights for individual articles remain with the authors, as indicated. All chapters are distributed under the terms of the Creative Commons Attribution License, which permit unrestricted use, distribution, and reproduction in any medium, provided the original author and source are credited.

Notice

Contributors, whose names have been given on the book cover, are not associated with the Publisher. The editors and the Publisher have attempted to trace the copyright holders of all material reproduced in this publication and apologise to copyright holders if permission has not been obtained. If any copyright holder has not been acknowledged, please write to us so we may rectify.

Reasonable efforts have been made to publish reliable data. The views articulated in the chapters are those of the individual contributors, and not necessarily those of the editors or the Publisher. Editors and/or the Publisher are not responsible for the accuracy of the information in the published chapters or consequences from their use. The Publisher accepts no responsibility for any damage or grievance to individual(s) or property arising out of the use of any material(s), instruction(s), methods or thoughts in the book.

Genetics: Analysis of Genes and Genomes

ISBN: 978-1-78154-773-1

British Library Cataloguing in Publication Data
A CIP record for this book is available from the British Library

Printed in the United Kingdom

Exclusively distributed by CBS Publishers & Distributors Pvt. Ltd.

Sales & Distribution Rights only for India, Pakistan, Bangladesh, Sri Lanka, Nepal and Bhutan. This book is not to be sold outside these territories.

Contents

List of Abbreviations .... *vii*

List of Contributors... ....*ix*

Preface... ....*xxiii*

Chapter 1   **A Preliminary Genetic Analysis of Complement 3 Gene and Schizophrenia** 1

Jianliang Ni, Shuangfei Hu, Jiangtao Zhang, Wenxin Tang, Weihong Lu, Chen Zhang

Chapter 2   **Multidimensional Gene Set Analysis of Genomic Data** 15

David Montaner, Joaquı́n Dopazo

Chapter 3   **Comparative Genomics of Mycoplasma: Analysis of Conserved Essential Genes and Diversity of the PanGenome** 41

Wei Liu, Liurong Fang, Mao Li, Sha Li, Shaohua Guo, Rui Luo, Zhixin Feng, Bin Li, Zhemin Zhou, Guoqing Shao, Huanchun Chen, Shaobo Xiao

Chapter 4   **Robust Gene-Gene Interaction Analysis in Genome Wide Association Studies** 61

Yongkang Kim, Taesung Park

Chapter 5   **Dynamic Pathway Analysis of Genes Associated with Blood Pressure Using Whole Genome Sequence Data** 83

Pingzhao Hu and Andrew D Paterson

Chapter 6   **Network Analysis of Gene Essentiality in Functional Genomics Experiments** 93

Peng Jiang, Hongfang Wang, Wei Li, Chongzhi Zang, Bo Li, Yinling J. Wong, Cliff Meyer, Jun S. Liu, Jon C. Aster and X. Shirley Liu

Chapter 7   **Castor Bean Organelle Genome Sequencing and Worldwide Genetic Diversity Analysis** 115

Maximo Rivarola, Jeffrey T. Foster, Agnes P. Chan, Amber L. Williams, Danny W. Rice, Xinyue Liu, Admasu Melake-Berhan, Heather Huot Creasy, Daniela Puiu, M. J. Rosovitz, Hoda M. Khouri, Stephen M. Beckstrom-Sternberg, Gerard J. Allan, Paul Keim, Jacques Ravel, Pablo D. Rabinowicz

Chapter 8   **Analysis of East Asia Genetic Substructure Using Genome-Wide SNP Arrays**.. 135

Chao Tian, Roman Kosoy, Annette Lee, Michael Ransom, John W. Belmont, Peter K. Gregersen, Michael F. Seldin

Chapter 9   **Assembly of Inflammation-Related Genes for Pathwayfocused Genetic Analysis** 153

Matthew J. Loza, Charles E. McCall, Liwu Li, William B. Isaacs, Jianfeng Xu, Bao-Li Chang

Chapter 10   **Analysis of Gene Expression Data Using Biclustering Algorithms** 175

Fadhl M. Al-Akwaa

Chapter 11   **Genome Sequence and Genetic Diversity of the Common Carp, Cyprinus Carpio**.. 195

Peng Xu, Xiaofeng Zhang, Xumin Wang, Jiongtang Li, Guiming Liu, Youyi Kuang, Jian Xu, Xianhu Zheng, Lufeng Ren, Guoliang Wang, Yan Zhang, Linhe Huo, Zixia Zhao, Dingchen Cao, Cuiyun Lu, Chao Li, Yi Zhou, Zhanjiang Liu, Zhonghua Fan, Guangle Shan, Xingang Li, Shuangxiu Wu, Lipu Song, Guangyuan Hou, Yanliang Jiang, Zsigmond Jeney, Dan Yu, Li Wang, Changjun Shao, Lai Song, Jing Sun, Peifeng Ji, Jian Wang, Qiang Li, Liming Xu, Fanyue Sun, Jianxin Feng, Chenghui Wang, Shaolin Wang, Baosen Wang, Yan Li, Yaping Zhu, Wei Xue, Lan Zhao, Jintu Wang, Ying Gu, Weihua Lv, Kejing Wu, Jingfa Xiao, Jiayan Wu, Zhang Zhang, Jun Yu & Xiaowen Sun

Chapter 12   **Comparative Genomic Analysis of Soybean Flowering Genes**.......... 219

Chol-Hee Jung, Chui E. Wong Mohan B. Singh , Prem L. Bhalla

Chapter 13   **Genome-Wide Analysis of the Nadk Gene Family in Plants** 251

Wen-Yan Li, Xiang Wang, Ri Li, Wen-Qiang Li, Kun-Ming Chen

Citations 283

Index.. 287

List of Abbreviations

AIM	Ancestry informative markers
AP	Alternative pathway
CHA	Chinese Americans
CRY2	Cryptochrome2
EASTASAIMS	East Asian substructure ancestry informative markers
FDR	False discovery rate
FGF	Fibroblast growth factor
GBM	Glioblastoma
GG	Gene-gene
GMDR	Generalized Multifactor Dimensionality Reduction
GO	Gene Ontology
GSEA	Gene Set Enrichment Analysis
GSEA	Gene Set Enrichment Analysis
GWAS	Genome-wide association studies
HMM	Hidden Markov model
KEGG	Kyoto Encyclopedia of Genes and Genomes
LD	Linkage disequilibrium
LGT	Lateral gene transfer
MDR	Multifactor Dimensionality Reduction), 1
NIAID	National Institute of Allergy and Immunologic Diseases
OG	Orthologue groups
OV	Ovarian cancer
PCA	Principal Components Analyses
PCA	Principal-component analysis
PCR	Polymerase chain reaction
SBP	Simulated quantitative trait
SBP	Systolic blood pressure
SLE	Ystem lupus erythamatosus
SNP	Single nucleotide polymorphism
SNPQ	Single nucleotide polymorphisms
TR	Transcription rate
TSG	Tumor suppressor genes
TSGD	Teleost-specific genome duplication
Tsnp	Tagging single nucleotide polymorphism
WGD	Whole-genome duplication

List of Contributors

Jianliang Ni
Tongde Hospital of Zhejiang Province, Zhejiang, China

Shuangfei Hu
Zhejiang Provincial People's Hospital, Zhejiang, China

Jiangtao Zhang
Tongde Hospital of Zhejiang Province, Zhejiang, China

Wenxin Tang
Hangzhou Seventh People's Hospital, Zhejiang, China

Weihong Lu
Schizophrenia Program, Shanghai Mental Health Center, Shanghai Jiao Tong University School of Medicine, Shanghai, China

Chen Zhang
Schizophrenia Program, Shanghai Mental Health Center, Shanghai Jiao Tong University School of Medicine, Shanghai, China

David Montaner
Department of Bioinformatics and Genomics, Centro de Investigacio'n Pri'ncipe Felipe (CIPF), Valencia, Spain
Functional Genomics Node (INB), Centro de Investigacio'n Pri'ncipe Felipe (CIPF), Valencia, Spain

Joaqui'n Dopazo
Department of Bioinformatics and Genomics, Centro de Investigacio'n Pri'ncipe Felipe (CIPF), Valencia, Spain
Functional Genomics Node (INB), Centro de Investigacio'n Pri'ncipe Felipe (CIPF), Valencia, Spain
CIBER de Enfermedades Raras (CIBERER), Valencia, Spain

Wei Liu
Division of Animal Infectious Diseases, State Key Laboratory of Agricultural Microbiology, College of Veterinary Medicine, Huazhong Agricultural University, Wuhan, People's Republic of China

Liurong Fang
Division of Animal Infectious Diseases, State Key Laboratory of Agricultural Microbiology, College of Veterinary Medicine, Huazhong Agricultural University, Wuhan, People's Republic of China

Mao Li
Division of Animal Infectious Diseases, State Key Laboratory of Agricultural Microbiology, College of Veterinary Medicine, Huazhong Agricultural University, Wuhan, People's Republic of China

Sha Li
Division of Animal Infectious Diseases, State Key Laboratory of Agricultural Microbiology, College of Veterinary Medicine, Huazhong Agricultural University, Wuhan, People's Republic of China

Shaohua Guo
Division of Animal Infectious Diseases, State Key Laboratory of Agricultural Microbiology, College of Veterinary Medicine, Huazhong Agricultural University, Wuhan, People's Republic of China

Rui Luo
Division of Animal Infectious Diseases, State Key Laboratory of Agricultural Microbiology, College of Veterinary Medicine, Huazhong Agricultural University, Wuhan, People's Republic of China

Zhixin Feng
Institute of Veterinary Medicine, Jiangsu Academy of Agricultural Sciences, Nanjing, People's Republic of China

Bin Li
Institute of Veterinary Medicine, Jiangsu Academy of Agricultural Sciences, Nanjing, People's Republic of China

Zhemin Zhou
Environmental Research Institute, University College Cork, Cork, Ireland

Guoqing Shao
Institute of Veterinary Medicine, Jiangsu Academy of Agricultural Sciences, Nanjing, People's Republic of China

Huanchun Chen
Division of Animal Infectious Diseases, State Key Laboratory of Agricultural Microbiology, College of Veterinary Medicine, Huazhong Agricultural University, Wuhan, People's Republic of China

Shaobo Xiao
Division of Animal Infectious Diseases, State Key Laboratory of Agricultural Microbiology, College of Veterinary Medicine, Huazhong Agricultural University, Wuhan, People's Republic of China

Yongkang Kim
Department of Statistics, Seoul National University, Seoul, South Korea

Taesung Park
Department of Statistics, Seoul National University, Seoul, South Korea
Interdisciplinary Program in Bioinformatics, Seoul National University, Seoul, 151–741, South Korea

Pingzhao Hu
The Centre for Applied Genomics, The Hospital for Sick Children, 686 Bay Street, Toronto, ON, M5G 0A4, Canada
Department of Biochemistry and Medical Genetics and George and Fay Yee Centre for Healthcare Innovation, University of Manitoba,745 Bannatyne Avenue, Winnipeg, MB, R3E 0W3, Canada

Andrew D Paterson
The Centre for Applied Genomics, The Hospital for Sick Children, 686 Bay Street, Toronto, ON, M5G 0A4, Canada
rogram in Genetics and Genome Biology, The Hospital for Sick Children, 686 Bay Street, Toronto, ON, M5G 0A4, Canada
Dalla Lana School of Public Health, University of Toronto, Health Sciences Building, 155 College St, Toronto, ON, M5T 3M7, Canada

Peng Jiang
Department of Biostatistics and Computational Biology, Dana-Farber Cancer Institute, Harvard T.H. Chan School of Public Health, Boston, MA 02215, USA

Hongfang Wang
Department of Pathology, Brigham and Women's Hospital, Boston, MA 02115, USA

Wei Li
Department of Biostatistics and Computational Biology, Dana-Farber Cancer Institute, Harvard T.H. Chan School of Public Health, Boston, MA 02215, USA

Chongzhi Zang
Department of Biostatistics and Computational Biology, Dana-Farber Cancer Institute, Harvard T.H. Chan School of Public Health, Boston, MA 02215, USA

Bo Li
Department of Biostatistics and Computational Biology, Dana-Farber Cancer Institute, Harvard T.H. Chan School of Public Health, Boston, MA 02215, USA

Yinling J. Wong
Department of Pathology, Brigham and Women's Hospital, Boston, MA 02115, USA

Cliff Meyer
Department of Biostatistics and Computational Biology, Dana-Farber Cancer Institute, Harvard T.H. Chan School of Public Health, Boston, MA 02215, USA

Jun S. Liu
Department of Statistics, Harvard University, Cambridge 200092, USA

Jon C. Aster
Department of Pathology, Brigham and Women's Hospital, Boston, MA 02115, USA

X. Shirley Liu
Department of Biostatistics and Computational Biology, Dana-Farber Cancer Institute, Harvard T.H. Chan School of Public Health, Boston, MA 02215, USA
School of Life Science and Technology, Tongji University, Shanghai, MA 02138, China

Maximo Rivarola
Institute for Genome Sciences, University of Maryland School of Medicine, Baltimore, Maryland, United States of America

Jeffrey T. Foster
Center for Microbial Genetics and Genomics, Northern Arizona University, Flagstaff, Arizona, United States of America

Agnes P. Chan
J. Craig Venter Institute, Rockville, Maryland, United States of America

Amber L. Williams
Department of Biological Sciences, Environmental Genetics and Genomics Laboratory, Northern Arizona University, Flagstaff, Arizona, United States of America

Danny W. Rice
Department of Biology, Indiana University, Bloomington, Indiana, United States of America

Xinyue Liu
Institute for Genome Sciences, University of Maryland School of Medicine, Baltimore, Maryland, United States of America

Admasu Melake-Berhan
J. Craig Venter Institute, Rockville, Maryland, United States of America

Heather Huot Creasy
Institute for Genome Sciences, University of Maryland School of Medicine, Baltimore, Maryland, United States of America

Daniela Puiu
J. Craig Venter Institute, Rockville, Maryland, United States of America

M. J. Rosovitz
J. Craig Venter Institute, Rockville, Maryland, United States of America

Hoda M. Khouri
Institute for Genome Sciences, University of Maryland School of Medicine, Baltimore, Maryland, United States of America

Stephen M. Beckstrom-Sternberg
Center for Microbial Genetics and Genomics, Northern Arizona University, Flagstaff, Arizona, United States of America
Pathogen Genomics Division, Translational Genomics Research Institute, Phoenix, Arizona, United States of America

Gerard J. Allan
Department of Biological Sciences, Environmental Genetics and Genomics Laboratory, Northern Arizona University, Flagstaff, Arizona, United States of America

Paul Keim
Center for Microbial Genetics and Genomics, Northern Arizona University, Flagstaff, Arizona, United States of America

Jacques Ravel
Institute for Genome Sciences, University of Maryland School of Medicine, Baltimore, Maryland, United States of America
Department of Microbiology and Immunology, University of Maryland School of Medicine, Baltimore, Maryland, United States of America

Pablo D. Rabinowicz
Institute for Genome Sciences, University of Maryland School of Medicine, Baltimore, Maryland, United States of America
J. Craig Venter Institute, Rockville, Maryland, United States of America
Department of Biochemistry and Molecular Biology, University of Maryland School of Medicine, Baltimore, Maryland, United States of America

Chao Tian
Rowe Program in Human Genetics, Departments of Biochemistry and Medicine, University of California Davis, Davis, California, United States of America

Roman Kosoy
Rowe Program in Human Genetics, Departments of Biochemistry and Medicine, University of California Davis, Davis, California, United States of America

Annette Lee
The Robert S. Boas Center for Genomics and Human Genetics, Feinstein Institute for Medical Research, North Shore LIJ Health System, Manhasset, New York, United States of America

Michael Ransom
Rowe Program in Human Genetics, Departments of Biochemistry and Medicine, University of California Davis, Davis, California, United States of America

John W. Belmont
Department of Molecular and Human Genetics, Baylor College of Medicine, Houston, Texas, United States of America

Peter K. Gregersen
The Robert S. Boas Center for Genomics and Human Genetics, Feinstein Institute for Medical Research, North Shore LIJ Health System, Manhasset, New York, United States of America

Michael F. Seldin
Rowe Program in Human Genetics, Departments of Biochemistry and Medicine, University of California Davis, Davis, California, United States of America

Matthew J. Loza
Center for Human Genomics, Department of Internal Medicine, Wake Forest University School of Medicine, Winston-Salem, North Carolina, United States of America

Charles E. McCall
Department of Internal Medicine, Wake Forest University School of Medicine, Winston-Salem, North Carolina, United States of America

Liwu Li
Department of Biology, Virginia Polytechnic Institute and State University, Blacksburg, Virginia, United States of America

William B. Isaacs
Department of Urology, Johns Hopkins University Medical Institutions, Baltimore, Maryland, United States of America
Department of Oncology, Johns Hopkins University Medical Institutions, Baltimore, Maryland, United States of America

Jianfeng Xu
Center for Human Genomics, Department of Epidemiology and Prevention, Wake Forest University School of Medicine, Winston-Salem, North Carolina, United States of America

Bao-Li Chang
Center for Human Genomics, Department of Pediatric Medicine, Wake Forest University School of Medicine, Winston-Salem, North Carolina, United States of America

Fadhl M. Al-Akwaa
Biomedical Eng. Dept., Univ. of Science & Technology, Sana'a, Yemen

Peng Xu
Centre for Applied Aquatic Genomics, Chinese Academy of Fishery Sciences, Beijing, China
These authors contributed equally to this work. Correspondence should be addressed to X.S

Xiaofeng Zhang
Heilongjiang River Fisheries Research Institute, Chinese Academy of Fishery Sciences, Harbin, China
These authors contributed equally to this work. Correspondence should be addressed to X.S

Xumin Wang
Chinese Academy of Sciences Key Laboratory of Genome Sciences and Information, Beijing Institute of Genomics, Chinese Academy of Sciences, Beijing, China
These authors contributed equally to this work. Correspondence should be addressed to X.S

Jiongtang Li
Centre for Applied Aquatic Genomics, Chinese Academy of Fishery Sciences, Beijing, China
These authors contributed equally to this work. Correspondence should be addressed to X.S

Guiming Liu
Chinese Academy of Sciences Key Laboratory of Genome Sciences and Information, Beijing Institute of Genomics, Chinese Academy of Sciences, Beijing, China

These authors contributed equally to this work. Correspondence should be addressed to X.S

Youyi Kuang
Heilongjiang River Fisheries Research Institute, Chinese Academy of Fishery Sciences, Harbin, China
These authors contributed equally to this work. Correspondence should be addressed to X.S

Jian Xu
Centre for Applied Aquatic Genomics, Chinese Academy of Fishery Sciences, Beijing, China
These authors contributed equally to this work. Correspondence should be addressed to X.S

Xianhu Zheng
Heilongjiang River Fisheries Research Institute, Chinese Academy of Fishery Sciences, Harbin, China
These authors contributed equally to this work. Correspondence should be addressed to X.S

Lufeng Ren
Chinese Academy of Sciences Key Laboratory of Genome Sciences and Information, Beijing Institute of Genomics, Chinese Academy of Sciences, Beijing, China

Guoliang Wang
Chinese Academy of Sciences Key Laboratory of Genome Sciences and Information, Beijing Institute of Genomics, Chinese Academy of Sciences, Beijing, China

Yan Zhang
Centre for Applied Aquatic Genomics, Chinese Academy of Fishery Sciences, Beijing, China

Linhe Huo
Chinese Academy of Sciences Key Laboratory of Genome Sciences and Information, Beijing Institute of Genomics, Chinese Academy of Sciences, Beijing, China

Zixia Zhao
Centre for Applied Aquatic Genomics, Chinese Academy of Fishery Sciences, Beijing, China

Dingchen Cao
Heilongjiang River Fisheries Research Institute, Chinese Academy of Fishery Sciences, Harbin, China

Cuiyun Lu
Heilongjiang River Fisheries Research Institute, Chinese Academy of Fishery Sciences, Harbin, China

Chao Li
Heilongjiang River Fisheries Research Institute, Chinese Academy of Fishery Sciences, Harbin, China

Yi Zhou
Stem Cell Program, Division of Hematology and Oncology, Boston Children's Hospital and Dana-Farber Cancer Institute, Harvard Medical School, Boston, Massachusetts, USA

Zhanjiang Liu
Centre for Applied Aquatic Genomics, Chinese Academy of Fishery Sciences, Beijing, China
Fish Molecular Genetics and Biotechnology Laboratory, Department of Fisheries and Allied Aquacultures, Auburn University, Auburn, Alabama, USA

Zhonghua Fan
Chinese Academy of Sciences Key Laboratory of Genome Sciences and Information, Beijing Institute of Genomics, Chinese Academy of Sciences, Beijing, China

Guangle Shan
Chinese Academy of Sciences Key Laboratory of Genome Sciences and Information, Beijing Institute of Genomics, Chinese Academy of Sciences, Beijing, China

Xingang Li
Chinese Academy of Sciences Key Laboratory of Genome Sciences and Information, Beijing Institute of Genomics, Chinese Academy of Sciences, Beijing, China

Shuangxiu Wu
Chinese Academy of Sciences Key Laboratory of Genome Sciences and Information, Beijing Institute of Genomics, Chinese Academy of Sciences, Beijing, China

Lipu Song
Chinese Academy of Sciences Key Laboratory of Genome Sciences and Information, Beijing Institute of Genomics, Chinese Academy of Sciences, Beijing, China

Guangyuan Hou
Centre for Applied Aquatic Genomics, Chinese Academy of Fishery Sciences, Beijing, China

Yanliang Jiang
Centre for Applied Aquatic Genomics, Chinese Academy of Fishery Sciences, Beijing, China

Zsigmond Jeney
Research Institute for Fisheries, Aquaculture and Irrigation, Szarvas, Hungary

Dan Yu
Chinese Academy of Sciences Key Laboratory of Genome Sciences and Information, Beijing Institute of Genomics, Chinese Academy of Sciences, Beijing, China

Li Wang
Chinese Academy of Sciences Key Laboratory of Genome Sciences and Information, Beijing Institute of Genomics, Chinese Academy of Sciences, Beijing, China

Changjun Shao
Chinese Academy of Sciences Key Laboratory of Genome Sciences and Information, Beijing Institute of Genomics, Chinese Academy of Sciences, Beijing, China

Lai Song
Chinese Academy of Sciences Key Laboratory of Genome Sciences and Information, Beijing Institute of Genomics, Chinese Academy of Sciences, Beijing, China

Jing Sun
Chinese Academy of Sciences Key Laboratory of Genome Sciences and Information, Beijing Institute of Genomics, Chinese Academy of Sciences, Beijing, China

Peifeng Ji
Centre for Applied Aquatic Genomics, Chinese Academy of Fishery Sciences, Beijing, China

Jian Wang
Centre for Applied Aquatic Genomics, Chinese Academy of Fishery Sciences, Beijing, China

Qiang Li
Centre for Applied Aquatic Genomics, Chinese Academy of Fishery Sciences, Beijing, China

Liming Xu
Centre for Applied Aquatic Genomics, Chinese Academy of Fishery Sciences, Beijing, China

Fanyue Sun
Fish Molecular Genetics and Biotechnology Laboratory, Department of Fisheries and Allied Aquacultures, Auburn University, Auburn, Alabama, USA

Jianxin Feng
Henan Academy of Fishery Science, Zhengzhou, China

Chenghui Wang
College of Fisheries and Life Science, Shanghai Ocean University, Shanghai, China

Shaolin Wang
Department of Psychiatry and Neurobiology Science, University of Virginia, Charlottesville, Virginia, USA

Baosen Wang
Centre for Applied Aquatic Genomics, Chinese Academy of Fishery Sciences, Beijing, China

Yan Li
Centre for Applied Aquatic Genomics, Chinese Academy of Fishery Sciences, Beijing, China

Yaping Zhu
Centre for Applied Aquatic Genomics, Chinese Academy of Fishery Sciences, Beijing, China

Wei Xue
Centre for Applied Aquatic Genomics, Chinese Academy of Fishery Sciences, Beijing, China

Lan Zhao
Centre for Applied Aquatic Genomics, Chinese Academy of Fishery Sciences, Beijing, China

Jintu Wang
Centre for Applied Aquatic Genomics, Chinese Academy of Fishery Sciences, Beijing, China

Ying Gu
Heilongjiang River Fisheries Research Institute, Chinese Academy of Fishery Sciences, Harbin, China

Weihua Lv
Heilongjiang River Fisheries Research Institute, Chinese Academy of Fishery Sciences, Harbin, China

Kejing Wu
Chinese Academy of Sciences Key Laboratory of Genome Sciences and Information, Beijing Institute of Genomics, Chinese Academy of Sciences, Beijing, China

Jingfa Xiao
Chinese Academy of Sciences Key Laboratory of Genome Sciences and Information, Beijing Institute of Genomics, Chinese Academy of Sciences, Beijing, China

Jiayan Wu
Chinese Academy of Sciences Key Laboratory of Genome Sciences and Information, Beijing Institute of Genomics, Chinese Academy of Sciences, Beijing, China

Zhang Zhang
Chinese Academy of Sciences Key Laboratory of Genome Sciences and Information, Beijing Institute of Genomics, Chinese Academy of Sciences, Beijing, China

Jun Yu
Chinese Academy of Sciences Key Laboratory of Genome Sciences and Information, Beijing Institute of Genomics, Chinese Academy of Sciences, Beijing, China

Xiaowen Sun
Centre for Applied Aquatic Genomics, Chinese Academy of Fishery Sciences, Beijing, China
Heilongjiang River Fisheries Research Institute, Chinese Academy of Fishery Sciences, Harbin, China

Chol-Hee Jung
Plant Molecular Biology and Biotechnology Laboratory, ARC Centre of Excellence for Integrative Legume Research, Melbourne School of Land and Environment, The University of Melbourne, Parkville, Victoria, Australia

Chui E. Wong Mohan B. Singh
Plant Molecular Biology and Biotechnology Laboratory, ARC Centre of Excellence for Integrative Legume Research, Melbourne School of Land and Environment, The University of Melbourne, Parkville, Victoria, Australia

Prem L. Bhalla
Plant Molecular Biology and Biotechnology Laboratory, ARC Centre of Excellence for Integrative Legume Research, Melbourne School of Land and Environment, The University of Melbourne, Parkville, Victoria, Australia

Wen-Yan Li
State Key Laboratory of Crop Stress Biology in Arid Areas, College of Life Sciences, Northwest A&F University, Yangling, Shaanxi, China

Xiang Wang
State Key Laboratory of Crop Stress Biology in Arid Areas, College of Life Sciences, Northwest A&F University, Yangling, Shaanxi, China

Ri Li
State Key Laboratory of Crop Stress Biology in Arid Areas, College of Life Sciences, Northwest A&F University, Yangling, Shaanxi, China

Wen-Qiang Li
State Key Laboratory of Crop Stress Biology in Arid Areas, College of Life Sciences, Northwest A&F University, Yangling, Shaanxi, China

Kun-Ming Chen
State Key Laboratory of Crop Stress Biology in Arid Areas, College of Life Sciences, Northwest A&F University, Yangling, Shaanxi, China

David Montaner
Department of Bioinformatics and Genomics, Centro de Investigacio´n Prı´ncipe Felipe (CIPF), Valencia, Spain,
Functional Genomics Node (INB), Centro de Investigacio´n Prı´ncipe Felipe (CIPF), Valencia, Spain,

Joaquı´n Dopazo
Department of Bioinformatics and Genomics, Centro de Investigacio´n Prı´ncipe Felipe (CIPF), Valencia, Spain,
Functional Genomics Node (INB), Centro de Investigacio´n Prı´ncipe Felipe (CIPF), Valencia, Spain,
CIBER de Enfermedades Raras (CIBERER), Valencia, Spain

Preface

Genetic analysis is the overall process of studying and researching in fields of science that involve genetics and molecular biology. The text *Genetics Analysis of Genes and Genomes* provides a clear, balanced, and comprehensive introduction to genetics and genomics. Genomics is a discipline in genetics that applies recombinant DNA, DNA sequencing methods, and bioinformatics to sequence, assemble, and analyze the function and structure of genomes. A preliminary genetic analysis of complement three gene and schizophrenia has been presented in first chapter. In second chapter, we present a multidimensional logistic model that allows studying the relationship of gene modules with different genome-scale. Third chapter employs comparative evolutionary analysis of twenty mycoplasma genomes to gain an improved understanding of essential genes. Fourth chapter proposes a robust GMDR estimation method (based on the L-estimator and M-estimator estimation methods) in an attempt to reduce the effects caused by outlying traits. In fifth chapter, we evaluate the associations between rare and common genetic variants and the simulated quantitative trait (SBP) measured at three time points at the gene and pathway levels. Sixth chapter provides a general method for gene essentiality analysis in functional genomic experiments. In seventh chapter, we report the compete castor bean chloroplast and mitochondrion genome sequences generated from a whole genome shotgun (WGS) sequencing project of the cultivar. The analysis of East Asia genetic substructure using genome-wide SNP arrays has been presented in eighth chapter. In ninth chapter, we review various phases of inflammation responses, including the development of immune cells, sensing of danger, influx of cells to sites of insult, activation and functional responses of immune and non-immune cells, and resolution of the immune response. Tenth chapter focuses on analysis of gene expression data using biclustering algorithms. Genome sequence and genetic diversity of the common carp have been discussed in eleventh chapter. Twelfth chapter provides a framework for the soybean flowering pathway and insights into the relationship and evolution of flowering genes between a short-day soybean and the long-day plant, Arabidopsis. Last chapter provides a comparative genomic analysis that identified 74 NADK gene homologs from 24 species representing the eight major plant lineages within the supergroup Plantae: glaucophytes, rhodophytes, chlorophytes, bryophytes, lycophytes, gymnosperms, monocots and eudicots. Phylogenetic and structural analysis classified these NADK genes into four well-conserved subfamilies with considerable variety in the domain organization and gene structure among subfamily members.

Chapter 1

A PRELIMINARY GENETIC ANALYSIS OF COMPLEMENT 3 GENE AND SCHIZOPHRENIA

Jianliang Ni[1], Shuangfei Hu[2], Jiangtao Zhang[1], Wenxin Tang[3], Weihong Lu[4], Chen Zhang[4]

[1] Tongde Hospital of Zhejiang Province, Zhejiang, China

[2] Zhejiang Provincial People's Hospital, Zhejiang, China

[3] Hangzhou Seventh People's Hospital, Zhejiang, China

[4] Schizophrenia Program, Shanghai Mental Health Center, Shanghai Jiao Tong University School of Medicine, Shanghai, China

ABSTRACT

Complement pathway activation was found to occur frequently in schizophrenia, and complement 3 (C3) plays a major role in this process. Previous studies have provided evidence for the possible role of C3 in the development of schizophrenia. In this study, we hypothesized that the gene encoding C3 (C3) may confer susceptibility to schizophrenia in Han Chinese. We analyzed 7 common single nucleotide polymorphisms (SNPs) of C3 in 647 schizophrenia patients and 687 healthy controls. Peripheral *C3* mRNA expression level was measured in 23 drug-naïve patients with schizophrenia and 24 controls. Two SNPs (rs1047286 and rs2250656) that deviated from Hardy-Weinberg equilibrium were excluded for further analysis. Among the remaining 5 SNPs, there was no significant difference in allele and genotype frequencies between the patient and control groups. Logistic regression analysis showed no significant SNP-gender interaction in either dominant model or recessive model. There was no significant difference in the level of peripheral *C3* expression between the drug-naïve schizophrenia patients and healthy controls. In conclusion, the results of this study do not support *C3* as a major genetic susceptibility factor in schizophrenia. Other factors in AP may have critical roles in schizophrenia and be worthy of further investigation.

INTRODUCTION

Schizophrenia is a chronic, severe and disabling brain disorder that affects approximately 1% of worldwide population. In the past decades, schizophrenia has been regarded as a neurodevelopment disorder. Early literature reported that adverse conditions may result in abnormal brain development during the perinatal period, whilst schizophrenic symptoms appear in later life after the synaptic pruning process [1,2]. However, the pathophysiology of schizophrenia remains unknown [3].

Although heritability estimates for schizophrenia reach 80%, twin concordance is around 50% [4]. Hence, non-genetic factors also play an important role in this disorder [5]. It has been well-documented that maternal virus infection is one of the most consistently identified environmental risk factors for schizophrenia [6]. On the other hand, clinical observations indicated that schizophrenia and certain autoimmune diseases share some key clinical, epidemiological and genetic features [7]. Such findings suggested that immune abnormalities may be implicated with the pathophysiology of schizophrenia [8].

Complement acts as a rapid and efficient immune surveillance system that serves to protect the body against the invasion and proliferation of various microorganisms [9,10]. Complement pathway activation was reported to occur frequently in schizophrenia, in which complement 3 (C3) regulates the process [11]. C3 is a protein of the immune system that plays a central role in the complement cascade and contributes to innate immunity. Recent observations have demonstrated that C3 is a critical mediator for synaptic refinement and plasticity in neurodevelopment [12,13]. In comparison with healthy controls, Hakobyan et al. [14] observed a significant higher level of C3 protein in schizophrenia patients. As such, the above findings provide interesting clues for the potential role of C3 in schizophrenia.

At the molecular level, the gene encoding C3 (*C3*) is located at chromosome 19, which has been reported to be a genetic schizophrenia susceptibility region [15]. However, few genetic studies have been carried out to investigate the association of *C3* with schizophrenia and yielded inconsistent results [16,17,18]. It is known that *C3* contains 41 exons and spreads over 41kb. One weakness for the early genetic studies is too few polymorphisms tested. In the present study, we aimed to examine whether the region of *C3* is associated with schizophrenia. A total of 7 polymorphisms were selected for a better coverage of this region. As a secondary aim, prior study reported that *C3* has a gender-specific effect [19], which may underlie differential susceptibility to schizophrenia [20]. So we attempted to examine whether there was any gender difference in the association of *C3* with schizophrenia. Data has shown that *C3*polymorphisms result in alternations in its protein function. To validate previous findings, we opted to measure the serum *C3* expression level among drug-naïve schizophrenia patients and healthy controls.

METHODS

Subjects

All subjects provided written informed consent prior to performing any of the procedures related to this study. All procedures were reviewed and approved by the ethical committees at Tongde Hospital of Zhejiang Province and Hangzhou Seventh People's Hospital, and performed in strict accordance with the Declaration of Helsinki, and other relevant national and international regulations.

For the genetic analysis, a total of 647 schizophrenia patients recruited from Tongde Hospital of Zhejiang Province and The Seventh People's Hospital of Hangzhou. The inclusion criteria for this study were according to our previous ones [20,21,22]. All patients (1) met the Diagnostic and Statistical Manual of Mental Disorders, Fourth Edition (DSM-IV) criteria for schizophrenia; (2) were not first-episode; (3) had no chronic physical disease or other psychiatric disorder aside from schizophrenia. Prior to analysis, all diagnosis and review of psychiatric case records were independently checked and verified by two senior psychiatrists. The control group comprised of 687 Han Chinese enrolled from the local community in Hangzhou. Before sampling, the volunteers self-reported that they were in good physical health and have no family history of psychiatric disorders. Those who have medical illnesses or drug and alcohol abuse/dependence were excluded. Demographic and clinical characteristics were presented inS1 Table.

For the expression analysis, twenty-three drug-naïve patients with first-episode schizophrenia were recruited from Tongde Hospital of Zhejiang Province. The patients were diagnosed according to the DSM-IV criteria for schizophrenia and had no physical disease. Twenty-four healthy subjects from Hangzhou city were also recruited for control group. Basic blood and urine tests were performed prior to recruitment in order to exclude any current physical illness. Patients and controls did not significantly differ for age, gender, BMI and smoking status. Detailed information was presented in S2 Table.

SNP Selection

We retrieved CHB data from the HapMap database (http://www.hapmap. org) and defined linkage disequilibrium (LD) blocks using Haploview

4.2 (Broad Institute, Cambridge, MA, USA) to set inclusion criteria for tagging SNPs. Haplotype-tagging single nucleotide polymorphisms (htSNPs) with R^2 cutoff>0.8 and minor allele frequency (MAF)>0.1 were selected. In total, three tag SNPs of C3 were selected for genotyping. Four potential C3 functional SNPs (rs7951, rs2230199, rs2250656 and rs11672613) [23,24,25,26] were also examined in this study (S3 Table).

Genotyping

Genomic DNA of all participants was extracted from peripheral blood using a Tiangen DNA Isolation Kit (Tiangen Biotech, Beijing, China). All 7 SNPs were amplified independently via polymerase chain reaction (PCR) and then genotyped via direct sequencing on an ABI PRISM 3730 Genetic Analyzer (Perkin-Elmer Applied Biosystems). S4 Table detailed the primers information. Genotyping was carried out according to the methods described in our previous studies [27,28]. PCR amplification was performed in a volume of 25 µL containing primer pair for each SNP. PCR primers were also used for sequencing. Sequencing results were handled using DNAStar package (DNA Star Inc., USA), and the original sequencing chromatograms of each sample were then manually checked.

Quantitative Real-Time Polymerase Chain Reaction (qRT-PCR)

We carried out the C3 mRNA expression analysis using qRT-PCR as previously described [29,30]. Peripheral blood was collected and mononuclear cells were separated by Ficoll-Paque PLUS density gradient centrifugation (GE Healthcare, Amersham, NJ, USA) within 2 hour, placed in TRIzol (Invitrogen, Carlsbad, CA, USA) and stored at -80°C. The total RNA was isolated from peripheral blood mononuclear cells according to the manufacturer's protocol, and 2 µg total RNA was re-transcribed into complementary DNA with reverse transcription (ReverTra Ace, Toyobo, Osaka, Japan) according to the manufacturer's instruction. RelativeC3 mRNA expression levels were assessed by real-time PCR with commercially available TaqMan gene expression assays for target gene C3 and glyceraldehydes-3-phosphate dehydrogenase (GAPDH) as reference gene (Applied Biosystems, CA, USA). All experiments were conducted in optical 384-well reaction microtiter plates on an ABI Prism 7900HT Sequence Detection System (Applied Biosystems, CA, USA). PCR was performed in a total volume of 10µL containing 1×TaqMan Universal Master Mix with AmpErase UNG, 1×Assay Mix (Applied Biosystems, CA, USA) and complementary DNA template at cycle conditions: 95°C for 15 min, followed by 40 cycles at 95°C for 15s and 60°C for 60s. All reactions were run in triplicate. In each sample, the expression of C3 was normalized to the expression of the reference gene. Results were reported in fold change using $2^{-\Delta\Delta Ct}$.

Statistical Analysis

The Hardy-Weinberg equilibrium testing and individual SNP association analyses were conducted using SHEsis (http://analysis.bio-x.cn). The odds ratio (OR) and corresponding 95% confidence interval (CI) were calculated with the major allele as reference. Pairwise linkage disequilibrium of all pairs of htSNPs was performed using HaploView 4.2 (Broad Institute, Cambridge, MA, USA), and the extent of linkage disequilibrium (LD) was measured by the standardized D' and R^2. Referring to the previous report [19], logistic regression was performed with SNP-gender interaction to adjust the effect of gender on SNPs. For the expression analysis, ANCOVA was carried out with age, gender, smoking status and BMI as covariates controlled in the model, to minimize the potential effect of these factors on the expression levels of $C3$ mRNA. The ANCOVA analysis was performed using SPSS 17.0 (SPSS Inc., Chicago, IL, USA). To adjust for multiple testing, the level of significance was corrected via Bonferroni correction. Power calculations were carried out using Quanto 1.2.3 (http://hydra.usc.edu/GxE).

Results

For the genetic analysis, there was no significant difference between the schizophrenia and control groups in term of age and gender. Seven SNPs were genotyped to investigate the association of $C3$ with schizophrenia. Two SNPs (rs1047286 and rs2250656) that deviated from Hardy-Weinberg equilibrium were excluded for the further analysis. Among the remaining 5 SNPs, no deviation from the Hardy-Weinberg was observed in genotype distribution. Table 1 showed that there was no significant difference in allele and genotype frequencies between the patient and control groups. After calculating LD for all pairs of SNPs, we found a low R^2 in $C3$(S1 Fig), indicating that no specific haplotype block could be identified. We further investigate whether there was any gender difference in the association of $C3$ with schizophrenia. Logistic regression analysis showed no significant SNP-gender interaction in either dominant model or recessive model (Tables 2 and 3). A total of 23 drug-naïve patients with schizophrenia and 24 well-matched healthy controls were recruited for the $C3$ expression study. As shown in Fig 1, there was no significant difference in the level of peripheral $C3$ mRNA expression between the drug-naïve schizophrenia patients and healthy controls. On the basis of the genotype data, the statistical power of the 5 SNPs within $C3$ was more than 85% (= 0.05) for schizophrenia samples under the assumption of a moderate effect size (OR = 1.5), a log additive model, and the prevalence of schizophrenia (≈1%).

Table 1: Comparison of genotype and allele frequencies of 5 *C3* SNPs in 647 schizophrenia patients and 687 healthy controls

		Genotype, n (%)					Allele, n (%)					
rs2277984	N	G/G	G/A	A/A	P^a	P^b	N	G	A	OR (95%CI)	P^a	P^c
Case	647	132 (20.4)	336 (51.9)	179 (27.7)	0.95	0.26	1294	600 (46.4)	694 (53.6)	1.02 (0.88–1.19)	0.76	
Control	687	137 (19.9)	355 (51.7)	195 (28.4)		0.28	1374	629 (45.8)	745 (54.2)			
rs7951		T/T	T/C	C/C				T	C			
Case	647	7 (1.1)	122 (18.9)	518 (80.1)	0.42	0.95	1294	136 (10.5)	1158 (89.5)	0.91 (0.71–1.16)	0.45	
Control	687	5 (0.7)	147 (21.4)	535 (77.9)		0.13	1374	157 (11.4)	1217 (88.6)			
rs11672613		C/C	C/T	T/T				C	T			
Case	647	98 (15.1)	307 (47.4)	242 (37.4)	0.22	0.97	1294	503 (38.9)	791 (61.1)	0.90 (0.77–1.05)	0.17	
Control	687	109 (15.9)	352 (51.2)	226 (32.9)		0.15	1374	570 (41.5)	804 (58.5)			
rs2230205		A/A	A/G	G/G				A	G			
Case	647	141 (21.8)	346 (53.5)	160 (24.7)	0.09	0.07	1294	628 (48.5)	666 (51.5)	1.18 (1.01–1.37)	0.035	0.175
Control	687	127 (18.5)	357 (52.0)	203 (29.5)		0.17	1374	611 (44.5)	763 (55.5)			
rs2230199		G/G	G/C	C/C				G	C			
Case	647	0 (0.0)	10 (1.5)	637 (98.5)	0.50	0.84	1294	10 (0.8)	1284 (99.2)	0.76 (0.33–1.71)	0.50	
Control	687	0 (0.0)	14 (2.0)	673 (98.0)		0.79	1374	14 (1.0)	1360 (99.0)			

[a] P values were not adjusted by Bonferroni correction
[b] P values were calculated for Hardy-Weinberg equilibrium
[c] P values were adjusted by Bonferroni correction.

Table 2: Logistic regression analysis of *C3* SNPs×gender interaction in dominant model

Variables	B	S.E	Wals	OR (95%CI)	P^a	P^b
rs2277984×gender	0.54	0.22	5.87	1.71 (1.11–2.65)	0.015	0.075
rs7951×gender	-0.40	0.24	2.78	0.67 (0.42–1.07)	0.10	
rs11672613×gender	-0.23	0.21	1.26	0.79 (0.53–1.19)	0.26	
rs2230205×gender	0.21	0.21	0.95	1.23 (0.81–1.87)	0.33	
rs2230199×gender	-0.21	0.27	0.62	0.81 (0.47–1.38)	0.43	

[a] P values were not adjusted by Bonferroni correction
[b] P values were adjusted by Bonferroni correction.

Table 3: Logistic regression analysis of *C3* SNPs×gender interaction in recessive model

Variables	B	S.E	Wals	OR (95%CI)	P^a
rs2277984×gender	0.04	0.24	0.02	1.04 (0.65–1.64)	0.88
rs7951×gender	0.35	1.25	0.08	1.42 (0.12–16.26)	0.78
rs11672613×gender	-0.17	0.26	0.43	0.84 (0.60–1.41)	0.79
rs2230205×gender	-0.14	0.23	0.37	0.87 (0.55–1.37)	0.55
rs2230199×gender	NA	NA	NA	NA	NA

[a] P values were not adjusted by Bonferroni correction
NA, Not applicable.

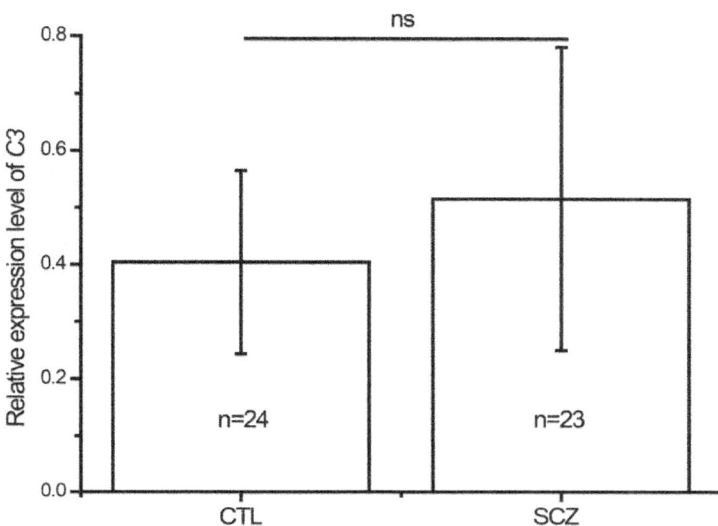

Figure 1: Expression levels of *C3* mRNA in peripheral blood in drug-naïve schizophrenia patients and healthy subjects.

C3 mRNA was normalized to that of *GAPDH*. CTL, control subjects (n = 24); SCZ, schizophrenia patients (n = 23); ns, no significance.

DISCUSSION

As a key component of innate immunity, accumulating evidence has indicated that abnormalities in the complement system are implicated in the etiology of schizophrenia [23]. We have examined the association of schizophrenia with the gene encoding C4-binding protein (*C4BPB/C4BPA*), a potent circulating soluble inhibitor of the classical and lectin pathways of complement. However, our results did not support the involvement of *C4BPB/C4BPA* in schizophrenia [22]. Here, we aimed to investigate the association of *C3*, another critical factor in complement system, with schizophrenia in Chinese Han population.

In this study, we did not observe any significant difference of allele and genotype frequencies between the schizophrenia patients and healthy controls. The statistical power of our study was also enough to detect an association between the variants and schizophrenia. Although Liu et al. [19] reported a SNP-gender interaction in *C3*, we did not have such findings in our sample. These results demonstrated that there is no genetic association between *C3* and schizophrenia, at least in Han Chinese. However, a recent study showed that increased levels of C3, acting as activation of complement system, can be found in schizophrenia patients when compared with healthy controls [31]. In contrast, Wong et al. [32] found a lower level of C3 in schizophrenia patients than that in controls. We noticed that the patients in both studies were those with chronic schizophrenia [31,32]. The aforementioned inconsistent results prompted us to determine the expression of *C3* in drug-naïve patients with first-episode schizophrenia. Our results showed no significant difference in the level of *C3* expression between schizophrenia patients and healthy controls. Therefore, our findings suggested that *C3* may not confer susceptibility to schizophrenia in Han Chinese.

Recently, Li et al. [33] performed a label-free quantitative proteomics analysis to identify 27 proteins as being schizophrenia related proteins, and found dysregulation of the alternative complement pathway in schizophrenia patients. The alternative pathway (AP) is one of three complement pathways, which is initiated by the spontaneous hydrolysis of C3. A number of molecules are involved in the occurrence of AP. Even though no association of *C3* with schizophrenia was found in this study, we could not exclude possible role of AP in the development of schizophrenia.

On the other side, cytokines are believed to play a vital role in coordinating immunologic and inflammatory responses in physiological and pathological conditions [34]. Therefore, cytokines may be critical mediators of the cross-talk between immune system and neuropsychiatric disorders [35]. Miller et al. [36] meta-analyzed 40 studies on cytokines and schizophrenia, and observed significant alternations of cytokine network in schizophrenia. Therefore, imbalance of cytokine network may be involved in the pathophysiology of schizophrenia. Prior literature indicated that complement activation products, such as C3a and C3a desArg, may enhance cytokine synthesis and inhibit the systemic synthesis of proinflammatory cytokines [37]. It is known that schizophrenia results from the cumulative impact of multiple common small-effect genetic variants and interactions between genes with small effect may contribute a larger heritable proportion to the overall risk of this disorder [38]. Therefore, we assumed that interaction of *C3* with genes encoding cytokines may be more sensitive to account for its susceptibility to schizophrenia. There is a need for further investigations to validate this hypothesis.

This study has some limitations that should be noted. First, the lack of a significant association may be caused by the modest sample size, possibly resulting in a type II error. Second, we did not psychiatrically screen the control subjects. Third, the principal hypothesis underlying this study is that common SNPs within *C3* may confer susceptibility to schizophrenia. Therefore, we did not sequence the *C3* to assess the influence of rare variants on schizophrenia, and this prevented us to detect their active role in the development of this disorder. Fourth, the case-control association analyses have the potential for population stratification, although all participants were ethnically matched in our sample. Finally, Wong et al. [32] reported lower level of C3 protein in schizophrenia patients in comparison to healthy controls only in male subjects, suggesting that there might be interesting to test *C3* expression separately in male and female subjects. However, we recruited only 47 individuals with or without schizophrenia in the *C3* expression study. The small sample size limited us to further analyze the gender-specific effect of *C3* expression on schizophrenia. Meanwhile, this also limited us to detect the association of studied *C3* SNPs with *C3* mRNA.

In conclusion, the results of this study do not support *C3* as a major genetic susceptibility factor in schizophrenia. Other factors in AP may have critical roles in schizophrenia and be worthy of further investigation.

ACKNOWLEDGMENTS

We are deeply grateful to all participants. We thank the anonymous reviewers for their insightful comments.

AUTHOR CONTRIBUTIONS

Conceived and designed the experiments: JN CZ. Performed the experiments: JN JZ. Analyzed the data: JN SH CZ. Contributed reagents/materials/analysis tools: JN JZ WT WL CZ. Wrote the paper: JN CZ.

REFERENCES

1. Weinberger DR (1996) On the plausibility of "the neurodevelopmental hypothesis" of schizophrenia. Neuropsychopharmacology 14: 1S–11S. pmid:8866738 doi: 10.1016/0893-133x(95)00199-n

2. Schmitt A, Malchow B, Hasan A, Falkai P (2014) The impact of environmental factors in severe psychiatric disorders. Front Neurosci 8: 19. doi: 10.3389/fnins.2014.00019. pmid:24574956

3. Zhang C, Fang Y, Xie B, Cheng W, Du Y, Wang D, et al. (2009) DNA methyltransferase 3B gene increases risk of early onset schizophrenia. Neurosci Lett 462: 308–311. doi: 10.1016/j.neulet.2009.06.085. pmid:19576953

4. Cannon TD, Kaprio J, Lonnqvist J, Huttunen M, Koskenvuo M (1998) The genetic epidemiology of schizophrenia in a Finnish twin cohort. A population-based modeling study. Arch Gen Psychiatry 55: 67–74. pmid:9435762 doi: 10.1001/archpsyc.55.1.67

5. Avramopoulos D, Pearce BD, McGrath J, Wolyniec P, Wang R, Eckart N, et al. (2015) Infection and inflammation in schizophrenia and bipolar disorder: a genome wide study for interactions with genetic variation. PLoS One 10: e0116696. doi: 10.1371/journal.pone.0116696. pmid:25781172

6. Schmitt A, Malchow B, Hasan A, Falkai P (2014) The impact of environmental factors in severe psychiatric disorders. Frontiers in Neuroscience 8. doi: 10.3389/fnins.2014.00019

7. Benros ME, Nielsen PR, Nordentoft M, Eaton WW, Dalton SO, Mortensen PB. (2011) Autoimmune Diseases and Severe Infections as Risk Factors for Schizophrenia: A 30-Year Population-Based Register Study. Am J Psychiatry 168: 1303–1310. doi: 10.1176/appi.ajp.2011.11030516. pmid:22193673

8. Leza JC, Bueno B, Bioque M, Arango C, Parellada M, Do K, et al. (2015) Inflammation in schizophrenia: A question of balance. Neurosci Biobehav Rev 55: 612–626. doi: 10.1016/j.neubiorev.2015.05.014. pmid:26092265

9. Ricklin D, Hajishengallis G, Yang K, Lambris JD (2010) Complement: a key system for immune surveillance and homeostasis. Nat Immunol 11: 785–797. doi: 10.1038/ni.1923. pmid:20720586

10. Tichaczek-Goska D (2012) Deficiencies and excessive human complement system activation in disorders of multifarious etiology. Adv Clin Exp Med 21: 105–114. pmid:23214307

11. Severance EG, Gressitt KL, Halling M, Stallings CR, Origoni AE, Vaughan C, et al. (2012) Complement C1q formation of immune complexes with milk caseins and wheat glutens in schizophrenia. Neurobiol Dis 48: 447–453. doi: 10.1016/j.nbd.2012.07.005. pmid:22801085

12. Fourgeaud L, Boulanger LM (2007) Synapse remodeling, compliments of the complement system. Cell 131: 1034–1036. pmid:18083091 doi: 10.1016/j.cell.2007.11.031

13. Michailidou I, Willems JG, Kooi EJ, van Eden C, Gold SM, Geurts JJ, et al. (2015) Complement C1q-C3 associated synaptic changes in multiple sclerosis hippocampus. Ann Neurol 77: 1007–1026. doi: 10.1002/ana.24398. pmid:25727254

14. Hakobyan S, Boyajyan A, Sim RB (2005) Classical pathway complement activity in schizophrenia. Neurosci Lett 374: 35–37. pmid:15631892 doi: 10.1016/j.neulet.2004.10.024

15. Francks C, Tozzi F, Farmer A, Vincent JB, Rujescu D, St Clair D, et al. (2010) Population-based linkage analysis of schizophrenia and bipolar case-control cohorts identifies a potential susceptibility locus on 19q13.

Mol Psychiatry 15: 319–325. doi: 10.1038/mp.2008.100. pmid:18794890

16. Rudduck C, Beckman L, Franzen G, Lindstrom L (1985) C3 and C6 complement types in schizophrenia. Hum Hered 35: 255–258. pmid:4029965 doi: 10.1159/000153555

17. Fananas L, Moral P, Panadero MA, Bertranpetit J (1992) Complement genetic markers in schizophrenia: C3, BF and C6 polymorphisms. Hum Hered 42: 162–167. pmid:1511994 doi: 10.1159/000154060

18. Blackwood DHR, Muir WJ, Stephenson A, Wentzel J, Adhiah A, Walker MJ, et al. (1996) Reduced expression of HLA-B35 in schizophrenia. Psychiatr Genet 6: 51–59. pmid:8840390 doi: 10.1097/00041444-199622000-00004

19. Liu K, Lai TYY, Chiang SWY, Chan VCK, Young AL, Tam PO, et al. (2014) Gender specific association of a complement component 3 polymorphism with polypoidal choroidal vasculopathy. Sci Rep 4: 7018. doi: 10.1038/srep07018. pmid:25388911

20. Tang WX, Cai J, Yi ZH, Zhang Y, Lu WH, Zhang C. (2014) Association study of common variants within the G protein-coupled receptor kinase 6 gene and schizophrenia susceptibility in Han Chinese. Hum Psychopharmacol 29: 100–103. doi: 10.1002/hup.2375. pmid:24302161

21. Zhu YL, Wang ZL, Ni JL, Zhang Y, Chen MJ, Cai J, et al. (2015) Genetic variant in NDUFS1 gene is associated with schizophrenia and negative symptoms in Han Chinese. J Hum Genet 60: 11–16. doi: 10.1038/jhg.2014.94. pmid:25354934

22. Wang SH, Lu HQ, Ni JL, Zhang JT, Tang WX, Lu WH, et al. (2015) An evaluation of association between common variants in C4BPB/C4BPA genes and schizophrenia. Neurosci Lett 590: 189–192. doi: 10.1016/j.neulet.2015.02.005. pmid:25660618

23. Miyagawa H, Yamai M, Sakaguchi D, Kiyohara C, Tsukamoto H, Kimoto Y, et al. (2008) Association of polymorphisms in complement component C3 gene with susceptibility to systemic lupus erythematosus. Rheumatology 47: 158–164. doi: 10.1093/rheumatology/kem321. pmid:18174230

24. Yu QQ, Yao Y, Zhu J, Bao X, Xie TH, Sun C, et al. (2015) Nonsynonymous single nucleotide polymorphisms in the complement component 3 gene are associated with risk of age-related macular degeneration: A meta-analysis. Gene 561: 249–255. doi: 10.1016/j.gene.2015.02.039. pmid:25688879

25. Phillips CM, Goumidi L, Bertrais S, Ferguson JF, Field MR, Kelly ED, et al. (2009) Complement component 3 polymorphisms interact with

polyunsaturated fatty acids to modulate risk of metabolic syndrome. Am J Clin Nutr 90: 1665–1673. doi: 10.3945/ajcn.2009.28101. pmid:19828715

26. Rhodes B, Hunnangkul S, Morris DL, Hsaio LC, Graham DSC, Nitsch D, et al. (2009) The heritability and genetics of complement C3 expression in UK SLE families. Genes Immun 10: 525–530. doi: 10.1038/gene.2009.23. pmid:19387462

27. Zhang Y, Chen MJ, Wu ZG, Chen J, Yu SY, Fang Y, et al. (2013) Association Study of Val66Met Polymorphism in Brain-Derived Neurotrophic Factor Gene with Clozapine-Induced Metabolic Syndrome: Preliminary Results. PLoS One 8: e72652. doi: 10.1371/journal.pone.0072652. pmid:23967328

28. Zhang Y, Chen M, Chen J, Wu Z, Yu S, Fang Y, et al. (2014) Metabolic syndrome in patients taking clozapine: prevalence and influence of catechol-O-methyltransferase genotype. Psychopharmacology (Berl) 231: 2211–2218. doi: 10.1007/s00213-013-3410-4

29. Zhang C, Wang Z, Hong W, Wu Z, Peng D, Fang Y. (2015) ZNF804A Genetic Variation Confers Risk to Bipolar Disorder. Mol Neurobiol. doi: 10.1007/s12035-015-9193-3.

30. Zhang C, Wu Z, Hong W, Wang Z, Peng D, Chen J, et al. (2014) Influence of BCL2 gene in major depression susceptibility and antidepressant treatment outcome. J Affect Disord 155: 288–294. doi: 10.1016/j.jad.2013.11.010. pmid:24321200

31. Soria LD, Gubert CD, Cereser KM, Gama CS, Kapczinski F (2012) Increased serum levels of C3 and C4 in patients with schizophrenia compared to eutymic patients with bipolar disorder and healthy. Revista Brasileira De Psiquiatria 34: 119–120. pmid:22392401 doi: 10.1590/s1516-44462012000100022

32. Wong CT, Tsoi WF, Saha N (1996) Acute phase proteins in male Chinese schizophrenic patients in Singapore. Schizophrenia Research 22: 165–171. pmid:8958601 doi: 10.1016/s0920-9964(96)00037-0

33. Li Y, Zhou KJ, Zhang Z, Sun LY, Yang JL, Zhang M, et al. (2012) Label-free quantitative proteomic analysis reveals dysfunction of complement pathway in peripheral blood of schizophrenia patients: evidence for the immune hypothesis of schizophrenia. Mol Biosyst 8: 2664–2671. doi: 10.1039/c2mb25158b. pmid:22797129

34. Deleidi M, Jaggle M, Rubino G (2015) Immune aging, dysmetabolism, and inflammation in neurological diseases. Front Neurosci 9: 172. doi: 10.3389/fnins.2015.00172. pmid:26089771

35. Potvin S, Stip E, Sepehry AA, Gendron A, Bah R, Kouassi E. (2008) Inflammatory cytokine alterations in schizophrenia: a systematic quantitative review. Biol Psychiatry 63: 801–808. pmid:18005941 doi: 10.1016/j.biopsych.2007.09.024

36. Miller BJ, Buckley P, Seabolt W, Mellor A, Kirkpatrick B (2011) Meta-analysis of cytokine alterations in schizophrenia: clinical status and antipsychotic effects. Biol Psychiatry 70: 663–671. doi: 10.1016/j. biopsych.2011.04.013. pmid:21641581

37. Takabayashi T, Vannier E, Clark BD, Margolis NH, Dinarello CA, Burke JF, et al. (1996) A new biologic role for C3a and C3a desArg: regulation of TNF-alpha and IL-1 beta synthesis. J Immunol 156: 3455–3460. pmid:8617973

38. McClellan JM, Susser E, King MC (2007) Schizophrenia: a common disease caused by multiple rare alleles. Br J Psychiatry 190: 194–199. pmid:17329737 doi: 10.1192/bjp.bp.106.025585

Chapter 2

MULTIDIMENSIONAL GENE SET ANALYSIS OF GENOMIC DATA

David Montaner[1,2], Joaquín Dopazo[1,2,3]

[1] Department of Bioinformatics and Genomics, Centro de Investigación Príncipe Felipe (CIPF), Valencia, Spain

[2] Functional Genomics Node (INB), Centro de Investigación Príncipe Felipe (CIPF), Valencia, Spain

[3] CIBER de Enfermedades Raras (CIBERER), Valencia, Spain

ABSTRACT

Understanding the functional implications of changes in gene expression, mutations, etc., is the aim of most genomic experiments. To achieve this, several functional profiling methods have been proposed. Such methods study the behaviour of different gene modules (e.g. gene ontology terms) in response to one particular variable (e.g. differential gene expression). In spite to the wealth of information provided by functional profiling methods, a common limitation to all of them is their inherent unidimensional nature. In order to overcome this restriction we present a multidimensional logistic model that allows studying the relationship of gene modules with different genome-scale measurements (e.g. differential expression, genotyping association, methylation, copy number alterations, heterozygosity, etc.) simultaneously. Moreover, the relationship of such functional modules with the interactions among the variables can also be studied, which produces novel results impossible to be derived from the conventional unidimensional functional profiling methods. We report sound results of gene sets associations that remained undetected by the conventional one-dimensional gene set analysis in several examples. Our findings demonstrate the potential of the proposed approach for the discovery of new cell functionalities with complex dependences on more than one variable.

INTRODUCTION

The development of new genomic technologies, such as microarrays of gene expression, genotyping or array-CGH, along with the new next-generation sequencing techniques is increasing the volume of data throughput amazingly. As a direct consequence of this, the bottleneck in functional genomics has shifted from the data production phase to the data analysis steps. In particular, the necessity for providing a functional interpretation at molecular level that accounts for the genome-scale experimental designs has promoted the development of different methods for the functional analysis of this type of data in the last years [1], [2].

It is widely accepted that most of the biological functionality of the cell arises from complex interactions among their molecular components that define operational interacting entities or modules [3]. Functions collectively performed by such modules have conceptually been represented in different ways. Gene ontology (GO) [4] and KEGG pathways [5] are the most popular and widely used module definitions although many other are available in different repositories (e.g., Reactome [6], Biocarta, etc.) For practical purposes, functional modules are henceforth defined as sets of genes sharing functional annotations extracted from any of these repositories. Functional profiling methods exploit different definitions of modules in an attempt of understanding the functional basis of high-throughput experimental results [7]. Initially, functional enrichment methods, in different implementations [7], [8], have been used for this purpose. More sensitive approaches, generically known as gene-set analysis (GSA) methods, pioneered by the Gene Set Enrichment Analysis (GSEA) [9], were later proposed [1], [10]. In the original formulation, GSA methods aimed to directly detect sets of functionally related genes (modules) with a coordinate and significant over- or under-expression across the complete list of genes ranked according to their differential expression [9], [11], [12], [13], [14],[15]. GSA methods can detect such modules even if their gene components are not significantly differentially expressed when tested individually. GSA has been successfully applied to the analysis of microarray experiments and has contributed to the adoption of a systems-biology perspective in distinct fields such as cancer [16]. Recent findings, brought about by the application of GSA methods on microarray experiments [17] are consistent with the idea that pathways, rather than individual genes, appear to govern the course of tumorigenesis [18]. The use of GSA has been extended to other areas beyond transcriptomics, such as evolution [19], QTL analysis [20] or genotyping [21].

Nevertheless, the different versions of GSA published to date [1],[2],[10] are inherently one-dimensional. Its application to the analysis of genomic datasets is at present limited to the study of a unique variable measured for the genes. The experimental conditions studied, even if corrected by other variables (e.g. age, gender, treatments, etc.), are typically summarized into a unique value for each gene (e.g., differential expression in a case-control, risk in the case of survival analysis, etc.) which is used to rank them accordingly.

Nowadays, the extensive use of different high-throughput methodologies allows the obtention of different measurements for the genes such as methylation status, splicing variants, linkage to diseases, etc., in a straightforward manner. As an illustration of this, a pilot study by The Cancer Genome Atlas (http://cancergenome.nih.gov/) consortium on glioblastomas has recently been published [22]. In it, different types of transcriptomic and genomic profiling were obtained and analyzed in an example of application of different genomic methodologies that would become routine soon. In addition, different measurements of the same type in different experimental contexts can easily be done. For instance, gene expression measurements in case-controls of different, but mechanistically related experimental conditions, phenotypes, diseases, treatments, etc. can be easily obtained. In such scenario, more than one measurement could be obtained to rank the genes involved in the study. Under the conventional GSA paradigm the different ranked lists of genes could be analyzed one at a time and still a good deal of information might be obtained. Nevertheless, by taking this approach any list of ranked genes is considered independent from each other and, consequently, behaviour of functional modules which are dependent on the combination of the studied ranking variables will, most likely, remain undetected.

Here we focus on a conceptually different strategy for GSA by extending the gene set based functional analysis to a multidimensional scenario in which more than one variable or genomic measurement is available for all genes in the study. Logistic regression allows for fitting models that include more than one variable. We show here, by means of several examples, how the application of the multidimensional GSA (MD-GSA) uncovers biological processes activated by different combinations of parameters (measured for all the genes and derived from microarray of other experiments) that would have remained undetected if the parameters would have been analysed one at a time, independently.

RESULTS

Gene-Set Activation Dependent on the Transcription Rates and MRNA Activities In Yeast

Gene expression is a process that involves two steps of synthesis which end when the appropriate level of protein required for performing a given function is reached. Some processes in the cell require of a quick activation and/or deactivation, while others remain in activity for longer periods and their activation processes do not involve any urgency. Thus, it is expectable different cell functionalities will use different strategies of gene and protein expression and degradation. Measurements of these parameters can be found in a recent genome-wide analysis on common gene expression strategies in yeast [23]. Using these data, we have studied two relevant and opposite biological processes that account for the steady-state mRNA level in the cell: transcription and stability [24]. The authors used a functional enrichment strategy [25] to test the GO terms associated to the parameters measured and to their correlations. Essentially, they used quintiles as cut-off values and tested for enrichments in the genes showing a high or low correlation in rates (transcription and translation) or abundances (mRNA and protein copy number), finding a total of 22 GO terms significantly over-represented at different combinations of rates and abundances. Nevertheless, other interesting situations in which the measurements are not correlated (e.g. transcription rate and mRNA stability) could not be analysed with this approach that, in addition, has the disadvantage of requiring an arbitrary threshold.

Here we analysed the dependences of GO terms on two measurements, transcription rate (TR) and mRNA stability (RS), as well as on the interaction between them. When the logistic model was applied to the mRNA stability and to the transcription rate independently, we obtained 170 and 80 GO terms significantly associated to extreme values of these variables (see Table S1). This increase in the number of GO terms found was due to the well known fact that GSA strategies are much more sensitive than threshold-based functional enrichment strategies [1],[10]. Actually, similar results were obtained when other equivalent GSA strategies were used (data not shown) [11], [19].

Nevertheless, the most interesting aspect of this study is the analysis of the interaction between both variables. Table 1 shows 18 GO terms which were significantly associated to the interaction between transcription rate and mRNA stability. Figure S1 depicts the GO terms within the GO hierarchy. Nine of these GO terms could only be detected when the model takes into account simultaneously both parameters. In most of the cases, the GO was associated to both low transcription rate and mRNA stability (pattern *q3i*, see methods for an explanation of the patterns) such as *sister chromatid segregation* (Figure 1 top) in a subtle way that can only be detected when both parameters are included in the model. On the other hand, other processes, such as *DNA packaging, Chromatin assembly* (Figure 1 bottom), *Chromatin assembly or disassembly* and *Establishment and/or maintenance of chromatin architecture*(which are related terms, see File S1), or *protein-DNA complex assembly* are associated to high transcription rates but low mRNA stability (pattern *q4i,* see methods). This last strategy, opposite to the first one, suggest a transient necessity of these processes, whose genes are produced at a fast rate but quickly discarded after their functions have been carried out.

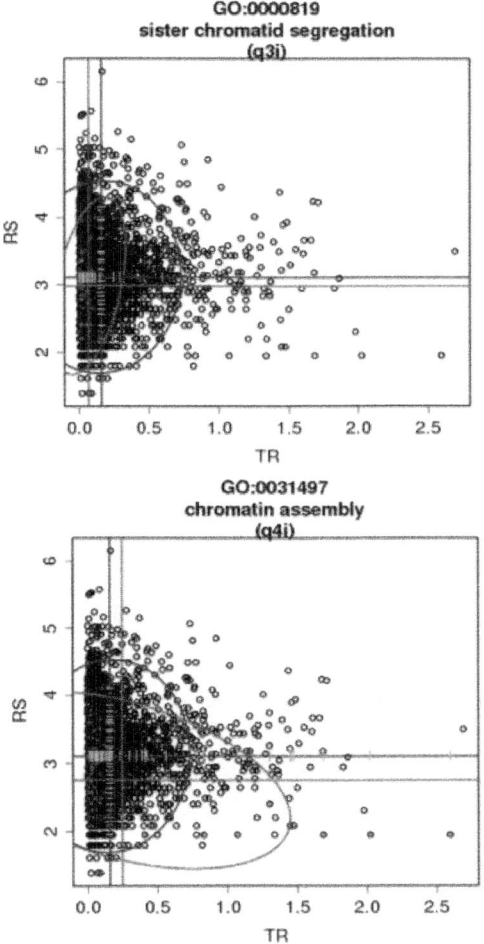

Figure 1: Combined analysis of transcription rates and mRNA stability in yeast with the logistic model.

RS (mRNA stability) is represented in vertical axis and TR (transcription rate) is represented in the horizontal axis for GO terms sister chromatid segregation (top) and chromatin assembly (bottom). Blue lines intersect in the mean of the distribution of all the values and red lines intersect in the mean of the distribution of values of the genes corresponding to the GO term analysed. Blue ellipse delimits the confidence interval for all the values and red ellipse delimits the confidence interval for the GO term analysed. The red ellipse marks the trend of the relationship between both variables. MD-GSA assigns patterns q3i and q4i respectively to these functional modules.

Table 1: Significant GO terms when transcription rate and mRNA stability are taken into account in the model

GO Id	Log odds ratio (model coefficients)			Adjusted p-value			pattern	new	GO name
	TR	RS	inter	TR	RS	inter			
GO:0019953	−11.87	−0.82	3.29	0.04	0.01	0.02	q3i	yes	sexual reproduction
GO:0051704	−11.98	−0.69	3.23	0.04	0.02	0.02	q3i	yes	multi-organism process
GO:0000819	−30.49	−0.87	7.1	0.02	0.03	0.02	q3i	yes	sister chromatid segregation
GO:0006260	−20.35	−0.97	4.99	0	0	0.01	q3i	no	DNA replication
GO:0006261	−25.15	−1.31	6.28	0	0	0.01	q3i	no	DNA-dependent DNA replication
GO:0022613	−4.69	−1.78	1.61	0.08	0	0.03	q3i	no	ribonucleoprotein complex biogenesis and assembly
GO:0042254	−5.05	−1.91	1.75	0.09	0	0.03	q3i	no	ribosome biogenesis
GO:0000746	−11.48	−0.73	3.17	0.06	0.02	0.03	q3i	yes	conjugation
GO:0000747	−11.39	−0.74	3.16	0.06	0.02	0.03	q3i	yes	conjugation with cellular fusion
GO:0042221	−6.65	−0.12	2.05	0.02	0.6	0.01	q3i	yes	response to chemical stimulus
GO:0000070	−30.23	−0.78	7.01	0.03	0.07	0.03	q3i	yes	mitotic sister chromatid segregation
GO:0019725	−9.13	−0.38	2.71	0.02	0.15	0.01	q3i	yes	cellular homeostasis
GO:0042592	−8.75	−0.3	2.59	0.02	0.27	0.01	q3i	yes	homeostatic process
GO:0006325	8.01	−0.47	−3.09	0	0.03	0.01	q4i	no	establishment and/or maintenance of chromatin architecture
GO:0065004	12.12	−0.49	−4.6	0	0.21	0.02	q4i	no	protein-DNA complex assembly
GO:0006323	12.63	−0.48	−4.96	0	0.15	0.01	q4i	no	DNA packaging
GO:0006333	12.44	−0.4	−4.84	0	0.23	0.01	q4i	no	chromatin assembly or disassembly
GO:0031497	12.51	−0.44	−4.84	0	0.2	0.01	q4i	no	chromatin assembly

A total of 18 GO terms were found as significant at FDR-adjusted p<0.05, nine of them were also found by the multivariate analysis. Column new indicates if the term as been found only because of the interaction factor (yes) or if it was found also in the univariate analysis in one or both dimensions independently.
doi:10.1371/journal.pone.0010348.t001

Different strategies of production and degradation, corresponding to different biological requirements of the cell, can be thus detected by the combined analysis of these parameters.

Gene-Set Dependences on Differential Expression and Splicing Index

Recent studies have shown that more that 70% of the multi-exon genes, corresponding to about 50% of all human genes, are predicted to be alternatively spliced [26]. It is well known that alternative splicing participates in many pathways and processes. Also alterations in splicing function has been implicated in many diseases, including neuropathological conditions such as Alzheimer disease, cystic fibrosis, defects in growth and development, and many human cancers [27].

The magnitude of the alterations in the splicing process can be studied through the splicing index. This index accounts for changes at the exon level that are relative to the expression of the gene. In particular, the intensity value of an exon's probeset is divided by an estimate of the expression level of the transcript cluster to which the exon belongs to. In this way, a gene-level-normalized intensity that can be compared across samples or conditions is created. Changes in this value between case and control samples provide a quantitative measure of alternative splicing between the two conditions [28]. Thus each gene in the data set can be studied both in terms of its differential expression and its alternative splicing. Our multidimensional logistic model can be used to explore this two dimensional gene space.

Here we reanalyze data obtained using Affymetrix exon arrays [29] in which human breast cancer cell lines are compared to non tumorigenic human breast epithelial cell lines. The parameters studied by means of the multidimensional logistic model are: differential gene expression estimates obtained upon the application of a t-test for the above mentioned comparison and a splicing index, that accounts for changes at the exon level that are relative to the expression of the gene [30].

A total of 141 GO terms were found to be significantly associated to high values of the differential gene expression dimension (pattern yh, yl; see methods section). These terms are equivalent to those that would be found by conventional one-dimensional GSA methods and, as expected, GO definitions related to cell proliferation, cell signalling, apoptosis, cellular adhesion, etc., were found among them. One significant GO term, *regulation of viral reproduction*, was significant in the splicing index dimension alone. The trend of the enrichment was towards the positive values of the splicing index (pattern *xh*; see methods section) meaning that genes in the GO term are "subordinately" more spliced in the tumour than in the normal tissue (see File S2A).

Another 12 terms were found by the MD-GSA (see Table 2), whose relationships within the GO hierarchy is depicted in Figure S2. The processes discovered here were related (but yet undetected) to other processes already detected by the conventional analysis of differential expression (see File S2A). For example, *positive regulation of cell adhesion* and its parent*regulation of cell adhesion* are descendants of *cell adhesion*, and two sister processes (*cell-matrix adhesion* and *cell-substrate adhesion*) were found by the model when the two variables were taken into account, and would have remained undetected if a conventional, unidimensional GSA approach would have been used. The patterns for these terms are bimodal in the two dimensional space (pattern *b24*, see methods section) indicating that the genes annotated to them behave as if they were in two sub-modules. For example, *positive regulation of cell adhesion* and its parent processes *regulation of cell adhesion*, which are known to be related to cancer, show a bimodal pattern towards the quadrants 2 and 4 (pattern*b24*). This means that part of the annotated genes are more spliced but underexpressed in the tumour samples while the other part is more spliced but underexpressed in the control samples (see Figure 2).

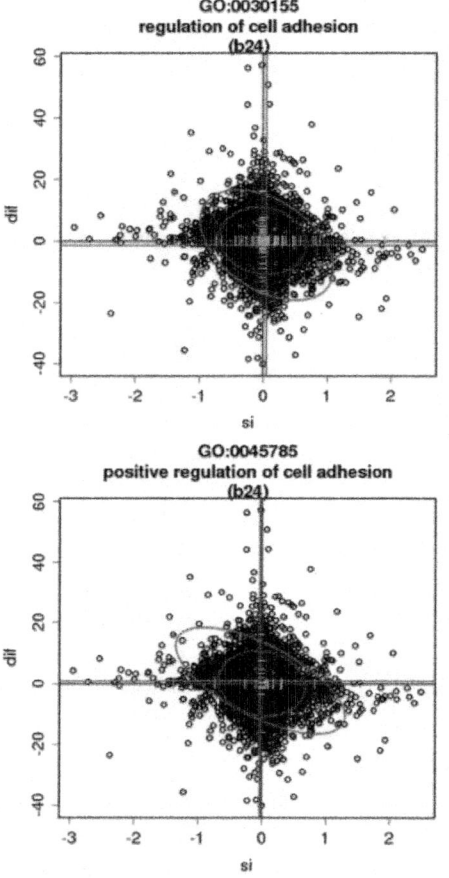

Figure 2: Combined analysis of differential gene expression and splicing index with the logistic model.

Differential expression is represented in vertical axis and splicing index is represented in the horizontal axis for GO terms positive regulation of cell adhesion (bottom) and its parent processes regulation of cell adhesion (top). Blue lines intersect in the mean of the distribution of all the values and red lines intersect in the mean of the distribution of values of the genes corresponding to the GO term. Blue ellipse delimits the confidence interval for all the values and red ellipse delimits the confidence interval for the GO term analysed. The red ellipse marks the trend of the relationship between both variables. MD-GSA assigns a bimodal pattern b24 to these functional modules.

doi:10.1371/journal.pone.0010348.g002

Table 2: Significant GO terms when differential expression and splicing index are taken into account in the model

GO id	Log odds ratio (model coefficients)			Adjusted p-value			pattern	GO name
	splicing	diff.exp	inter	splicing	diff.exp	inter		
GO:0006767	0.15	−0.15	0.14	1	0.61	0.04	b13	water-soluble vitamin metabolic process
GO:0045216	0.29	−0.04	0.17	1	0.95	0.02	b13	cell-cell junction assembly and maintenance
GO:0007043	0.38	−0.03	0.18	1	0.97	0.02	b13	cell-cell junction assembly
GO:0048706	0.2	0.08	0.17	1	0.89	0.03	b13	embryonic skeletal development
GO:0007034	0.32	−0.18	0.17	1	0.65	0.02	b13	vacuolar transport
GO:0007041	0.32	−0.1	0.18	1	0.86	0.01	b13	lysosomal transport
GO:0048704	0.23	0.12	0.19	1	0.84	0.02	b13	embryonic skeletal morphogenesis
GO:0048705	0.17	0.1	0.17	1	0.85	0.02	b13	skeletal morphogenesis
GO:0016197	0.08	0.1	0.15	1	0.79	0.02	b13	endosome transport
GO:0030155	0.01	−0.16	−0.15	1	0.43	0.01	b24	regulation of cell adhesion
GO:0045785	−0.04	0.06	−0.18	1	0.94	0.02	b24	positive regulation of cell adhesion
GO:0030032	−0.16	−0.17	−0.18	1	0.72	0.03	b24	lamellipodium biogenesis

A total of 12 GO terms were found as significant in the interaction at FDR-adjusted p<0.05.
doi:10.1371/journal.pone.0010348.t002

An equivalent analysis for KEGG can be found in File S2B.

Gene-Sets Differentially Activated In Related Diseases: A Case Study with Psoriasis and Dermatitis.

The study of gene expression at genomic level in both psoriasis [31] and dermatitis [32] and further functional analysis reveals a considerable number of deregulated pathways when both diseases are compared to their corresponding healthy samples. Thus, when the multivariate logistic model was applied to gene lists arranged by differential expression 172 GO terms were found to be significant only for dermatitis (patterns *xh*, *xl*; see methods section) and 202 only for psoriasis (patterns *yh, yl*). Another 77 GO terms were found to be significant in both, dermatitis and psoriasis but did not show an interaction effect (patterns *q1f, q2f, q3f, q4f*) Most of this terms will also be found by the independent unidimensional analysis of the dermatitis dataset and the psoriasis dataset. In the case of dermatitis, terms related to signalling, cell proliferation, immune system and development of epidermis were found, among others (see Files S3A and S3B). Similar terms can be found in psoriasis with some variations (see Files S3A and S3B). A detailed comparative functional analysis of these diseases is beyond the scope of this manuscript and we will only focus on the results obtained when both diseases are simultaneously analysed.

Table 3 shows the GO terms that are significant when both diseases are taken into account in the logistic model (column labelled with "inter"). Figure S3 shows the GO terms within the GO hierarchy. The GO terms *M phase of mitotic cell cycle* (and their parent terms *M phase* and *cell cycle phase*) and *cell division* where associated to both diseases in their main effects and also in their interaction effect (pattern *q1i*, seemethods) reinforcing their relevance in the biological mechanisms underlying both skin syndromes. Some other GO terms are only significant in the interaction effect. Their genes show a bimodal behaviour as if the functional module was composed of two sub-units (pattern *b13, b24*; see methods). For instance, GO terms*phosphoinositide-mediated signaling* and *response to reactive oxygen species* have a positive interaction coefficient, which means that some of the genes of the module are being coordinately over-expressed in both diseases while the remaining genes in the GO term are under-expressed also in both diseases. In a symmetric way, *negative regulation of lymphocyte proliferation* (and the parent process *negative regulation of mononuclear cell proliferation*) shows a negative interaction. Part of the genes in these modules increase their expression in dermatitis but decrease it in psoriasis while the rest of them present the opposite behaviour. The reduced cutaneous IFNalpha2 transcription which has been described as a differential characteristic of dermatitis with respect to psoriasis [32] could be causing this effect detectable in the analysis when the two variables are included in the model. All this bimodal terms highlight antagonistic effect, detectable only trough the combined analysis of both diseases.

Table 3: Significant GO terms when differential expression of dermatitis and psoriasis are taken into account in the model

GO id	Log odds ratio (model coefficients)			Adjusted p-value			pattern	GO name
	dermatitis	psoriasis	inter	dermatitis	psoriasis	inter		
GO:0022403	−0.13	0.36	0.11	0.11	0	0.01	q1i	cell cycle phase
GO:0000279	−0.06	0.37	0.12	0.55	0	0.03	q1i	M phase
GO:0051301	−0.1	0.25	0.15	0.36	0	0	q1i	cell division
GO:0000087	−0.11	0.4	0.12	0.32	0	0.05	q1i	M phase of mitotic cell cycle
GO:0048015	0.08	0.07	0.16	0.72	0.68	0.05	b13	phosphoinositide-mediated signaling
GO:0000302	0.24	−0.06	0.29	0.59	0.85	0	b13	response to reactive oxygen species
GO:0032945	0.43	0.33	−0.79	0.26	0.39	0	b24	negative regulation of mononuclear cell proliferation
GO:0050672	0.43	0.33	−0.79	0.26	0.39	0	b24	negative regulation of lymphocyte proliferation
GO:0048589	−0.19	−0.06	−0.59	0.53	0.91	0.04	b24	developmental growth
GO:0007028	0.21	−0.11	−0.75	0.47	0.83	0	b24	cytoplasm organization and biogenesis
GO:0007043	0.07	−0.5	−0.91	0.86	0.22	0	b24	cell-cell junction assembly
GO:0045216	0.12	−0.26	−0.86	0.75	0.59	0	b24	cell-cell junction assembly and maintenance

A total of 12 GO terms were found as significant in the interaction at FDR-adjusted p<0.05.
doi:10.1371/journal.pone.0010348.t003

Combined analysis of several genomic measurements: a case study with genotyping, gene expression and copy number alterations in breast cancer

It is known that mutations or alteration in copy number are related to cancer and tumour development [33], [34]. Current microarray technologies allow for the measurement of SNP variation and copy number estimation at the same time [35], [36] and have been used to gain insights into breast cancer [37], [38], [39], among other diseases.

Using the multidimensional logistic model proposed we have re-analyzed here data from several separated studies previously collected by us in an integrative analysis of breast cancer disease [38]. In particular we provide a combined description of GO and KEGG relationship to different parameters such as SNP association, copy number alteration and differential gene expression in connection to disease outcome (all the data were taken from the additional information of the above mentioned study, see methods).

When analyzing SNP association data and copy number in luminal B tumours by the proposed MD-GSA, *basal cell carcinoma* KEGG pathway raised up (File S4B) showing a bimodal pattern towards quadrants 1 and 3 (*b13*, see methods). This indicates that the genes in the pathway highly associated to disease are also increased in their copy number, and that genes not associated to disease do not have an increased copy number (they may even have a reduced copy number what would fit with the no association or even protection of the SNPs to disease). Most probably, the SNPs are markers associated either to regions undergoing copy number alterations or to other mutations that affect the *basal cell carcinoma* pathway, which obviously underlies breast cancer disease. The same analysis using the GO reported some negative bimodal terms (Table 4 and File S4B) like *L-amino acid transport* which is known to be involved proliferation processes [40]. A similar analysis with GO terms can be found in File S4A. Figure S4 displays the GO terms in Table 4 within the GO hierarchy.

Table 4: Significant GO terms when copy number and gene association to the disease (see text) are taken into account in the model

GO id	Log odds ratio (model coefficients)			Adjusted p-value			pattern	GO name
	association	copy number	inter	association	copy number	inter		
GO:0015807	−0.09	−0.85	−0.59	0.98	0.46	0.04	b24	L-amino acid transport
GO:0032228	−0.63	−1.21	−0.68	0.65	0.24	0.01	b24	regulation of synaptic transmission, GABAergic
GO:0050805	−0.94	−1.24	−0.63	0.22	0.24	0.04	b24	negative regulation of synaptic transmission
GO:0051932	−0.82	−1.35	−0.67	0.49	0.17	0.02	b24	synaptic transmission, GABAergic
GO:0042398	−0.77	−0.02	0.12	0.04	0.99	1	xl	amino acid derivative biosynthetic process
GO:0042401	−0.93	0.12	0.2	0.01	0.98	1	xl	biogenic amine biosynthetic process
GO:0030216	0.2	0.41	−0.03	0.8	0.03	1	yh	keratinocyte differentiation
GO:0031424	0.29	0.59	−0.01	0.81	0	1	yh	keratinization

A total of 8 GO terms were found as significant at FDR-adjusted p<0.05.
doi:10.1371/journal.pone.0010348.t004

We also applied the MD-GSA to the variables prognosis and differential expression in tumours. In the representation (File S5A), high values in the differential expression dimension indicate under-expression in tumour while low values indicate over-expression. Conversely, high values in the prognosis dimension indicate bad prognosis (if the gene is expressed) while low values in the prognosis dimension indicate good prognosis (if the gene is expressed).

Table 5 (more details in File S5A) show results obtained from the application of the MD-GSA using modules defined with GO terms. The relationships among them within the GO hierarchy are depicted in Figure S5. Most of the GO terms related to *cell division* and *cell cycle* show a$q2i$ pattern (see methods) indicating a significant convergence of their genes in the prognosis and differential expression dimensions. From the relatively high prognosis value associated to the genes annotated to this GO terms we know that, if over expressed they indicate bad prognosis. From the low values in the t-statistic we know these GO terms are enriched in the tumours samples. Hence the multivariate logistic model is pointing out those modules which are dangerous to the patient if they are activated, and, that are certainly know to be activated in luminal B tumours. This extended functional analysis provides the researcher not only with a quick an easy interpretation of the combined data but also with the additional information of the interaction term in the model. It is worth pointing out here that better and more detailed results are obtained by combining both datasets under the proposed methodology than by applying independently the univariant methodology to any of the datasets and summing up the results obtained. The equivalent MD-GSA for KEGG pathways can be found in File S5B.

Table 5: Significant GO terms when differential expression and prognosis are taken into account in the model

GO id	Log odds ratio (model coefficients)			Adjusted p-value			pattern	GO name
	diff.exp	prognosis	inter	diff.exp	prognosis	inter		
GO:0000087	-0.45	-0.08	-0.42	0.01	0.81	0	q2i	M phase of mitotic cell cycle
GO:0000279	-0.53	-0.07	-0.38	0.04	0.85	0	q2i	M phase
GO:0000910	-0.27	-0.09	-0.57	0.01	0.95	0	q2i	cytokinesis
GO:0007067	-0.47	-0.07	-0.4	0.04	0.9	0	q2i	mitosis
GO:0022618	-0.22	-0.33	-0.42	0.03	0.21	0	q2i	ribonucleoprotein complex assembly
GO:0051301	-0.38	0	-0.38	0.01	0.99	0	q2i	cell division
GO:0051726	-0.01	0.05	-0.22	0.03	0.91	0.01	q2i	regulation of cell cycle
GO:0045638	0.09	-0.35	-0.6	0.01	0.65	0.04	q4i	negative regulation of myeloid cell differentiation
GO:0000226	-0.08	0.16	-0.31	0.11	0.47	0.02	b24	microtubule cytoskeleton organization and biogenesis
GO:0000278	-0.34	0.04	-0.28	0.11	0.94	0	b24	mitotic cell cycle
GO:0007346	-0.3	-0.08	-0.39	0.07	0.9	0	b24	regulation of mitotic cell cycle
GO:0022403	-0.42	0	-0.31	0.09	0.99	0	b24	cell cycle phase
GO:0042254	-0.4	-0.45	-0.42	0.19	0.1	0.01	b24	ribosome biogenesis
GO:0006412	0.06	-0.28	-0.2	0.02	0.01	0.07	q4f	translation
GO:0006414	0.45	-1.12	-0.43	0	0	0.28	q4f	translational elongation
GO:0042312	0.45	0.08	-0.51	0.03	0.97	0.22	xh	regulation of vasodilation
GO:0000209	-0.25	0.55	0.13	0.94	0.01	1	yh	protein polyubiquitination
GO:0006066	0.08	0.2	-0.02	0.97	0.02	1	yh	alcohol metabolic process
GO:0010033	0.05	0.29	0	0.99	0.02	1	yh	response to organic substance
GO:0032944	-0.17	-0.7	0.06	0.97	0.02	1	yl	regulation of mononuclear cell proliferation
GO:0042098	-0.18	-0.61	0.08	0.95	0.04	1	yl	T cell proliferation
GO:0042110	0.03	-0.38	0.14	0.75	0.03	0.86	yl	T cell activation
GO:0042129	-0.33	-0.74	-0.02	0.99	0.05	1	yl	regulation of T cell proliferation
GO:0045321	-0.04	-0.28	0.06	0.92	0.03	1	yl	leukocyte activation
GO:0046649	-0.06	-0.33	0.07	0.89	0.02	1	yl	lymphocyte activation
GO:0046651	-0.19	-0.49	-0.05	0.99	0.05	1	yl	lymphocyte proliferation
GO:0050670	-0.17	-0.7	0.06	0.97	0.02	1	yl	regulation of lymphocyte proliferation
GO:0051249	-0.06	-0.44	0.24	0.52	0.04	0.71	yl	regulation of lymphocyte activation

Terms were significant at FDR-adjusted $p<0.05$.
doi:10.1371/journal.pone.0010348.t005

Advantages and Limitations of the Logistic Regression Methodology

The major advantage of the logistic regression methodology is it flexibility. It can be used in any genomic context in which certain biological characteristic of a gene is measured using a numerical scale. This numerical scale may be a continuous "ranking statistic" as described previously [41] or in this paper, but it may also be a categorical variable [42].

Moreover, many modifications of the logistic model with potential applications in biology are already statistically developed and can be used straight forward. Here, for instance we showed how to extend the methodology to study not one but two gene characteristics at a time. It is also straightforward to include the interaction in the model as we showed here. A unidimensional binary logistic model can be used instead the conventional 2×2 contingency table alternative because the logistic model easily allows for weighting genes [42]. This simplicity of extension is not at all intrinsic to most other GSA approaches, what makes the logistic model worth to be explored.

Another advantage of the method is that it does not start from the original observed data set (gene expression matrix for instance) but from a ranking statistic that already summarizes the relevant characteristic under study. This makes the methodology useful in many genomic contexts beyond the microarray paradigm. One example of ranking statistic we have discussed is the classical t-test which, perhaps with some modification, is underneath most GSA methodologies. For each gene, this statistic measures the biological characteristic of "how much" the gene is differentially expressed in a particular biological experiment. But we also exemplified how the ranking statistic can be a hazard ratio form a Cox model or other gene-wise variable[19]. In the case of the hazard ratio, the biological characteristic measured for each gene by the statistic is the association of expression and risk disease. The GSA for this second example can be directly carried out using the logistic methodology and software. On the contrary, most GSA approaches will require major modifications of their methods and software to be applied in a case other than differential gene expression in a class comparison experiment.

Virtually any gene-wise variable can be explored from a GSA perspective using the logistic regression model. In this paper we presented examples for the analysis of transcription rates, mRNA stabilities, splicing, SNP association to disease and copy number estimation. The analysis of other measurements is possible, including the evolutionary selective pressure in the human genome or a study of gene connectivity in the interactome [19]. Other publications also discuss on the advantage of a methodology that starts form a single ranking statistic and not from the whole expression data matrix [42], [43].

Having said that, some remarks and warnings should be given related mainly with the null hypothesis that underpin the method and p-value computation.

In Sator's logistic regression approach [41] and in the extension we are proposing here, the distribution of the ranking statistic within each module is compared to that of its complement. Thus, following Goeman's nomenclature they are "competitive" tests [10]. Also, the way p-values are computed in the logistic model make of this approach a "gene sampling model" methodology [10].

It has been shown that, in general contexts of gene expression, where gene measurements are correlated within modules, GSA approaches that test "competitive" hypothesis based on "gene sampling models" are anticonservative [10]. This undesirable property also applies to the main effects of the bivariate logistic model as we could confirm in simulation studies (only in the case of internal correlation in the gene sets, which is the case of gene expression but not of the rest of the measurements used in this study). Interestingly, the consequence of gene correlation over the interaction effect, which is the main contribution of the proposed methodology, was the opposite and makes the method more conservative (see File S6). One way to avoid the bias of the particular context of gene expression would be to compute p-values based on a subject sampling permutation.

Care should be taken also when interpreting p-values from the method proposed here due to its "competitive" nature and the fact that it starts from a ranking statistic instead of the original data. Consequently, p-values test whether the distribution of the ranking statistic within each module is different to that of the whole genome. Therefore p-values do not extrapolate directly to the individual level class comparison which was done in order to compute the ranking statistic.

DISCUSSION

Functional annotations, such as GO or KEGG pathways, have been used for the definition of modules of genes, carrying out common functional roles, in functional profiling methods [1], [2]. All these methods, including the most recent versions, such as the GSA, can only deal with data that have been preselected or arranged by a unique variable (e.g. differential gene expression between cases and controls, etc.) The approach we are presenting here constitutes a novel and conceptually different proposal for the functional analysis of genomic experiments. It allows the simultaneous analysis of several variables, which can account for different properties of the genes.

This approach can detect interactions between these variables that account for functional roles dependent on several genomic properties or measurements.

We have used for this purpose a logistic model. It has recently been shown that the application of the logistic model to one single variable (differential gene expression in this case) produces results conceptually similar to the outcome of any conventional GSA method [41]. The aim here is not to improve the one dimensional detection of gene modules related to the measurement, but to look for gene modules that have complex dependences on several genomic variables or measurements. Thus, in the first example we show how some functional GO categories depend on particular combinations of their transcription rates and mRNA stabilities. Different strategies can be used by the cellular machinery to ensure, for example, a rapid activation or a long lasting of a particular team of genes that cannot be explained with only one variable. Thus, combinations of several variables (e.g. a rapid transcription rate and a low mRNA stability can be useful for a rapid release and a rapid deactivation of a transient function) are on the root of many biological processes. The variables used can be properties of the genes or can be also measurements of behaviours such as their expression in a given condition. In the second case example we have analyzed a combination of gene property (splicing index) and gene behaviour (differential gene expression). The MD-GSA was able of detecting biological processes that depend on combinations of both variables and would remain undetected if the variables were independently analyzed. Finally, we applied the same concept to the same type of measurement (differential gene expression) in two different but related scenarios: a case control of dermatitis and another case-control of psoriasis. In this example we were able of finding common and distinctive altered functionalities of both related diseases that remained otherwise undetected with the conventional one-dimensional GSA. The combination of measurements that can be studied under this framework and their biological relevance is unimaginable. Thus the relation of biological roles to combinations of different parameters of different types, such as gene intrinsic properties (e.g. mRNA stability), gene behaviours (e.g. level of expression) or gene states (e.g. methylation, SNPs, copy number), etc., can be easily be studied using this approach.

Summarizing, MD-GSA constitutes a novel approach to the functional profiling of genome scale experiments that paves the way for a higher level understanding of the behaviour of functional modules in the cell.

MATERIALS AND METHODS

Datasets and Data Preprocessing

Transcription Rates and mRNA Stabilities in Yeast

Genome-wide values for the transcription rates (TR) and mRNA stabilities (RS) of the genes of yeast used in the first sub-section of results can be found in the supplementary material of the manuscript by Garcia-Martinez et al. [23].

Gene Expression and Splicing Index.

Okoniewski & Miller [44] used exon arrays to compare breast cancer cell line MCF7 (fetal calf serum) to non tumorgenic breast epithelial cell line MCF10A (horse serum). They estimated differential gene expression using standard t-statistics and alternative splicing using the splicing index described in [30]. Since the splicing index is defined for each exon, we have used here median values to provide splicing measurements at a gene level. Thus, we have two numerical variables recorded for each gene in the study. The first one assesses the variation in the general expression level. The second one quantifies the change in splicing pattern of the gene, independently of its expression levels.

Differential Expression in Psoriasis and Dermatitis

Expression data from two separated case control experiments where combined in this analysis. The first experiment consisted of the comparison of lessional and non lessional skin samples in atopic dermatitis patients [32] (data were obtained from the GEO database, accession: GSE5667). The second experiment compared affected and unaffected skin in psoriatic patients[31] (GEO database, accession: GSE6710). Separated gene expression analyses of these two datasets were performed using standard methods: RMA algorithm [45] was used to normalize data within each of the experiments. The limma package [46] from Bioconductor [47] was used to estimate, separately for each of the studies, differential gene expression between diseased and non-diseased skin. Hence, two experimental measurements (limma t-statistics) where generated for each gene and used in the analysis: a first measurement of differential gene expression in dermatitis and a second measurement of differential gene expression in psoriasis.

Combined Analysis of Several Breast Cancer Genomic Measurements

Data used in the combined analysis of genomic measurements, in the results section, were taken from the supplementary material of [38]. SNP association to disease was measured using Odds Ratio (OR) of their corresponding minor allele frequencies. Then, the magnitude of the association of each gene to the disease was obtained as the value of association of the SNP more associated to the disease among all the SNPs mapping in the gene (or near the gene and being in linkage disequilibrium) [21], [38]. Differences in gene expression between tumour and normal breast tissues where estimated using t-statistics. Cox regression models where used to correlate survival time and gene expression, yielding a "prognosis" value for each gene (genes with "high" hazard ratios in the Cox model are associated to poor prognosis; genes with "low" hazard ratios associated to good prognosis). Another genomic measurements used was the average copy number for each gene in luminal B tumours, obtained from the hybridization intensity of the probesets corresponding to each gene (taken from the additional material of our study [38]).

Annotation Data

Functional modules are defined according the annotations of the GO [4] and the KEGG Pathway [48] repositories. Functional modules of more than 500 genes where considered to be too general to be informative so they where filtered out. Functional modules having less than 10 genes annotated to them where considered to be too small to be properly fitted by the multivariate logistic model and where also discarded.

Multi dimensional GSA (MD-GSA) using a logistic model that considers more than one variable

Logistic regression is a well established statistical methodology used to model the probability of occurrence of a binary event as a function of some other independent variables [49]. In the context of genomic studies, univariate logistic models have been shown to be suitable to perform gene set enrichment analysis [41].

Modelling functional class membership in terms of some measurement, X, of differential gene expression between two conditions as follows:

$$\ln\left(\frac{P(g \in F)}{P(g \notin F)}\right) = K + \alpha X \tag{1}$$

we can call the gene set F enriched in one of the conditions a significant estimate of the α coefficient is obtained [41].

In this paper we extend the use logistic models to perform a multidimensional gene set enrichment analysis. Our model describes the probability of a gene belonging to a functional class as a function of not one, but several experimental measurements. For two of those measurements the model will be as follows:

$$\ln\left(\frac{P(g\in F)}{P(g\notin F)}\right) = K + \alpha X + \beta Y + \gamma XY$$

(2)

where α and β are the main effects and γ is the interaction effect.

In a case-control study measuring, for instance, gene expression and genotype, we could model the probability of genes being annotated to a GO term as a function of both, differential gene expression (X) and allelic association to disease (Y).

Modelling not only the additive effects but also the interaction term, we accurately describe how the genes in a gene set are related to both measurements X and Y together, allowing for the detection of enrichment patterns which will remain unnoticed in two independent univariate analyses.

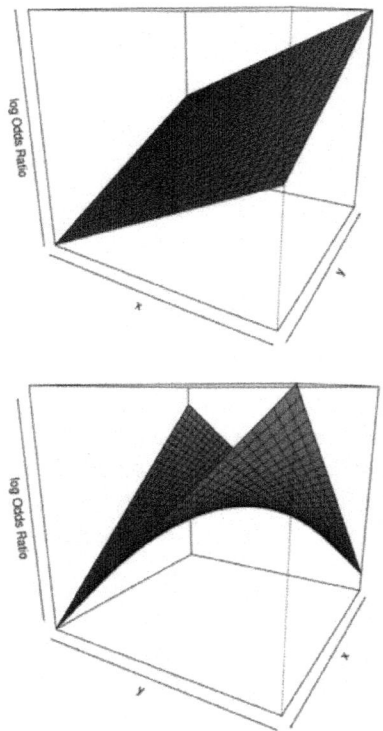

Figure 3: Surfaces described by the logistic model.

The model in equation (2) describes the log odds ratio of a gene being annotated to functional module F as a function of two variables, X and Y. The shape of this surface when embedded in a 3D space is that of a plane if the interaction coefficient γ is zero (Figure 3, top), or a hyperbolic paraboloid, also called saddle surface, when the estimate of γ is different from zero (Figure 3, bottom). Hence, from the sign and significance of the fitted coefficients, we can find the direction in the two dimensional space XY in which the genes annotated to the function F are more likely to be found.

The surface described by the logistic model is a plane when the interaction term (γ) is 0 (top) and a hyperbolic paraboloid when the interaction term (γ) is not zero (bottom).

When γ is zero the sign of the coefficients α and β describe the slopes of the plane and therefore, the direction towards which the probability of genes being annotated is greater.Figure 4 describes the areas where genes belonging to a functional module are more likely to be found, depending on the estimated α and β coefficients of the logistic model (2) and provided that the estimate of γ is not significantly different from zero.

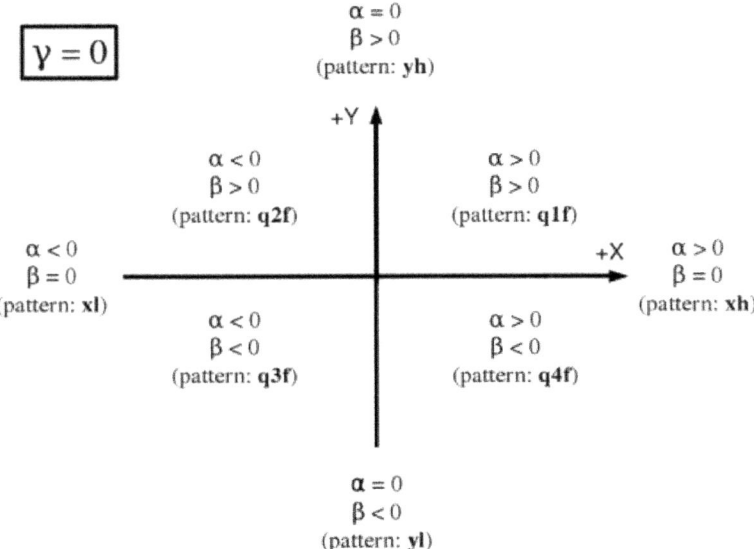

Figure 4: Location of the areas where genes are more likely to be annotated to the function F depending on the coefficients of the fitted model.

When γ=0 the fitted surface is a plane which slope grows towards the area.

When γ is different from zero the interaction dominates the growth of the log odds ratio while the saddle point in the surface has the coordinates $(-\beta/\gamma, -\alpha/\gamma)$. If for instance, for a particular functional module F, all estimated coefficients are positive, then, the saddle point of the hyperbolic paraboloid will be in the third quadrant and the surface will grow to the infinite in the first quadrant. As the surface represents how likely we are to find genes annotated to module F in the plane XY, we will conclude that the module F is located towards the firs quadrant. Moreover, as the interaction effect is positive we know that the evidence of this localization is greater than the one we will get from separated analysis of each one of the dimensions X and Y on their own (following equation 1). Then, biological interpretation can be done recalling the meaning of the X and Y quantities. Figure 5 (top) describes the areas where genes belonging to a functional module are more likely to be found, depending on the estimates of α, β and γ and when γ is estimated to be different from zero.

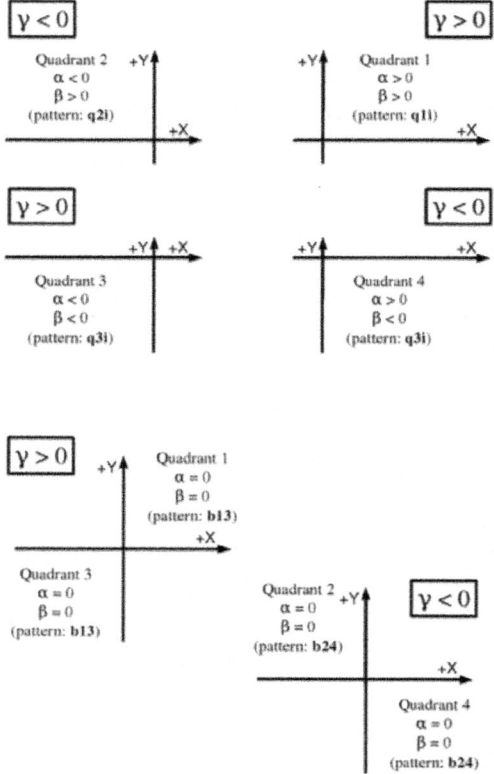

Figure 5: Location of the areas where genes are more likely to be annotated to the function F depending on the coefficients of the fitted model.

If $\gamma \neq 0$ the fitted surface is a hyperbolic paraboloid, when $\alpha \neq 0$ and $\beta \neq 0$ (top part) the most likely area to find genes annotated to F is the quadrant opposite to the saddle point of the surface. When $\alpha = 0$ and $\beta = 0$ (bottom part) the saddle point of the surface is in the (0,0) and the genes annotated to the function F are more likely to be found in two opposite quadrants, reflecting the bimodality of the function F.

If it was the case that just the interaction coefficient γ would be different from zero, then the saddle point will be the (0, 0) and the genes annotated to functional module F will be allocated to opposite quadrants of the XY space; the first and the third quadrant if $\gamma > 0$; the second and the fourth quadrants if $\gamma < 0$. In this latest case we will call the functional module F bimodal and the biological interpretation will be that, genes in F are effectively spited up in two groups of opposite patterns. Figure 5 (bottom) describes the areas where genes belonging to a functional module are more likely to be found, if the estimates of α and β are zero.

Table 6 shows how to interpret all possible combinations of α, β and γ estimates.

Table 6: Interpretation of all relevant combinations of α, β and γ estimates

α	β	γ	pattern identifier	pattern	description
+	+	+	q1i	Quadrant 1 with interaction	F is allocated towards one of the quadrants and the evidence is greater than just the additive evidences from the univariate analysis.
+	0	+			
0	+	+			
-	-	+	q3i	Quadrant 3 with interaction	
-	0	+			
0	-	+			
-	+	-	q2i	Quadrant 2 with interaction	
-	0	-			
0	+	-			
+	-	-	q4i	Quadrant 4 with interaction	
+	0	-			
0	-	-			
0	0	+	b13	Bimodal + (quadrants 1 and 3)	F is split in two opposite quadrants.
0	0	+	b24	Bimodal − (quadrants 2 and 4)	
+	+	0	q1f	Quadrant 1 flat	F is allocated towards one of the quadrants and the evidence is similar to the additive evidences from the univariate analysis.
-	-	0	q3f	Quadrant 3 flat	
-	+	0	q2f	Quadrant 2 flat	
+	-	0	q4f	Quadrant 4 flat	
+	0	0	xh	X high (+) values	F is enriched just in the first condition.
-	0	0	xl	X low (−) values	
0	+	0	yh	Y high (+) values	F is enriched just in the second condition.
0	-	0	yl	Y low (−) values	

doi:10.1371/journal.pone.0010348.t006

Wald statistics to test the main effect coefficients and the interaction effects [41]. Other approaches like likelihood ratio tests could also have been used.

As one logistic regression model needs to be fit for each functional module in the analysis, multiple testing occurs and p-value correction must be performed. In this paper we use Benjamini and Hochberg [50] approach to correct all p-values of the same parameter of the model α, β or γ.

Implementation

The proposed algorithm has been implemented as an R library available athttp://bioinfo.cipf.es/supplementary/multidimensional_GSA, released under the GPL license.

AUTHOR CONTRIBUTIONS

Conceived and designed the experiments: DM JD. Performed the experiments: DM. Analyzed the data: DM. Wrote the paper: JD.

REFERENCES

1. Dopazo J (2009) Formulating and testing hypotheses in functional genomics. Artif Intell Med 45: 97–107.

2. Huang DW, Sherman BT, Lempicki RA (2008) Bioinformatics enrichment tools: paths toward the comprehensive functional analysis of large gene lists. Nucleic Acids Res 37: 1–13.

3. Hartwell LH, Hopfield JJ, Leibler S, Murray AW (1999) From molecular to modular cell biology. Nature 402: C47–52.

4. Ashburner M, Ball CA, Blake JA, Botstein D, Butler H, et al. (2000) Gene ontology: tool for the unification of biology. The Gene Ontology Consortium. Nat Genet 25: 25–29.

5. Kanehisa M, Goto S, Kawashima S, Okuno Y, Hattori M (2004) The KEGG resource for deciphering the genome. Nucleic Acids Res 32: D277–280.

6. Vastrik I, D'Eustachio P, Schmidt E, Joshi-Tope G, Gopinath G, et al. (2007) Reactome: a knowledge base of biologic pathways and processes. Genome Biol 8: R39.

7. Dopazo J (2006) Functional interpretation of microarray experiments. Omics 10: 398–410.

8. Khatri P, Draghici S (2005) Ontological analysis of gene expression data: current tools, limitations, and open problems. Bioinformatics 21: 3587–3595.

9. Mootha VK, Lindgren CM, Eriksson KF, Subramanian A, Sihag S, et al. (2003) PGC-1alpha-responsive genes involved in oxidative

phosphorylation are coordinately downregulated in human diabetes. Nat Genet 34: 267–273.

10. Goeman JJ, Buhlmann P (2007) Analyzing gene expression data in terms of gene sets: methodological issues. Bioinformatics 23: 980–987.

11. Al-Shahrour F, Diaz-Uriarte R, Dopazo J (2005) Discovering molecular functions significantly related to phenotypes by combining gene expression data and biological information. Bioinformatics 21: 2988–2993.

12. Goeman JJ, van de Geer SA, de Kort F, van Houwelingen HC (2004) A global test for groups of genes: testing association with a clinical outcome. Bioinformatics 20: 93–99.

13. Subramanian A, Tamayo P, Mootha VK, Mukherjee S, Ebert BL, et al. (2005) Gene set enrichment analysis: a knowledge-based approach for interpreting genome-wide expression profiles. Proc Natl Acad Sci U S A 102: 15545–15550.

14. Clark AG, Eisen MB, Smith DR, Bergman CM, Oliver B, et al. (2007) Evolution of genes and genomes on the Drosophila phylogeny. Nature 450: 203–218.

15. Kim SY, Volsky DJ (2005) PAGE: parametric analysis of gene set enrichment. BMC Bioinformatics 6: 144.

16. Kitano H (2004) Cancer as a robust system: implications for anticancer therapy. Nat Rev Cancer 4: 227–235.

17. Bentink S, Wessendorf S, Schwaenen C, Rosolowski M, Klapper W, et al. (2008) Pathway activation patterns in diffuse large B-cell lymphomas. Leukemia 22: 1746–1754.

18. Bardelli A, Velculescu VE (2005) Mutational analysis of gene families in human cancer. Curr Opin Genet Dev 15: 5–12.

19. Al-Shahrour F, Arbiza L, Dopazo H, Huerta-Cepas J, Minguez P, et al. (2007) From genes to functional classes in the study of biological systems. BMC Bioinformatics 8: 114.

20. Wu C, Delano DL, Mitro N, Su SV, Janes J, et al. (2008) Gene set enrichment in eQTL data identifies novel annotations and pathway regulators. PLoS Genet 4: e1000070.

21. Medina I, Montaner D, Bonifaci N, Pujana MA, Carbonell J, et al. (2009) Gene set-based analysis of polymorphisms: finding pathways or biological processes associated to traits in genome-wide association studies. Nucleic Acids Res 37: W340–344.

22. McLendon R, Friedman A, Bigner D, Van Meir EG, Brat DJ, et al. (2008) Comprehensive genomic characterization defines human glioblastoma genes and core pathways. Nature.

23. Garcia-Martinez J, Gonzalez-Candelas F, Perez-Ortin JE (2007) Common gene expression strategies revealed by genome-wide analysis in yeast. Genome Biol 8: R222.

24. Perez-Ortin JE (2007) Genomics of mRNA turnover. Brief Funct Genomic Proteomic 6: 282–291.

25. Al-Shahrour F, Diaz-Uriarte R, Dopazo J (2004) FatiGO: a web tool for finding significant associations of Gene Ontology terms with groups of genes. Bioinformatics 20: 578–580.

26. Johnson JM, Castle J, Garrett-Engele P, Kan Z, Loerch PM, et al. (2003) Genome-wide survey of human alternative pre-mRNA splicing with exon junction microarrays. Science 302: 2141–2144.

27. Faustino NA, Cooper TA (2003) Pre-mRNA splicing and human disease. Genes Dev 17: 419–437.

28. Srinivasan K, Shiue L, Hayes JD, Centers R, Fitzwater S, et al. (2005) Detection and measurement of alternative splicing using splicing-sensitive microarrays. Methods 37: 345–359.

29. Bitton DA, Okoniewski MJ, Connolly Y, Miller CJ (2008) Exon level integration of proteomics and microarray data. BMC Bioinformatics 9: 118.

30. Clark TA, Schweitzer AC, Chen TX, Staples MK, Lu G, et al. (2007) Discovery of tissue-specific exons using comprehensive human exon microarrays. Genome Biol 8: R64.

31. Reischl J, Schwenke S, Beekman JM, Mrowietz U, Sturzebecher S, et al. (2007) Increased expression of Wnt5a in psoriatic plaques. J Invest Dermatol 127: 163–169.

32. Plager DA, Leontovich AA, Henke SA, Davis MD, McEvoy MT, et al. (2007) Early cutaneous gene transcription changes in adult atopic dermatitis and potential clinical implications. Exp Dermatol 16: 28–36.

33. Wood LD, Parsons DW, Jones S, Lin J, Sjoblom T, et al. (2007) The genomic landscapes of human breast and colorectal cancers. Science 318: 1108–1113.

34. Pinkel D, Albertson DG (2005) Array comparative genomic hybridization and its applications in cancer. Nat Genet 37: SupplS11–17.

35. Bignell GR, Huang J, Greshock J, Watt S, Butler A, et al. (2004) High-resolution analysis of DNA copy number using oligonucleotide microarrays. Genome Res 14: 287–295.

36. Peiffer DA, Le JM, Steemers FJ, Chang W, Jenniges T, et al. (2006) High-resolution genomic profiling of chromosomal aberrations using Infinium whole-genome genotyping. Genome Res 16: 1136–1148.

37. Easton DF, Pooley KA, Dunning AM, Pharoah PD, Thompson D, et al. (2007) Genome-wide association study identifies novel breast cancer susceptibility loci. Nature 447: 1087–1093.

38. Bonifaci N, Berenguer A, Diez J, Reina O, Medina I, et al. (2008) Biological processes, properties and molecular wiring diagrams of candidate low-penetrance breast cancer susceptibility genes. BMC Med Genomics 1: 62.

39. Hunter DJ, Kraft P, Jacobs KB, Cox DG, Yeager M, et al. (2007) A genome-wide association study identifies alleles in FGFR2 associated with risk of sporadic postmenopausal breast cancer. Nat Genet 39: 870–874.

40. Singh RK, Rinehart CA, Kim JP, Tolleson-Rinehart S, Lawing LF, et al. (1996) Tumor cell invasion of basement membrane in vitro is regulated by amino acids. Cancer Invest 14: 6–18.

41. Sartor MA, Leikauf GD, Medvedovic M (2008) LRpath: A logistic regression approach for identifying enriched biological groups in gene expression data. Bioinformatics 25: 211–217.

42. Montaner D, Minguez P, Al-Shahrour F, Dopazo J (2009) Gene set internal coherence in the context of functional profiling. BMC Genomics 10: 197.

43. Pavlidis P, Qin J, Arango V, Mann JJ, Sibille E (2004) Using the gene ontology for microarray data mining: a comparison of methods and application to age effects in human prefrontal cortex. Neurochem Res 29: 1213–1222.

44. Okoniewski MJ, Miller CJ (2008) Comprehensive analysis of affymetrix exon arrays using BioConductor. PLoS Comput Biol 4: e6.

45. Irizarry RA, Hobbs B, Collin F, Beazer-Barclay YD, Antonellis KJ, et al. (2003) Exploration, normalization, and summaries of high density oligonucleotide array probe level data. Biostatistics 4: 249–264.

46. Smyth GK (2004) Linear models and empirical bayes methods for assessing differential expression in microarray experiments. Stat Appl Genet Mol Biol 3: Article3.

47. Gentleman RC, Carey VJ, Bates DM, Bolstad B, Dettling M, et al. (2004) Bioconductor: open software development for computational biology and bioinformatics. Genome Biol 5: R80.

48. Kanehisa M, Araki M, Goto S, Hattori M, Hirakawa M, et al. (2008) KEGG for linking genomes to life and the environment. Nucleic Acids Res 36: D480–484.

49. Agresti A (2002) Categorical data analysis. Hoboken, New Jersey: John Wiley and Sons.

50. Benjamini Y, Hochberg Y (1995) Controlling the false discovery rate: a practical and powerful approach to multiple testing. Journal of the Royal Statistical Society Series B 57: 289–300.

Chapter 3

COMPARATIVE GENOMICS OF MYCOPLASMA: ANALYSIS OF CONSERVED ESSENTIAL GENES AND DIVERSITY OF THE PANGENOME

Wei Liu[1] , Liurong Fang[1] , Mao Li[1] , Sha Li[1] , Shaohua Guo[1] , Rui Luo[1] , Zhixin Feng[2] , Bin Li[2] , Zhemin Zhou[3] , Guoqing Shao[2] , Huanchun Chen[1] , Shaobo Xiao[1]

[1] Division of Animal Infectious Diseases, State Key Laboratory of Agricultural Microbiology, College of Veterinary Medicine, Huazhong Agricultural University, Wuhan, People's Republic of China

[2] Institute of Veterinary Medicine, Jiangsu Academy of Agricultural Sciences, Nanjing, People's Republic of China

[3] Environmental Research Institute, University College Cork, Cork, Ireland

ABSTRACT

Mycoplasma, the smallest self-replicating organism with a minimal metabolism and little genomic redundancy, is expected to be a close approximation to the minimal set of genes needed to sustain bacterial life. This study employs comparative evolutionary analysis of twenty*Mycoplasma* genomes to gain an improved understanding of essential genes. By analyzing the core genome of mycoplasmas, we finally revealed the conserved essential genes set for mycoplasma survival. Further analysis showed that the core genome set has many characteristics in common with experimentally identified essential genes. Several key genes, which are related to DNA replication and repair and can be disrupted in transposon mutagenesis studies, may be critical for bacteria survival especially over long period natural selection. Phylogenomic reconstructions based on 3,355 homologous groups allowed robust estimation of phylogenetic relatedness among mycoplasma strains. To obtain deeper insight into the relative roles of molecular evolution in pathogen adaptation to their hosts, we also analyzed the positive selection pressures on particular sites and lineages. There appears to be an approximate correlation between the divergence of species and the level of positive selection detected in corresponding lineages.

INTRODUCTION

Mycoplasmas are widespread in nature as parasites of humans, mammals, reptiles, fish, arthropods, and plants [1]. As a conditional pathogenic organism, it associates with various diseases, including pneumonia, arthritis, meningitis and chronic urogenital tract disease [2]. Although they are the smallest self-replicating organisms, both commensal forms and pathogenic forms are diverse. With a minimal metabolism and little genomic redundancy, the genome of *Mycoplasma* is expected to be a close approximation to the minimal set of genes needed to sustain bacterial life [3]. An early projection proposed a minimal gene set composed of 206 genes based on the analysis of eight free-living and endosymbiotic bacterial genomes[4]. More recently, Glass *et al.* [5] performed a global transposon mutagenesis study and identified 100 putatively nonessential genes in *M. genitalium*. Logically, the remaining 387 genes presumably constitute the set of essential genes. However, these data greatly exceed theoretical projections of how many genes comprise a minimal genome, as proposed by Gil *et al.* [4].

Natural selection leads to the fixation of essential genes and can delete nonessential genes in a wide range of species [6], [7]. This process is similar, but more robust than manual mutagenesis studies. Through long term evolution form a more conventional progenitor in the *Firmicutes* taxon [8], *Mycoplasmas* have undergone a process of massive genome reduction[1]. These wall-less bacteria are obligate parasites that live in relatively unchanging niches requiring little adaptive capability. *M. genitalium*, a human urogenital pathogen, is the extreme manifestation of this genomic parsimony, having only 482 protein-coding genes and the smallest genome of any known free-living organism capable of being grown in axenic culture[9]. Thus, with little genomic redundancy and contingencies for different environmental conditions, *Mycoplasmas* are regarded as optimal microbes to perform genes essentiality studies [5]. Along with the burgeoning increase in *Mycoplasma* genome sequence data, this would appear to be the right time to explore gene essentiality from a comparative genomics perspective.

An enormous genetic diversity exists in the mycoplasmas, yet how much diversity is functional, and what are the important adaptations that serve to partition species into different niches? *M. hyopneumoniae* and *M. hyorhinis* are the causal agents of swine mycoplasmosis. The former causes a mild, chronic pneumonia of swine and results in deactivation of mucociliary functions[10]. This agent is infective for a single host species, but the mechanisms of host specificity are unknown. The latter is responsible for respiratory tract and arthritis disease in swine [1]. *M. hyorhinis* is generally considered a swine pathogen, yet is most commonly infect laboratory cell lines, implying that it can thrive among different species of cell lines [11]. A strong link between *M. hyorhinis* and human cancer was reported recently by Huang *et al.* [12], who used a monoclonal antibody against the unique *M. hyorhinis*–specific protein p37 to detect mycoplasma in over 600 carcinoma tissues from a variety of organs. The study indicated that up to 56% of gastric carcinoma and 55% of colon carcinoma biopsies were positive for *M. hyorhinis* [12]. With a similar genome size, *M. hyorhinis* and *M. hyopneumoniae* exhibit high levels of functional diversity. Interest has therefore shifted to questions of why *M. hyorhinis* can thrive among different species of cell lines.

This paper communicates the results of three major analyses. In the first analysis, we present the details of a comparative analysis of twenty *Mycoplasma* strains and investigate the conserved essential genes set for mycoplasma survival. For the second analysis, phylogenomic reconstructions based on 3,355 homologous groups allows robust estimation of phylogenetic relatedness among mycoplasma strains. The third analysis employs the branch-site method to assess positive selection pressures on particular sites and lineages. There appears to be an approximate correlation between the divergence of species and the level of positive selection detected in corresponding lineages.

RESULTS

Diversity of *Mycoplasmataceae* Family: Core Genome *vs.* Flexible Gene Pool

The number of protein coding genes per genome within the various strains and species of mycoplasmas is relatively similar (ranging from 475 to 1,037; Table 1), but the gene composition of these genomes is much more variable. Based on the gene content table (obtained as described in Materials and Methods; Table S1), three *M. hyopneumoniae* strains share about 95% of their genes, and three different species of mycoplasmas share only around 71% of their genes (Figure 1). This latter result appears to be independent of the particular strains or species involved in the comparison. Even with the inclusion of 20 genomes, the pan-genome size of *Mycoplasmas* appears not to be determined, and we estimate that the size probably surpasses 8,000 genes. This huge pan-genome size may be a reflection of their different lifestyles in distinct ecological niches. Within species, the pan-genome size also remains uncertain, although our estimates suggest that the pan-genome size of *M. hyopneumoniae* is smaller.

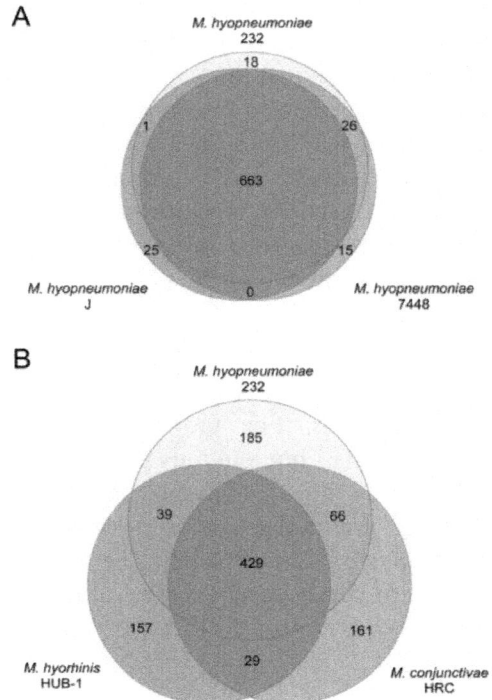

Figure 1: Venn diagram for two sets of three taxa.

Above are taxa of the same species and below are taxa of different species. The surfaces are approximately proportional to the number of genes.

Table 1: Bacterial Strains Used in This Study

Mollicute strains	Host	CDS	Genome size (bp)	Accession	Citation
U. urealyticum serovar 10 str. ATCC 33699	Human	646	874478	CP001184	-
U. parvum serovar 3 str. ATCC 700970	Human	611	751719	AF222894	[45]
U. parvum serovar 3 str. ATCC 27815	Human	609	751679	CP000942	-
M. synoviae 53	Bird	659	799476	NC_007294	[46]
M. pulmonis UAB CTIP	Rodent	-	963879	AL445566	[47]
M. pneumoniae M129	Human	688	816394	U00089	[48]
M. penetrans HF-2	Human	1037	1358633	BA000026	[49]
M. mycoides subsp. mycoides SC str. PG1	Ruminant	1016	1211703	BX293980	[50]
M. mobile 163K	Fish	635	777079	AE017308	[51]
M. hyorhinis HUB-1	Swine	658	839615	NC_014448	[52]
M. hyopneumoniae J	Swine	657	897405	NC_007295	[46]
M. hyopneumoniae 7448	Swine	657	920079	NC_007332	[46]
M. hyopneumoniae 232	Swine	691	892758	NC_006360	[53]
M. genitalium G37	Human	475	580076	NC_000908	[5]
M. gallisepticum str. R(low)	Bird	763	1012800	AE015450	[54]
M. crocodyli MP145	Crocodile	689	934379	CP001991	-
M. conjunctivae HRC	Sheep and goats	696	846214	FM864216	[55]
M. capricolum subsp. ATCC 27343	Ruminant	812	1010023	CP000123	-
M. arthritidis 158L3-1	Rats and mice	631	820453	NC_011025	[56]
M. agalactiae PG2	Sheep and goats	759	877438	CU179680	-

doi:10.1371/journal.pone.0035698.t001

The extent of the pan-genome is opposed to the core. Genes that are in common between the different species within the family *Mycoplasmataceae* comprised our core genome - the set of orthologous genes determined the common properties of this family. In this work, the tribeMCL program was used to cluster orthologous genes, and a total of 13,654 predicted proteins were grouped into 3,355 clusters, each cluster representing a group of putative orthologs. The 1,481 genes that are present in single genomes (Figure 2) represent lineage specific genes. In addition, the 196 genes shared by all the 20 strains comprised our core genome (Figure 3).

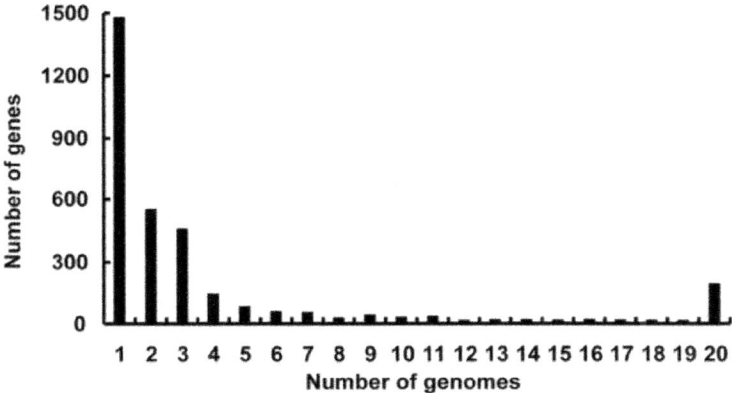

Figure 2: Frequency of genes within the 20 genomes included in this analysis.

Genes present in a single genome represent lineage specific genes, while at the opposite end of the scale, genes found in all 20 genomes represent the *Mycoplasma* core genome.

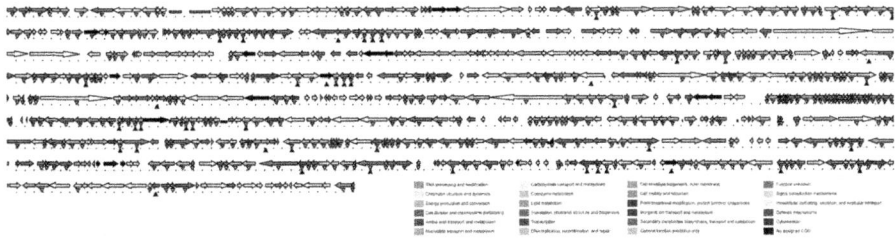

Figure 3: Viable core and pan genes distribution in the *M. hyorhinis* HUB-1 map.

The *M. hyorhinis* HUB-1 genome is shown at a scale of 100 kb per line. Colored arrows above the line indicate annotated genes. Genes are colored according to their functional category, as indicated in the key. rRNAs and tRNAs are denoted by red and purple rectangles, respectively. Red triangles above the line represent the core genes of*Mycoplasmataceae*. Blue triangles above the line represent the genes shared by all the species in clade III, but which are absent from other lineages. Black triangles below the line represent the disrupted genes documented in *M. genitalium* G37, mapped onto the*M. hyorhinis* HUB-1 genome [5].

Functional Characterization of the Core Genome

The use of the core genome concept has led to important insights into the evolution of bacterial species and identification of potentially important novel genes [13]. In terms of functional assignments according to COGs, almost half (42.3%) of the proteins observed from the core genome are devoted to translation, ribosomal structure, and biogenesis (Figure S1). Our results support the analysis of Ouzounis and Kyrpides [14], who demonstrated that genetic processes such as translation are conserved and close to the original form. Strikingly, 10.6% of the observed core genes have resisted functional assignments according to COGs classification (Table S2), which highlights the need for better functional characterization of these genes. Furthermore, by comparing functional categories of the core genome with the categories of the genome of *M. hyorhinis* HUB-1, we noticed that a large array of proteins devoted to amino acid, carbohydrate transport and metabolism, as well as defense mechanisms, were sharply reduced. Our results support the analysis of Fraser *et al.* [9] and Himmelreich *et al.* [15], who demonstrated that both *M. genitalium* and *M. pneumoniae* lost all the genes involved in amino acid synthesis, and their survival is totally dependent on an exogenous supply of the complete spectrum of amino acids. Beyond this, the pronounced reduction of those functional categories observed in the core genome might be the further genetic evidence for gene loss in *M. genitalium* and *M. pneumonia* [9], [15].

Persistent Nonessential Genes *vs.* Essential Genes

Identification of the core genome has important implications for a broad range of microbiological applications, such as determining the essentiality of genes derived from the core genome and deriving traits that correspond to a common ancestor (orthology) [4]. In this work, we classified the core genome into two classes according to persistence and essentiality: persistent nonessential genes and conserved essential genes (Figure 3). Glass *et al.* [5]performed a global transposon mutagenesis study and identified 100 putatively nonessential genes in *M. genitalium*. We mapped those nonessential genes onto the *M. hyorhinis* HUB-1 genome and 24 of them were persistent among *Mycoplasma* genomes.

Focusing on gene persistence, the essentiality of a gene is relative to a set of experimental conditions. It is quite different for a cell to survive in a laboratory setting, with plenty of supplied metabolites, compared to thriving in the wild, where it competes with other organisms for limited resources. Starvation or stresses are omnipresent, and the fitness effect of persistent genes may be essential for survival under transition from one environmental condition to another [16]. After transposon mutagenesis, disrupted genes may not be essential for growth in rich media, but their loss may lead to such a low fitness that its deletion will never be fixed in natural populations [16]. For example, Glass *et al.* [5] isolated six mutants involved in recombination and DNA repair: *recA*, *recU*, Holliday junction DNA helicases *uvrA* and *uvrB*, formamidopyrimidine-DNA glycosylase *mutM*, which excises oxidized urines from DNA, and a likely DNA damage inducible protein gene. Interestingly, we noticed that these six disrupted genes occur in the core genome set, which suggests that these disrupted persistent genes might be critical for growth in variable environments over long periods. The survival of these mutants is probably due to their tolerances of IS element within a relatively short period. According to Glass *et al.*, these six mutants grew more poorly after repeated passage, which indicates that those genes are also critical for bacterial survival [5]. Our results in this regard generally agree with the analysis by Glass *et al.* This analysis prompted us to explore gene essentiality by combining both the experimental approaches and comparative genomics analysis.

Genome-Based Reconstruction of *Mycoplasmataceae* Phylogeny

We constructed robust phylogenies for the family *Mycoplasmataceae* based on whole genome analysis. The supertree contains five major, distinct clades (Figure 4). In clade I, three*Ureaplasma* strains with a common host are clustered on a single branch. In clade II, six mycoplasma species (*M. pulmonis* etc.) with various lifestyles formed the cluster. In clade III, the strain HRC appears adjacent to strain HUB-1 and three isolates of *M. hyopneumoniae*strains are clustered closely, indicative of a recent common ancestry. In clade IV, *M. capricolum*and *M. mycoides* are clustered closely on a single branch and both of them are the agents of ruminant mycoplasmosis. In clade V, *M. genitalium* and *M. pneumoniae* are closely related, together with *M. penetrans* and *M. gallisepticum*, forming the cluster.

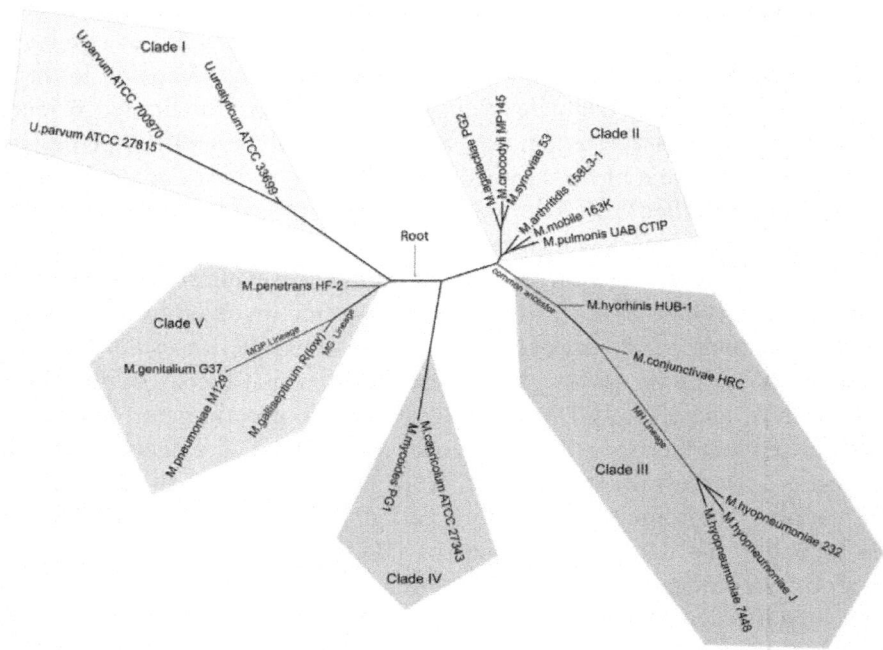

Figure 4: Phylogenetic tree of *Mycoplasmataceae*.

The phylogenetic relationship was estimated and tested in one thousand bootstrap samples using TREE-PUZZLE version 5.2 with a BIONJ model (see Materials and Methods). This supertree shows five major distinct clades. The four lineages that were used as foreground in the branch-site model positive selection test are highlighted in red.

Although single gene trees have been used extensively to estimate the species tree, evidence has shown that single gene trees may have particular difficulty in representing prokaryotic species phylogeny, because lateral gene transfer (LGT) occurs among prokaryotic genomes, and LGT may obscure the phylogenetic signal [17]. In this work, the 16S rRNA-based phylogeny has also been reconstructed for the family *Mycoplasmataceae* (Figure 5), which shows almost the same topology as the supertree. However, these two trees differ in the placement of *M. penetrans* HF-2 and *M. mobile* 163.

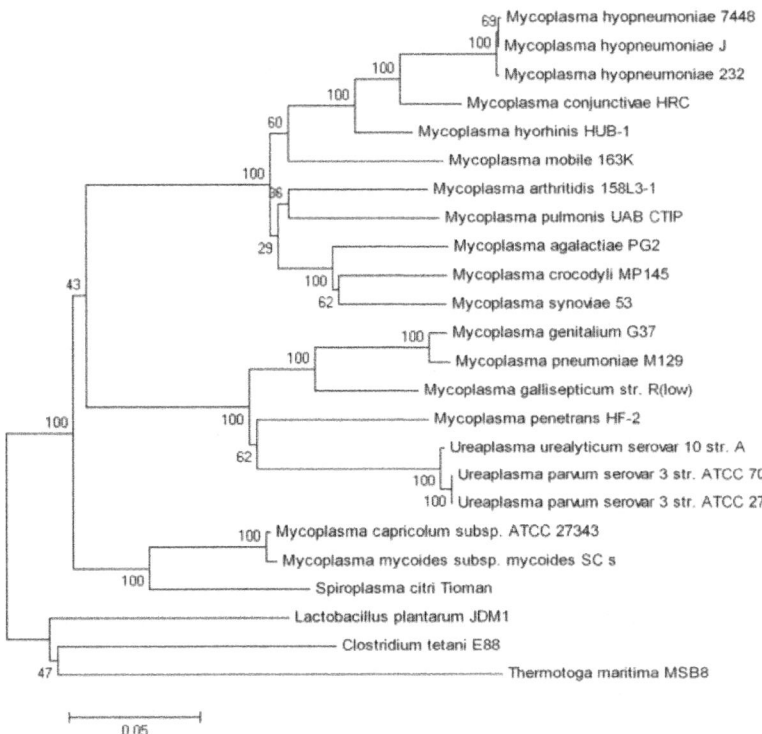

Figure 5: 16S rRNA tree of *Mycoplasmataceae*.

This consensus tree of 100 bootstrap replications was constructed based on 16S rRNA sequences using the Neighbor Joining (NJ) method implemented in MEGA 4.1. The bootstrap values are marked at the root of each branch.

A recent study proposed that the "Tree of Life" may be resolved by concatenation of 31 orthologs occurring in 191 species [18], and an analogous approach has been applied to infer*Mycoplasmas* phylogeny by random concatenation of 91 protein sequences shared by 16*mollicutes* [19]. The placement of *M. penetrans* in the concatenation-based phylogeny is consistent with the supertree (Figure 4).

Positive Selection Analysis

We employed the branch-site test of Yang and Nielsen [20], [21] implemented in the program HY-PHY (http://www.datam0nk3y.org/hyphy/doku.php) to assess positive selection [22], [23] at particular sites in particular lineages. This method compares synonymous and nonsynonymous substitution rates in protein coding genes and regards a nonsynonymous rate elevated above the synonymous rate as evidence for positive selection. For the *Mycoplasma* data set, both swine-infecting and human-infecting lineages were tested. The branches we tested are highlighted in red on Figure 4.

In the four swine-infected lineages (HUB-1 lineage, MC lineage, MH lineage, and the common ancestor of clade III), 661 orthologous groups shared by all species in clade III were tested. A total of 23 genes were identified to be under positive selection (Table 2). These genes were assigned to functional categories according to the COG database. We found that a large fraction of the genes subject to positive selection were connected to DNA replication, recombination, and repair. Successful genome replication is essential for growth and survival of an organism, and polymerase complexes often fail to complete this task [24]. Also, replication is thought to contribute to proliferation and efficiency of the colonization of hostile environments[25]. Interestingly, we detected that positive selection occurs in the both of the *dnaA* and *dnaN*genes, which comprise the *oriC* region in *M. hyorhinis*, and in several copies of proteins connected to replication, recombination and repair in both MC lineage and the common ancestor of clade III. Selection pressure on these genes could reflect constraints on efficient genome replication during colonization and proliferation in the hostile environment of the host[25].

Table 2: Genes under Positive Selection in Swine-infecting Lineages

Lineage	Gene	dN/dS[a]	Sequence %	COG(s)	Product
MH Lineage	mhp623	543.871	16.60%	COG1744R	ABC transporter
	mhp388	306.029	5.00%	COG3037S	ascorbate-specific PTS system enzyme IIC
	mhp603	4.42	0.00%	COG0195K	transcription elongation factor NusA
	mhp368	4.771	9.70%	COG0531E	putative membrane lipoprotein
	mhp595	515.577	13.40%	COG0266L	formamidopyrimidine-DNA glycosylase
	mhp480	1913.82	15.20%	-	hypothetical protein
MC Lineage	MCJ_005740	525.955	9.30%	COG1196D	putative ABC transporter ATP-binding protein P
	MCJ_003040	547.097	7.90%	COG0013J	alanyl-tRNA synthetase
	MCJ_007160	227.357	8.10%	COG2176L	DNA polymerase III PolC
	MCJ_002410	488.408	7.20%	COG4608E	oligopeptide ABC transporter ATP-binding protein
	MCJ_000340	316.719	5.10%	COG0556L	exinuclease ABC subunit B
HUB-1 Lineage	MHR_0001	2.122	7.10%	COG0593L	Chromosomal replication initiator protein dnaA
	MHR_0002	786.654	25.60%	COG0592L	DNA polymerase III beta subunit
	MHR_0009	8.155	5.50%	COG0525J	Valyl tRNA synthetase
	MHR_0132	306.824	7.30%	COG0060J	Isoleucyl tRNA synthetase
	MHR_0148	732.204	14.30%	COG0006E	Xaa-pro aminopeptidase
	MHR_0248	2.637	3.60%	COG0187L	DNA gyrase subunit B
	MHR_0318	180.115	16.40%	-	ABC transporter permease protein
	MHR_0377	6.345	9.70%	COG0202K	DNA-directed RNA polymerase subunit alpha
	MHR_0443	33.397	7.70%	COG0178L	Excinuclease ATPase subunit-like protein
	MHR_0609	14.416	16.70%	COG05440	Trigger factor
Common Ancester[b]	MHR_0128	133.038	23.40%	COG0322L	Excinuclease ABC subunit C
	MHR_0131	33.679	13.50%	COG0188L	Topoisomerase IV subunit A
	MHR_0310	754.512	15.80%	COG2274V	ABC transporter ATP-binding and permease protein
	MHR_0486	58.184	12.90%	COG0587L	DNA polymerase III alpha subunit
	MHR_0489	3.669	29.10%	COG0532J	Translation initiation factor IF-2
	MHR_0639	8298.46	31.60%	-	Lipoprotein
	MHR_0363	6343.51	14.00%	COG1164E	Oligoendopeptidase F
	MHR_0356	6343.51	14.00%	COG1164E	Oligoendopeptidase F

[a]Ratio of the nonsynonymous to the synonomous mutation rate (dN/dS) measures the strength of selection, where values >1 indicate positive selection, and larger values indicate stronger selection.
[b]In the common ancestor lineage: a single gene of *M. hyorhinis* HUB-1 was used to represent each ortholog group (Table S1). Genes of *M. conjunctivae* HRC and three *M. hyopneumoniae* strains in the same ortholog group are also under positive selection.
doi:10.1371/journal.pone.0035698.t002

In the case of human-infecting lineages, the lineage that stood out from the rest with regard to host specificity was *M. gallisepticum*, which is significantly associated with chronic respiratory disease in chickens [26]. Not surprisingly, this lineage was identified to be under the strongest selection pressure in clade V (Table S3). However, we failed to notice any selection pressure on the *dnaA* and *dnaN* genes in the MG lineage. A large number of genes related to DNA repair, RNA processing, Amino acid transport and metabolism were found under positive selection. Selection pressure on these genes may facilitate evolutionary flexibility in the MG lineage, hence its ability to adapt to new environments.

Discussion

Numerous global transposon mutagenesis studies of minimal genomes have been performed to identify essential genes [5], [27]. Long-term natural selection can also delete nonessential genes in a wide range of species [6], [7], which is similar to, but more robust than, manual mutagenesis studied. With a minimal metabolism and little genomic redundancy, mycoplasmas are regarded as optimal microbes for the identification of essential genes [1]. It is believed that the *Mycoplasmas* evolved from a more conventional progenitor in the *Firmicutes* taxon by a process of massive genome reduction [16]. We found that each of the species in*Mycoplasmataceae* has undergone a similar process. These species have undergone various natural selection pressures in different environments. Most of the core genes remaining for such a long time should be considered as the essential genes needed by all the species within this family. Generally speaking, our comparative analysis was highly consistent with the studies by Glass *et al.* (Figure 6); however, we identified more genes that may have been deleted due to natural selection. The six genes differing between the two studies are all key genes of DNA replication and DNA repair: *recA*, *recU*, Holliday junction DNA helicases *uvrA* and *uvrB*, *mutM*and a likely DNA damage inducible protein gene. Those genes are core genes, but were disrupted in transposon mutagenesis studies [5]. Interestingly, Glass *et al.* stated that these six mutants grew more poorly after repeated passage, probably due to an accumulation of cell damage over time. This indicates that these six disrupted core genes may be critical for bacterial survival, especially over long periods. Therefore, we classified these genes as truly essential genes.

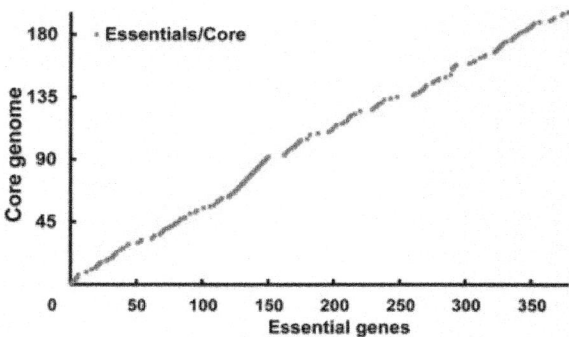

Figure 6: Similarity relationship between core and essential genes.

Genes on x-axis represent the essential genes documented in *M. genitalium* G37, while genes on y-axis represent the core genome identified in this study. Both the genes on x-axis and y-axis are distributed in terms of genomic coordinates. On-diagonal, genes that are both essentials and core are indicated as a series of colored dots.

Although transposon mutagenesis has proven to be a useful method and has been used extensively to determine the essentiality of genes [27], this method is highly dependent on environmental conditions. For the most part, mutagenesis studies performed under nutrient-rich conditions provide a substantial underestimate of the number of genes that are essential under host environmental conditions [28]. In reality, it is quite different for a cell to survive in a laboratory setting and to thrive in the wild. Transposon mutagenesis might misclassify nonessential genes that slow growth without arresting it but can also miss essential genes that tolerate transposon insertions [4]. However, comparative genomics analysis has also limitations, since it is likely to underestimate the core genome because it takes into account only the genes that have remained similar enough during the course of evolution to be recognized as true orthologues. Therefore, it will not include genes with a high rate of evolution, which may not show their relationship in comparisons of distant taxons [4]. Taken together, both the experimental approaches and comparative genomics analysis should be taken into account when addressing questions of essentiality. Besides, the core genes set proposed in the current study are only essential for most species within this family. We failed to estimate the conserved genes needed for single species. Each member of a particular species was maintained in a distinct ecological niche, in which some of the genes that were not present in core genome may also be important for the mycoplasma survival. These genes may be termed "lineage specific" essential genes. As more genome sequence becomes available in the future, there will be an opportunity to explore more properties of species special core genes set using comparative genomic tools.

To date, most mycoplasma phylogenies have been derived from single gene comparisons, or from concatenated alignments of a small number of genes. The increasing availability of mycoplasma genomes presents an opportunity to reconstruct evolutionary events using entire genomes [29], [30]. As a tool for future comparative phylogenetic studies, we used both supertrees and single gene alignments to infer relationships between 20 strains of the family*Mycoplasmataceae*. Our supertree and 16S rRNA phylogenies are consistent in most of their branches. However, there are conflicts regarding whether *M. penetrans* is clustered with the*Ureaplasma* lineage or with Clade V, as well as the placement of *M. mobile* 163. We also compared our trees with a recent study, in which the phylogeny of the *Mycoplasmas* was reconstructed by random concatenation of 91 protein sequences shared by 16 *mollicutes* [19]. The placement of *M. penetrans* in the concatenation-based phylogeny is consistent with the supertree, while the bootstrap value (=62) of the branch of *M. penetrans* and *Ureaplasma* in the 16S rRNA tree is low. Therefore, we placed *M. penetrans* into clade V. The location of *M. mobile* 163 is different among all three trees, which indicates a complicated phylogenic history for this strain, which may have involved recombination or other LGT events.

Phylogenetic reconstruction based upon concatenation of multiple orthologous genes can generate a more accurate tree than that done with a single gene [31], [32]. The supertree is even better than the concatenation-based tree, because it is immune to long-branch attraction artifacts [33], [34]. Thus, a robust supertree was constructed in this study to present the phylogeny of the family *Mycoplasmataceae*. The supertree was then used as a foreground for further analysis. Based on the supertree of *Mycoplasmataceae*, we classified these twenty strains into five different clades, between which the host specificity varies. All three strains in Clade I and three of the four strains in Clade V (except *M. gallisepticum*) were identified to be agents of human infection; therefore, these two clades form the human-infecting lineage. Both of the two sequenced strains in clade IV are the agent of ruminant infection, and thus represent the ruminant-infecting lineage. Four of the five strains in Clade III (except *M. conjunctivae*) are associated with swine mycoplasmosis and form the swine-infecting lineage. Briefly, most species with the same host specificity clustered together, forming a separate clade. There appears to be an approximate correlation between the divergence of species and the level of positive selection detected in different lineages. We suspect that host specificity was determined after the emergence of *Mycoplasma* species. Subsequent host jumping events may have been caused by a series of natural selection events during evolution.

To gain deeper insights into the molecular evolution events underlying natural selection, we employed the branch-site method to assess positive selection in swine-infecting and human-infect lineages. According to Petersen *et al.* [35], two categories of genes, immune-related and environmental adaptation related genes, are expected to show strong evidence for positive selection. Our analysis revealed that a number of genes related to DNA replication and repair (*dnaA, dnaN, gyrB, uvrA, polC, uvrB, uvrC, parC, dnaE*), show remarkably strong evidence for positive selection. These genes were unevenly distributed across HUB-1, the MC lineage and the common ancestor of clade III. Notably, both the *dnaA* and *dnaN* genes, which compose the *oriC* region, were identified to be under positive selection in the HUB-1 lineage. Previous studies have already demonstrated that replication may contribute to proliferation and efficiency of the colonization of hostile environments [25]. Therefore, we suspected that selection pressure on *oriC* may be one of the reasons why *M. hyorhinis* can thrive among different species of cell lines. This research provides a better insight into, and understanding of, persistent nonessential genes, and encourages exploration of essential genes by combining both the experimental approaches and comparative genomics analysis. This study also provides a comparative genomics method for addressing questions of essentiality. With the increasing number of genome sequences available for the same species in the future, this method will be useful for exploring species-specific essential genes.

MATERIALS AND METHODS

Bacterial Strains and Genome Sequences

M. hyorhinis strain HUB-1 was isolated from the respiratory tract of swine in China and confirmed to be an *M. hyorhinis* strain by verifying the 16S rRNA region. The main characteristics of 20 *Mycoplasmas* strains with freely available genomes at the time of the study are presented in Table 1. These genomes were used for comparative analysis.

Assignment of Orthologs and Phylogenetic Analysis

We analyzed *M. hyorhinis* HUB-1 and 19 other *Mycoplasmataceae* genomes from the NCBI databases. To ensure consistency, the annotations of all genomes were verified based on the similarity with *M. hyorhinis* HUB-1, using the tBLASTn algorithm [36]. The sets of orthologous protein-coding genes were defined as mutual fully transitive reciprocal BLASTP [37] hits (with E-value$<10^{-4}$) [38]. Co-ortholog groups were identified by the method similar to Inparanoid [39]and ortholog gene clusters were obtained using the tribeMCL program [40]. The nucleic acid sequence of each ortholog group was aligned using the CLUSTALW program version 1.83 [41]. For each data set, the phylogenetic relationship was estimated and tested in one thousand bootstrap samples using TREE-PUZZLE version 5.2 (general time reversible (GTR) $+\Gamma4+I$ model of evolution with a BIONJ starting tree) [42]. The bi-partitions with at least 70% support from the bootstrap test in each data set were recorded as "0/1" status and used to reconstruct the consensus sequence. The phylogenetic relationship of the consensus sequence was built using the SplitsTREE 4 with the BioNJ model.

Positive Selection Analysis

We employed the branch-site test of Yang and Nielsen [20], implemented in the program HY-PHY, to assess positive selection at particular sites and lineages. Briefly, the likelihood of a model that does not allow positive selection is compared to one allowing positive selection on some specified lineages. The model allowing positive selection is tested using a likelihood ratio test (LRT) [43] that is compared to a $\chi2$ statistic with two degrees of freedom. Likelihoods were estimated on the genes or species trees. For the *Mycoplasma* data set, both swine-infecting and human-infecting lineages were tested (Figure 4). To avoid the interference of recombination, only genes that support all four lineages in their gene trees (with $>70\%$ bootstrap support) were used. In total, 661 genes were tested. Finally, p values were corrected for multiple hypotheses testing using the Bonferroni method [44].

ACKNOWLEDGMENTS

We thank Dr. Maojun Liu (Jiangsu Academy of Agricultural Sciences, Nanjing, China), Dr. Lei Wang (Nankai University, Tianjin, China), Mr. Feng Li (Huazhong agricultural university, Wuhan, China) assistance with sequencing and analysis.

AUTHOR CONTRIBUTIONS

Conceived and designed the experiments: WL SX LF HC. Performed the experiments: WL ML SL SG ZF. Analyzed the data: WL ZZ GS RL BL. Contributed reagents/materials/analysis tools: WL GS RL BL. Wrote the paper: WL LF SX.

REFERENCES

1. Razin S, Yogev D, Naot Y (1998) Molecular biology and pathogenicity of mycoplasmas. Microbiol Mol Biol Rev 62: 1094–1156.

2. Waites KB, Katz B, Schelonka RL (2005) *Mycoplasmas* and *ureaplasmas* as neonatal pathogens. Clin Microbiol Rev 18: 757–89.

3. Morowitz HJ, Tourtellotte ME (1962) The smallest living cells. Sci Am 206: 117–126.

4. Gil R, Silva FJ, Pereto J, Moya A (2004) Determination of the core of a minimal bacterial gene set. Microbiol Mol Biol Rev 68: 518–537.

5. Glass JI, Assad-Garcia N, Alperovich N, Yooseph S, Lewis MR, et al. (2006) Essential genes of a minimal bacterium. Proc Natl Acad Sci U S A 103: 425–430.

6. Parkhill J, Sebaihia M, Preston A, Murphy LD, Thomson N, et al. (2003) Comparative analysis of the genome sequences of *Bordetella pertussis, Bordetella parapertussis*and *Bordetella bronchiseptica*. Nat Genet 35: 32–40.

7. Jin Q, Yuan ZH, Xu JG, Wang Y, Shen Y, et al. (2002) Genome sequence of *Shigella flexneri* 2a: insights into pathogenicity through comparison with genomes of *Escherichia coli* K12 and O157. Nucleic Acids Res 30: 4432–4441.

8. Wolf M, Muller T, Dandekar T, Pollack JD (2004) Phylogeny of Firmicutes with special reference to Mycoplasma (Mollicutes) as inferred from phosphoglycerate kinase amino acid sequence data. Int J Syst Evol Micr 54: 871–875.

9. Fraser CM, Gocayne JD, White O, Adams MD, Clayton RA, et al. (1995) The minimal gene complement of *Mycoplasma genitalium*. Science 270: 397–403.

10. DeBey MC, Ross RF (1994) Ciliostasis and loss of cilia induced by *Mycoplasma hyopneumoniae* in porcine tracheal organ cultures. Infect Immun 62: 5312–5318.

11. Kotani H, Butler GH, Tallarida D, Cody C, McGarrity GJ (1990) Microbiological cultivation of *Mycoplasma hyorhinis* from cell cultures. In Vitro Cell Dev Biol 26: 91–96.

12. Huang S, Li JY, Wu J, Meng L, Shou CC (2001) Mycoplasma infections and different human carcinomas. World J Gastroenterol 7: 266–269.

13. Lefebure T, Stanhope MJ (2007) Evolution of the core and pan-genome of*Streptococcus*: positive selection, recombination, and genome

composition. Genome Biol 8: R71.

14. Ouzounis C, Kyrpides N (1996) The emergence of major cellular processes in evolution. FEBS Lett 390: 119–123.

15. Himmelreich R, Hilbert H, Plagens H, Pirkl E, Li BC, et al. (1996) Complete sequence analysis of the genome of the bacterium *Mycoplasma pneumoniae*. Nucleic Acids Res 24: 4420–4449.

16. Fang G, Rocha E, Danchin A (2005) How essential are nonessential genes? Mol Biol Evol 22: 2147–2156.

17. Bapteste E, Boucher Y, Leigh J, Doolittle WF (2004) Phylogenetic reconstruction and lateral gene transfer. Trends Microbiol 12: 406–411.

18. Ciccarelli FD, Doerks T, von Mering C, Creevey CJ, Snel B, et al. (2006) Toward automatic reconstruction of a highly resolved tree of life. Science 311: 1283–1287.

19. Sirand-Pugnet P, Citti C, Barre A, Blanchard A (2007) Evolution of mollicutes: down a bumpy road with twists and turns. Res Microbiol 158: 754–766.

20. Zhang J, Nielsen R, Yang Z (2005) Evaluation of an improved branch-site likelihood method for detecting positive selection at the molecular level. Mol Biol Evol 22: 2472–2479.

21. Yang Z, Nielsen R (2002) Codon-substitution models for detecting molecular adaptation at individual sites along specific lineages. Mol Biol Evol 19: 908–917.

22. Kosiol C, Vinar T, da Fonseca RR, Hubisz MJ, Bustamante CD, et al. (2008) Patterns of positive selection in six Mammalian genomes. PLoS Genet 4: e1000144.

23. Shapiro BJ, Alm EJ (2008) Comparing patterns of natural selection across species using selective signatures. PLoS Genet 4: e23.

24. McGlynn P, Lloyd RG (2001) Rescue of stalled replication forks by RecG: simultaneous translocation on the leading and lagging strand templates supports an active DNA unwinding model of fork reversal and Holliday junction formation. Proc Natl Acad Sci U S A 98: 8227–8234.

25. Anisimova M, Bielawski J, Dunn K, Yang Z (2007) Phylogenomic analysis of natural selection pressure in *Streptococcus* genomes. BMC Evol Biol 7: 154.

26. Szczepanek SM, Tulman ER, Gorton TS, Liao X, Lu Z, et al. (2010) Comparative genomic analyses of attenuated strains of *Mycoplasma gallisepticum*. Infect Immun 78: 1760–1771.

27. Hutchison CA, Peterson SN, Gill SR, Cline RT, White O, et al. (1999) Global transposon mutagenesis and a minimal Mycoplasma genome. Science 286: 2165–2169.

28. Papp B, Pal C, Hurst LD (2004) Metabolic network analysis of the causes and evolution of enzyme dispensability in yeast. Nature 429: 661–664.

29. Snel B, Bork P, Huynen MA (1999) Genome phylogeny based on gene content. Nat Genet 21: 108–110.

30. Kunin V, Ahren D, Goldovsky L, Janssen P, Ouzounis CA (2005) Measuring genome conservation across taxa: divided strains and united kingdoms. Nucleic Acids Res 33: 616–621.

31. Daubin V, Gouy M, Perriere G (2002) A phylogenomic approach to bacterial phylogeny: evidence of a core of genes sharing a common history. Genome Res 12: 1080–1090.

32. Rokas A, Williams BL, King N, Carroll SB (2003) Genome-scale approaches to resolving incongruence in molecular phylogenies. Nature 425: 798–804.

33. Fitzpatrick DA, Logue ME, Stajich JE, Butler G (2006) A fungal phylogeny based on 42 complete genomes derived from supertree and combined gene analysis. BMC Evol Biol 6: 99.

34. Gadagkar SR, Rosenberg MS, Kumar S (2005) Inferring species phylogenies from multiple genes: concatenated sequence tree versus consensus gene tree. J Exp Zool B Mol Dev Evol 304: 64–74.

35. Petersen L, Bollback JP, Dimmic M, Hubisz M, Nielsen R (2007) Genes under positive selection in *Escherichia coli*. Genome Res 17: 1336–1343.

36. Iguchi A, Thomson NR, Ogura Y, Saunders D, Ooka T, et al. (2009) Complete genome sequence and comparative genome analysis of enteropathogenic *Escherichia coli* O127:H6 strain E2348/69. J Bacteriol 191: 347–354.

37. Altschul SF, Madden TL, Schaffer AA, Zhang J, Zhang Z, et al. (1997) Gapped BLAST and PSI-BLAST: a new generation of protein database search programs. Nucleic Acids Res 25: 3389–3402.

38. Zhaxybayeva O, Gogarten JP (2002) Bootstrap, Bayesian probability and maximum likelihood mapping: exploring new tools for comparative genome analyses. BMC Genomics 3: 4.

39. O'Brien KP, Remm M, Sonnhammer EL (2005) Inparanoid: a comprehensive database of eukaryotic orthologs. Nucleic Acids Res 33: D476–480.

40. Enright AJ, Van Dongen S, Ouzounis CA (2002) An efficient algorithm

for large-scale detection of protein families. Nucleic Acids Res 30: 1575–1584.

41. Thompson JD, Higgins DG, Gibson TJ (1994) CLUSTAL W: improving the sensitivity of progressive multiple sequence alignment through sequence weighting, position-specific gap penalties and weight matrix choice. Nucleic Acids Res 22: 4673–4680.

42. Schmidt HA, Strimmer K, Vingron M, von Haeseler A (2002) TREE-PUZZLE: maximum likelihood phylogenetic analysis using quartets and parallel computing. Bioinformatics 18: 502–504.

43. Yang Z (1998) Likelihood ratio tests for detecting positive selection and application to primate lysozyme evolution. Mol Biol Evol 15: 568–573.

44. Guilbaud O (2007) Bonferroni parallel Gatekeeping - Transparent generalizations, adjusted P-values, and short direct proofs. Biometrical J 49: 917–927.

45. Glass JI, Lefkowitz EJ, Glass JS, Heiner CR, Chen EY, et al. (2000) The complete sequence of the mucosal pathogen *Ureaplasma urealyticum*. Nature 407: 757–762.

46. Vasconcelos ATR (2005) Swine and poultry pathogens: the complete genome sequences of two strains of *Mycoplasma hyopneumoniae* and a strain of *Mycoplasma synoviae* (vol 187, pg 5568, 2005). J Bacteriol 187: 7548–7548.

47. Chambaud I, Heilig R, Ferris S, Barbe V, Samson D, et al. (2001) The complete genome sequence of the murine respiratory pathogen *Mycoplasma pulmonis*. Nucleic Acids Res 29: 2145–2153.

48. Dandekar T, Huynen M, Regula JT, Ueberle B, Zimmermann CU, et al. (2000) Re-annotating the *Mycoplasma pneumoniae* genome sequence: adding value, function and reading frames. Nucleic Acids Res 28: 3278–3288.

49. Sasaki Y, Ishikawa J, Yamashita A, Oshima K, Kenri T, et al. (2002) The complete genomic sequence of *Mycoplasma penetrans*, an intracellular bacterial pathogen in humans. Nucleic Acids Res 30: 5293–5300.

50. Westberg J, Persson A, Holmberg A, Goesmann A, Lundeberg J, et al. (2004) The genome sequence of *Mycoplasma mycoides* subsp. mycoides SC type strain PG1T, the causative agent of contagious bovine pleuropneumonia (CBPP). Genome Res 14: 221–227.

51. Jaffe JD, Stange-Thomann N, Smith C, DeCaprio D, Fisher S, et al. (2004) The complete genome and proteome of *Mycoplasma mobile*.

Genome Res 14: 1447–1461.

52. Liu W, Fang L, Li S, Li Q, Zhou Z, et al. (2010) Complete genome sequence of*Mycoplasma hyorhinis* strain HUB-1. J Bacteriol 192: 5844–5845.

53. Minion FC, Lefkowitz EJ, Madsen ML, Cleary BJ, Swartzell SM, et al. (2004) The genome sequence of *Mycoplasma hyopneumoniae* strain 232, the agent of swine mycoplasmosis. J Bacteriol 186: 7123–7133.

54. Papazisi L, Gorton TS, Kutish G, Markham PF, Browning GF, et al. (2003) The complete genome sequence of the avian pathogen *Mycoplasma gallisepticum* strain R(low). Microbiology 149: 2307–2316.

55. Calderon-Copete SP, Wigger G, Wunderlin C, Schmidheini T, Frey J, et al. (2009) The*Mycoplasma conjunctivae* genome sequencing, annotation and analysis. BMC Bioinformatics 10: Suppl 6S7.

56. Voelker LL, Dybvig K (1999) Sequence analysis of the *Mycoplasma arthritidis*bacteriophage MAV1 genome identifies the putative virulence factor. Gene 233: 101–107.

Chapter 4

ROBUST GENE-GENE INTERACTION ANALYSIS IN GENOME WIDE ASSOCIATION STUDIES

Yongkang Kim[1], Taesung Park[1,2]

[1] Department of Statistics, Seoul National University, Seoul, South Korea

[2] Interdisciplinary Program in Bioinformatics, Seoul National University, Seoul, 151–741, South Korea

ABSTRACT

Genome-wide association studies (GWAS) have successfully discovered hundreds of associations between genetic variants and complex traits. Most GWAS have focused on the identification of single variants. It has been shown that most of the variants that were discovered by GWAS could only partially explain disease heritability. The explanation for this missing heritability is generally believed to be gene-gene (GG) or gene-environment (GE) interactions and other structural variants. Generalized multifactor dimensionality reduction (GMDR) has been proven to be reasonably powerful in detecting GG and GE interactions; however, its performance has been found to decline when outlying quantitative traits are present. This paper proposes a robust GMDR estimation method (based on the L-estimator and M-estimator estimation methods) in an attempt to reduce the effects caused by outlying traits. A comparison of robust GMDR with the original MDR based on simulation studies showed the former method to outperform the latter. The performance of robust GMDR is illustrated through a real GWA example consisting of 8,577 samples from the Korean population using the Homeostasis Model Assessment of Insulin Resistance (HOMA-IR) level as a phenotype. Robust GMDR identified the KCNH1 gene to have strong interaction effects with other genes on the function of insulin secretion.

INTRODUCTION

A genome-wide association (GWA) study has become a common approach for testing the association between a single nucleotide polymorphism (SNP)

and a complex trait of interest [1]. There have been many successful results from genome-wide association studies (GWAS). However, SNPs that were identified by GWAS have been shown to explain only a small fraction of disease etiology, because the relatedness between complex diseases and multiple genes and/or their interactions are ignored. For this reason, the analysis of gene-gene (GG) and gene-environment (GE) interactions have been emphasized as a new alternative for understanding the etiology of common complex traits. However, GG and GE interactions are hard to detect and characterize by using traditional parametric statistical methods, for the following reasons [2]. First, high dimensional genetic data may be of a sparse nature. The issue of data sparseness can be addressed by using exponentially large sample sizes when parametric statistical methods are used for determining GG and GE interactions. Second, detecting GG and GE interactions using traditional procedures may lead to an increase in type II errors and a decrease in power. As a result, detecting GG and GE interactions has been a well-known challenge in statistics and data mining areas [3].

Ritchie et al. proposed a Multifactor Dimensionality Reduction (MDR) method for detecting GG interaction [4]. This method is not affected by data sparsity because it allows multilocus genotype combinations with very few or no data points. However, the original MDR method can only be used in association studies for qualitative traits. The MDR method was extended to more general types of traits, including quantitative traits, by Lou et al., who proposed the Generalized Multifactor Dimensionality Reduction (GMDR) method [5]. This method is quite flexible in handling various types of traits and is even able to account for individual covariates.

Since the initial introduction of the original MDR and GMDR methods, numerous extensions of MDR have been developed. Calle et al. proposed model-based MDR (MB-MDR) for the parametric extension of the MDR method [6]. Recently, Quantitative MDR (QMDR) was proposed for handling quantitative traits [7].

These MDR methods have brought many remarkable successes in GG interaction analysis. However, the effects of outlying trait observations on the results of MDR methods have not been studied. Outlying trait observations are those traits that diverge from the global distribution of traits. Bennett et al., Most et al., and Yang et al. discussed the treatment of outliers in GWAS and proposed simple approaches, such as removing outlying observations and transforming traits, including log-transformation or inverse-normalization [8, 9, 10]. The R-package QCGWAS includes a process for detecting outliers by skewness or kurtosis. However, these QC methods are very sensitive to the underlying distributional assumption and may lead to the wrong conclusions when the distributional assumption is not met. In this paper, we first demonstrate the effects of outlying trait observations on GMDR analysis. Next, we propose robust GMDR for reducing the effects caused by outlying traits. Robust GMDR uses robust estimation based on L-estimator and M-estimator. Simulation studies were used to compare the robust GMDR developed in this work to GMDR, and was shown to outperform GMDR. The performance of robust GMDR is illustrated through a real GWA example consisting of 8,577 samples from the Korean population.

MATERIALS AND METHODS

KARE Data

The Korea Association Resource (KARE) is a project for gathering large-scale GWA analyses in the Gyeonggi Province of South Korea [11]. Two community cohorts participated in the KARE project: the Ansung and Ansan cohorts. The number of participants in the Ansung cohort is 5,018 and that in Ansan cohort is 5,020. The age of the participants were between 40 and 69, and more than 260 traits have been extensively examined through epidemiological surveys, physical examinations, and laboratory tests. In this paper, the Homeostasis Model Assessment of Insulin Resistance (HOMA-IR) levels are used, which are widely employed to estimate insulin resistance. As HOMA-IR has a nonnegative skewness distribution, a gamma distribution is commonly assumed [12]. DNA samples were genotyped on the Affymetrix Genome-Wide Human SNP array 5.0, which is able to genotype 500,568 SNPs. The quality control processes are well described in a study by Cho et al. [13]. For the purposes of our work, samples without Body Mass Index (BMI) or HOMA-IR scores were disregarded. After removing the samples with the missing phenotypes, a total of 8,577 individuals and 327,872 SNPs remained in the study.

Review of MDR Method

The evaluation of genetic effects using the MDR method requires several steps. The first step involves selecting genetic factors according to their effects on the phenotype. If GG interactions of the nth order are required, n SNPs are selected and a 3^n contingency table is created. Each cell of the contingency table contains the number of cases and controls with the same genetic factor combinations. The second step calculates the ratio of the number of cases to the number of controls in each cell, and uses this ratio to determine whether a cell is "high risk" or "low risk". If the ratio of case to control in a cell exceeds the threshold (which is commonly specified as 1), then those cells are labeled "high risk"; otherwise they are labeled "low risk". This process enables each cell to be classified as either "high risk" or "low risk". In the third step, the ability of selected SNP sets to perform these classifications based on the case / control ratio are evaluated through accuracy and balanced accuracy (BA) measures, where BA is a measure for the prediction ability of the model and is defined as follows:

$$BA = \frac{1}{2}(sensitivity + specificity)$$

(1)

The larger the BA produced by the model, the more accurate the model's prediction ability [14]. These measures are computed from the 2×2 confusion table in which the first variable represents the case / control ratio and the second variable high risk and low risk. Then, the sensitivity is defined as the ratio of the number of cases to the number of high risk samples and the specificity is defined as the ratio of the number of controls to low risk samples. All these steps would be exhaustively performed for all SNP combinations.

The problem of over-fitting is usually avoided by performing a 10-fold cross-validation. It is also possible to use the cross-validation consistency (CVC) as a measure for selecting SNP combinations. This selection is based on the number of times the same SNP combination is selected as best out of 10 cross-validated samples. Alternatively, the SNP combination with the largest CVC measure can also be selected as the best SNP combination. BA and CVC measures have been widely used in MDR analysis

Review of GMDR Method

Although MDR has been used extensively in GG interaction analysis, it has several limitations. First, the MDR method is only capable of handling qualitative traits such as case and control data, because the method classifies each cell by using the case / control ratio. Second, it is not possible for MDR to adjust for other individual covariates such as environmental factors. Therefore, using MDR to analyze traits that are affected by environmental factors may produce biased results. In an attempt to overcome these limitations, Lou et al. proposed the Generalized Multifactor Dimensionality Reduction (GMDR) method [5], each step of which is very similar to those of the original MDR method. However, instead of using the counts of individuals, the GMDR method uses a residual-based score.

Let y_i denote the phenotype of individual i, which can be either quantitative or qualitative. Let $E(y_i) = \mu_i$. Consider the following generalized linear model (GLM):

$$l(\mu_i) = \alpha + x_i^T \beta + z_i^T \gamma$$

(2)

where $l(\mu_i)$ is the link function, α is the intercept, and x_i is a vector that expresses possible genotype combinations of interest. z_i is a vector representing environmental factors, and β and γ are coefficient vectors.

The first step when using GMDR is to fit the null model only with environmental factors under the null hypothesis, i.e., $\beta = 0$. The second step is to classify each SNP combination using the residuals from the null model. For each cell of the contingency table, the sum of the residuals is computed. Cells with a positive residual sum are classified as "high risk"; otherwise, they are classified as "low risk". Subsequent GMDR cross-validation steps are performed similar to those of MDR.

It is noteworthy that, in this GMDR procedure, the distinction between high and low classification is based on the sum of the residuals. Thus, a few outlying observations may have a large impact on the performance of GMDR, which provided the motivation for considering robust GMDR. Although GLM can handle both quantitative and qualitative traits, our efforts were focused on a quantitative trait with outlying observations. Thus, an identity link function was used with the supposition that y_i has an independent normal distribution with equal variance.

Proposed Robust GMDR Method

The effect of outlying observations can be reduced by considering robust scores such as an L-estimator and an M-estimator type score. The L-estimator is a linear combination of order statistics that have to be measured. A well-known example of the L-estimator is least trimmed square (LTS) [15]. LTS regression is formulated as follows.

$$\arg \min_\beta \sum_{i=1}^{k} |\gamma_{(i)}(\beta)|^2, \ where \ \gamma_i(\beta) = (y_i - f(x_i, \beta)), i = 1, \ldots, n.$$

(3)

Here, k is the number of samples that is actually used in the regression and n is the total sample size. $\gamma_i(\beta)$ is the ith residual for a fixed β that is ordered according to the absolute value of the residuals. Bickel and Lehmann considered trimmed expectations to be the only ones which are both robust and whose estimators have guaranteed high efficiency [16]. LTS is known to outperform other least square estimators. Using LTS regression, our L-estimator GMDR uses a studentized trimmed residual as a score for GMDR, where the trimmed studentized residual is defined as follows:

$$\dot{\gamma}_i(\beta) = \begin{cases} \dfrac{y_i - f(x_i, \beta)}{\sqrt{\mathrm{var}(y_i - f(\widehat{x_i, \beta}))}} & \text{if } \left| \dfrac{y_i - f(x_i, \beta)}{\sqrt{\mathrm{var}(y_i - f(\widehat{x_i, \beta}))}} \right| \leq T \\[4ex] 0 & \text{if } \left| \dfrac{y_i - f(x_i, \beta)}{\sqrt{\mathrm{var}(y_i - f(\widehat{x_i, \beta}))}} \right| > T \end{cases}$$

$$(4)$$

where T is the threshold value to determine whether the ith sample should be used, and it is possible to adjust T to determine the extent to which outliers would have to be trimmed.

The second robust GMDR uses an M-estimator type score, which is derived as the minima of the sums of the functions of the data. Least-squares estimators and many maximum-likelihood estimators are examples of M-estimators. The M-estimator was first derived for the purpose of introducing it into robust regression by Huber [17]. Tukey subsequently proposed a biweight function, which is a type of M-estimator for robust regression. The M-estimator using Tukey's biweight function is defined as follows.

$$\rho(\gamma_i(\beta)) = \begin{cases} \dfrac{1}{6}[1 - (1 - \gamma_i(\beta))^2)^3] & \text{if } |\gamma_i(\beta)| \leq 1 \\[3ex] \dfrac{1}{6} & \text{if } |\gamma_i(\beta)| > 1 \end{cases},$$

$$(5)$$

$$\arg \min_\beta \sum_{i=1}^{k} \rho(\gamma_i(\beta), \text{ where } \gamma_i(\beta) = (y_i - f(x_i, \beta)), i = 1, \ldots, n.$$

$$(6)$$

Solving the M-estimator requires differentiation of the ρ function, which usually requires selection of an appropriate shape for the biweight function.

Using the biweight function, our M-estimator GMDR defines the threshold residual score as:

$$\dot{\gamma}_i(\beta) = \begin{cases} \dfrac{y_i - f(x_i, \beta)}{\sqrt{\mathrm{var}(y_i - f(\widehat{x_i, \beta}))}} & \text{if } \left| \dfrac{y_i - f(x_i, \beta)}{\sqrt{\mathrm{var}(y_i - f(\widehat{x_i, \beta}))}} \right| \leq T \\[4ex] T \cdot sign(y_i - f(x_i, \beta)) & \text{if } \left| \dfrac{y_i - f(x_i, \beta)}{\sqrt{\mathrm{var}(y_i - f(\widehat{x_i, \beta}))}} \right| > T \end{cases}$$

$$(7)$$

In this formula, the threshold T is used to shrink all residuals exceeding T. As this function does not require differentiation; the biweight function is modified by simply reducing the weight caused by extreme loss.

RESULTS

Simulation Study

First, we simulated the phenotypes from the Normal distribution. In this case, samples were generated by using the following model.

$$y = envir_1 + \beta_1 SNP_1 + \beta_2 SNP_2 + \beta_{1 \times 2} SNP_1 SNP_2 + \varepsilon, \quad (8)$$
$$envir_1 \sim N(0,1), SNP_1, \ldots, SNP_p \sim Bin(2, MAF), \varepsilon \sim N(0,1).$$

The effects of MAFs were checked by considering two MAF values: 0.1 and 0.3. Next, outlying samples were generated by using the following model

$$y \sim N(0, 400) \quad (9)$$
$$SNP_1, \ldots, SNP_p \sim Bin(2, MAF).$$

A series of designs were created by using different settings for the purpose of data simulation, and the settings are summarized in Table 1. The first three columns of Table 1 denote the effect of the size of SNP1, SNP2, and the interaction in pure samples, respectively. The fourth column lists the mixed proportion, i.e., the proportion of outlying samples as generated by Model (9). Larger mixed proportions indicate that the particular simulation design generates more outlying observations. The fifth and sixth columns of Table 1 denote the heritability of each simulation setting when MAF = 0.1 and 0.3, respectively. The detailed equations for calculating the heritability are described in the Appendix.

Table 1: Models of simulation with Normal distribution

	SNP1	SNP2	interaction	mixed proportion	Heritability (MAF = 0.1)	Heritability (MAF = 0.3)
Design1	0.2	0.2	0.18	0.067	0.010	0.041
Design2	0.2	0.2	0.18	0.033	0.010	0.041
Design3	0.4	0.4	-0.3	0.067	0.022	0.027
Design4	0.4	0.4	-0.3	0.033	0.022	0.027
Design5	0.25	0.25	0.15	0.067	0.014	0.048
Design6	0.25	0.25	0.15	0.033	0.014	0.048
Design7	0.2	0.2	0.18	0	0.010	0.041

The first to third columns denote the effect sizes of SNP1, SNP2 and interaction in pure samples respectively. The fourth column of Table 1 indicates mixed proportion. The fifth and sixth columns of Table 1 indicates heritability and skewness respectively.

doi:10.1371/journal.pone.0135016.t001

Second, we performed simulation studies for the skewed distribution. The samples were generated from the gamma distribution given in (10). The MAF of SNPs was set to 0.3. The outlying observations were generated by the shifted Gamma distribution in (11). As the Gamma distribution has a long tail to the right, outlying observations were only generated for the right-hand side of the distribution. A series of designs are summarized in Table 2

$$y \sim Gamma(\alpha, \beta) \tag{10}$$

$$y \sim 0.1 + Gamma(\alpha, \beta) \tag{11}$$

$\alpha = (envir_1 + \beta_1 SNP_1 + \beta_2 SNP_2 + \beta_3 SNPS_1 NP_2)^2, \beta = 1/(envir_1 + \beta_1 SNP_1 + \beta_2 SNP_2 + \beta_3 SNPS_1 NP_2),$
$envir_1 \sim N(0,1), SNP_1, \ldots, SNP_p \sim Bin(2, 0.3)'.$

Table 2: Models of simulation with Gamma distribution

	SNP1	SNP2	interaction	mixed proportion	Heritability	Skewness
Design 1	0.6	0.6	0.3	0.067	0.117	0.770
Design 2	0.6	0.6	0.3	0.033	0.117	0.8
Design 3	1.2	0.4	0.3	0.067	0.197	0.617
Design 4	1.2	0.4	0.3	0.033	0.197	0.587
Design 5	0.8	-0.4	-0.2	0.067	0.074	0.965
Design 6	0.8	-0.4	-0.2	0.033	0.074	0.801
Design 7	0.6	0.6	0.3	0	0.117	0.602

The first column shows MAF of each design. Second to fourth columns denote the effect sizes of SNP1, SNP2 and interaction in pure samples respectively. The fifth and sixth columns of Table indicates heritability and skewness respectively.

doi:10.1371/journal.pone.0135016.t002

For these Normal and Gamma distributions, we considered two sample sizes of 1,000 and 3,000 to evaluate the effect of sample size, and assumed the number of SNPs to be 100 and 1000. Ordinary GMDR, L-estimator-based GMDR, and M-estimator-based GMDR were compared in terms of their power of detecting a true SNP pair. We used Eq (12) for calculating the power, where N refers to the total number of iterations.

$$\frac{1}{N}\sum_{i=1}^{N} I(\text{set } \{SNP_1, SNP_2\} \text{ is selected in } i\text{th iteration}). \tag{12}$$

The number of iterations was 1,000 when the number of SNPs is 100, while it was 100 when the number of SNPs is 1000. The best SNP pair was selected by using the average CVC through a 10-fold cross validation.

We also computed the false detection rate (FDER) [18], which was done by assuming that a targeted non-causal SNP pair, say SNP_1 and SNP_2, exists. To detect this pair, we simulated 500,000 null datasets by permuting the response variables. Because there were at most 100 SNPs in each dataset, the randomly selected rate of each SNP pair is at least 1/499,500. We counted how many times the pair of SNP_1 and SNP_2 was selected as the best model. We found all three methods selected the pair of SNP_1 and SNP_2 at most twice when the number of SNPs is 1000, which shows that the three methods have a very low FDER.

Fig 1 shows the results of the power comparison when SNPs are generated by Bin(2,0.3) and the phenotypes are obtained by Normal distributions.

Designs 1 to 6 show the results when outlying observations exist and Design 7 when outlying observations do not exist. For Design 7, the ordinary GMDR and M-estimator GMDR outperform L-estimator GMDR. As L-estimator GMDR uses a smaller number of samples than other methods, the power of L-estimator GMDR is less than that of the others. For all the other designs, the performance of L-estimator GMDR and M-estimator GMDR exceeds that of ordinary GMDR.

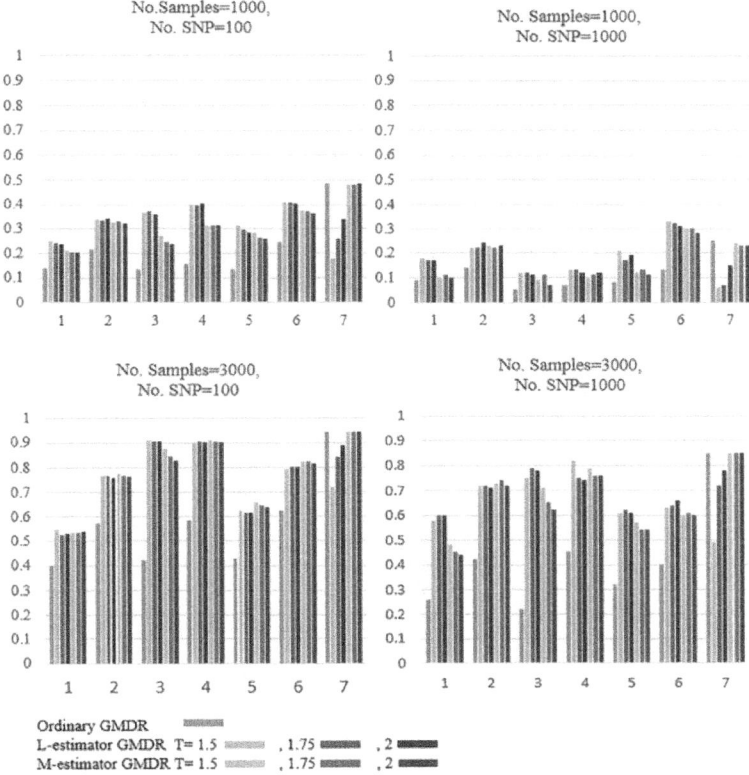

Figure 1: Results of Simulation: Normal distribution with MAF 0.3.

The upper and lower figures show the results when the number of samples is equal to 1,000 and 3,000, respectively. These four figures almost show the same patterns, but the simulation power is different.

doi:10.1371/journal.pone.0135016.g001

Fig 2 shows the results of the power comparison when SNPs are generated by Bin(2,0.1) and the phenotypes are obtained by Normal distributions. The patterns are quite similar to those inFig 1. For Designs 1 to 6, when outlying observations exist, the two robust GMDRs were found to outperform ordinary GMDR. For Design 7, when outlying observations do not exist, the performance of ordinary GMDR and M-estimator GMDR exceeds that of L-estimator GMDR.

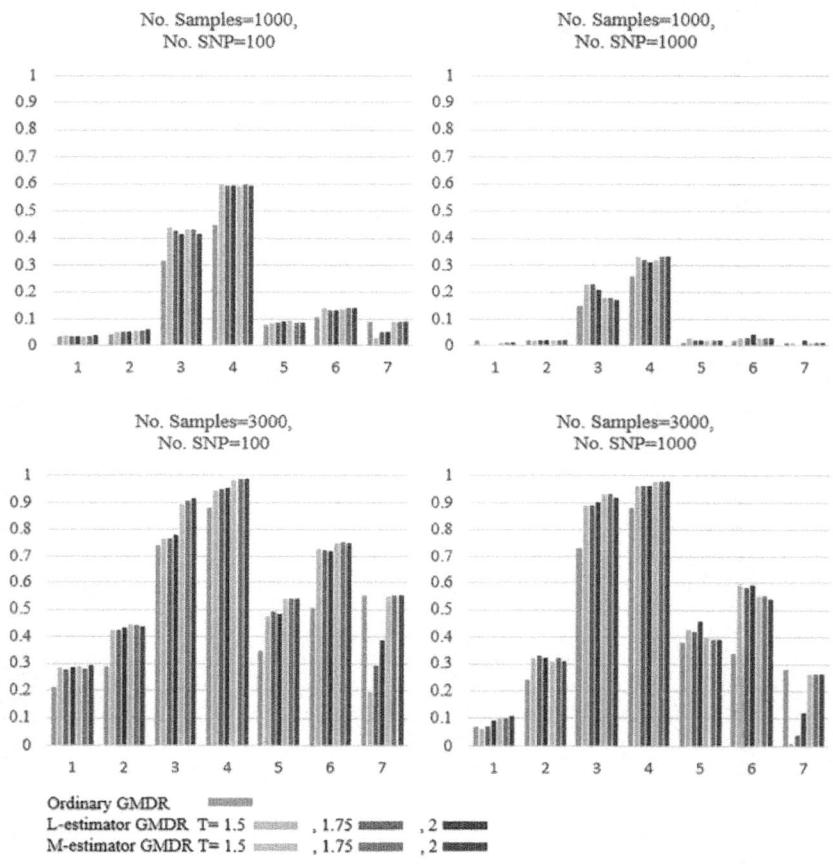

Figure 2: Results of Simulation: Normal distribution with MAF 0.1.

The upper figures show the results when the number of samples is equal to 1000 and the lower figures show the results when the number of samples equals 3,000. These four figures almost show the same patterns, but the simulation power is different.

Fig 3 shows the results of the power comparison when SNPs are generated by Bin(2,0.3) and the phenotypes are obtained by Gamma distributions. The residuals were calculated by generalized linear models under the Gamma distributional assumption. For Designs 1 to 6, when outlying observations exist, the two robust GMDRs were found to outperform ordinary GMDR to a large extent. In particular, L-estimator GMDR performed better than M-estimator GMDR. Note that the performance of L-estimator is highly dependent on the threshold values, whereas this is not the case for M-estimator. For Design 7, when outlying observations do not exist, all GMDRs perform similarly, probably due to the large SNP size effect.

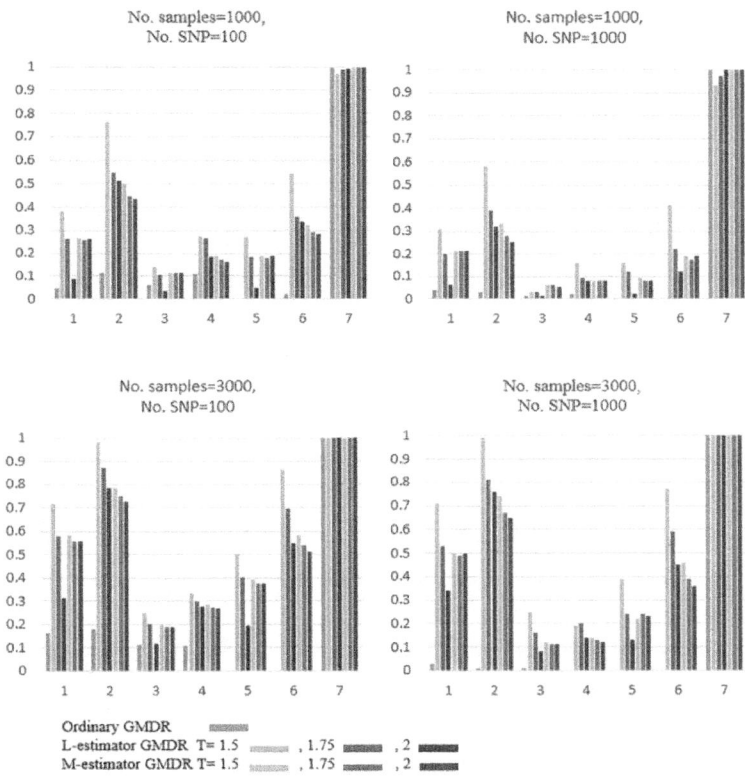

Figure 3: Results of Simulation: Gamma distribution with MAF 0.3.

The upper figures show the results when the number of samples is equal to 1000 and the lower figures show the results when the number of samples equals 3,000. These four figures almost show the same patterns, but the simulation power is different.

doi:10.1371/journal.pone.0135016.g003

Fig 4 shows the results of the power comparison when SNPs are generated by Bin(2,0.3) and the phenotypes obtained by Gamma distributions. The residuals were calculated by generalized linear models under the Normal distributional assumption (left panel) and under the Gamma distributional assumption (right panel). In general, the use of GMDR in combination with the correct distributional assumption is much more powerful than with an incorrect distributional assumption. For Designs 1 to 6, when outlying observations exist, two robust GMDRs perform much better than ordinary GMDR. In particular, L-estimator GMDR outperforms M-estimator GMDR.

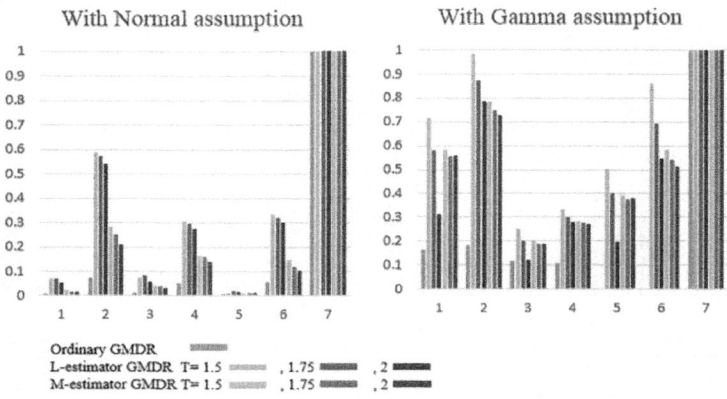

Figure 4: Results of Simulation: Power comparison between the residuals calculated under the Normal assumption and under the Gamma assumption.

The residuals were calculated by generalized linear models under the Normal distributional assumption (left panel) and under the Gamma distributional assumption (right panel).

The effect of threshold values were investigated by using the robust GMDR methods with various threshold values, i.e., T = 1.5, 1.75, and 2. As shown in our simulation results in Figs 1,2, 3 and 4, the dependence of L-estimator GMDR on T is much stronger than that of M-estimator GMDR. When the distribution of phenotypes is non-skewed (e.g., a Normal distribution), the effect of T is not large. On the other hand, when the distribution of phenotypes is skewed (e.g., a Gamma distribution), the effect of T is quite large. Although our simulation setting is limited, T = 1.5 consistently performed better than or equal to other threshold values in the presence of outlying observations. Thus, for practical applications, we recommend using T = 1.5.

Real Data Analysis

Our proposed robust GMDR was applied to KARE data with a HOMA-IR phenotype. Fig 5displays the boxplots of HOMA-IR without and with log-transformation. The skewed distribution of Homeostasis Model Assessment of

Insulin Resistance (HOMA-IR) levels prompted many researchers to perform a log-transform before conducting the regression analysis (shown in the upper panel of Fig 5) [12]. Thus, a log-transform was also performed in our work and its boxplot is given in the lower panel of Fig 5. In spite of performing a log-transformation, the distribution of HOMA-IR levels still remained skewed, with many samples apparently qualifying as outliers.Fig 6 shows the normal quantile-quantile (QQ) plots of the HOMA-IR levels: the left panel represents the case without log-transformation and the right panel the case with log-transformation. The distribution that was obtained after the log-transformation appears to be more symmetrical than without log-transformations. Thus, our GMDR analysis was performed using log-transformed HOMA-IR levels.

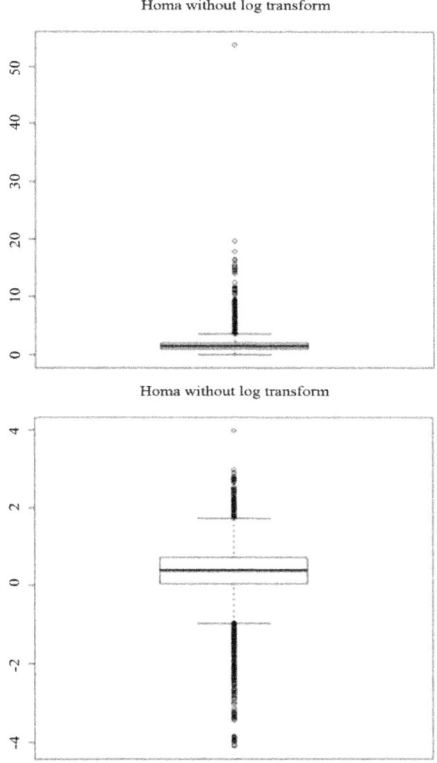

Figure 5: Boxplots of HOMA-IR.

Before log transformation to HOMA-IR, HOMA-IR has much skewed distribution. In contrasts, after log transformation, HOMA-IR has almost symmetric distribution.

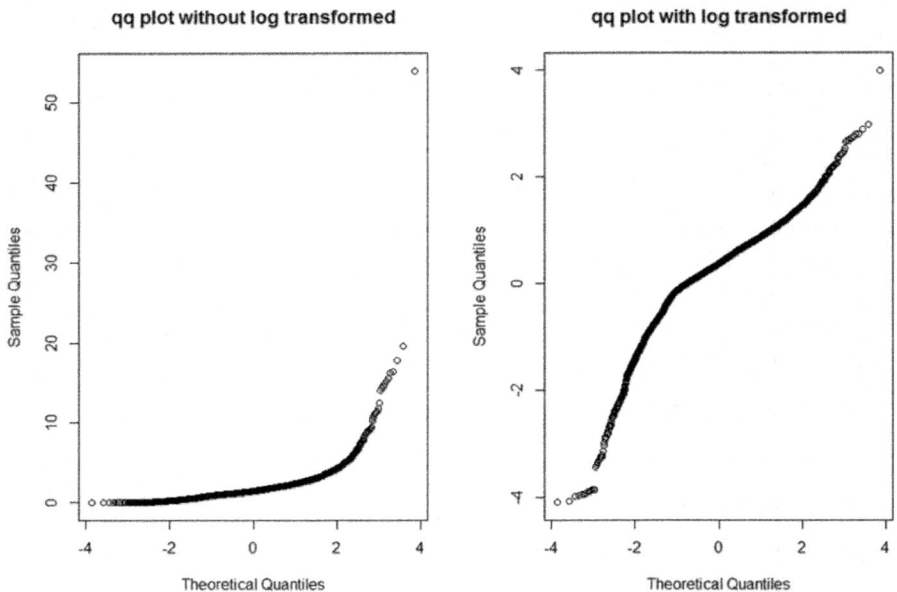

Figure 6: QQ-plot for HOMA-IR data.

This figure shows the QQ-plots for HOMA-IR data comparing with standard normal distribution. It is found that HOMA-IR with Log-transformation acts more like samples of normal distribution in right figure.

doi:10.1371/journal.pone.0135016.g006

Before the GMDR analysis, a single SNP analysis was first performed to reduce the computational burden on GMDR. A single SNP analysis was performed for all SNPs.

$$\log(HOMA - IR_i) = \alpha_0 + \alpha_1 SEX_i + \alpha_2 AGE_i + \gamma_3 AREA_i + \alpha_4 BMI_i + \alpha_{5j} SNP_{ij} + \varepsilon_i, \quad (13)$$

Noteworthy is that i (= 1,…,8577) represents individuals; whereas, j (= 1,…,327,872) represents SNPs. All SNPs are ranked in the order of ascending p-values. SNPs are filtered out again by the linkage disequilibrium (LD) coefficient r^2. Only one SNP was selected among the SNP pairs with $r^2 \geq 0.8$. Thus, the SNPs can be treated independently. After filtering SNPs via LD, we set the total number of selected SNPs as $n / \log(n) \approx 585$ by the sure independence screening (SIS) criterion for the purpose of performing the GMDR analysis [19].

The same model that was used for selecting SNPs was used for the GMDR analysis. Sex, age, area, and BMI were used as environmental covariates. The regression models for GMDR are given as follows:

$$\log(HOMA - IR_i) = \beta_0 + \beta_1 SEX_i + \beta_2 AGE_i + \beta_3 AREA_i + \beta_4 BMI_i + \varepsilon_i,$$

$$(14)$$

The residual scores were then calculated based on this model. The threshold value of T was assumed as 1.5, 1.75, or 2 in the L-estimator and M-estimator. Table 3 shows the results obtained with the three GMDR methods. Each method provided a different SNP combination. These three pairs of SNPs have not previously been reported in the literature in relation to HOMA-IR. When we performed the analysis using the real data, the computational times required by the M-estimator GMDR were similar to those required for ordinary GMDR analysis. However, the L-estimator required slightly more computational time, because it contains an additional step to compute residuals.

Table 3: GMDR Analyses for log-transformed HOMA-IR in KARE data

Method	Best Combination	CVC	Average Test BA
Ordinary GMDR	rs576563 rs2920792	4	0.512535
L-estimator GMDR	rs4915657 rs7500315	9	0.518679
M-estimator GMDR	rs11125090 rs9353581	9	0.522633

Table 3 shows the results of three GMDR methods. Each GMDR method provided different SNP combination.

doi:10.1371/journal.pone.0135016.t003

We performed the network analysis with the top 100 SNP pairs with the highest BAs. Each node represents an SNP and each edge connecting two nodes represents an SNP pair. We defined the hub genes as genes with at least four edges. Table 4 summarizes the list of hub genes for log-transformed HOMA-IR identified by robust GMDRs and ordinary GMDR. The KCNH1 gene, which has the largest number of edges, was reported to have the function of insulin secretion [20]. Both M-estimator GMDR and L-estimator GMDR successfully identified the KCNH1 gene as a hub gene regardless of the threshold values. In addition, M-estimator GMDR with a threshold of 1.5 and L-estimator GMDR with all three threshold values identified the RYR2 gene as a hub gene. This gene is reported to be related with diabetes [21]. The PBX1 was found by M-estimator GMDR with a threshold of 2 and an ordinary GMDR. It was also reported that the PBX1 gene is related to type 2 diabetes [22]. In summary, the network analysis that was performed with the top 100 SNP pairs showed supporting evidence that the robust GMDR methods identified a larger number of hub genes that were reported in the literature than the ordinary GMDR method.

Table 4: Hub-genes for log-transformed HOMA-IR in KARE data with several methods

Method	threshold = 1.5	No. nodes	threshold = 1.75	No. nodes	threshold = 2	No. nodes
L-estimator	KCNH1*	4	KCNH1*	9	KCNH1*	13
	RYR2*	7	RYR2*	5	RYR2*	9
	LAMB3	5	LAMB3	4	HPCAL4	4
	C1orf168	3	SMYD3	5	SMYD3	4
	LMX1A*	3				
M-estimator	KCNH1*	43	KCNH1*	52	KCNH1*	52
	RYR2*	6			PBX1*	4
	KIF26B	4				
	TXNRD2	3				
Ordinary GMDR	PBX1*	31				
	TMEM51	5				
	VSIG2	4				
	KIF26B	3				

Table 4 shows the hub-genes for log-transformed HOMA-IR with different threshold. KCNH1 and RYR2 are commonly detected in L-estimator and M-estimator GMDR. However, Ordinary GMDR could not detect those genes.

doi:10.1371/journal.pone.0135016.t004

For a selected SNP pair, we investigated the high /low risk prediction results of individual cells.Fig 7 shows the boxplots of residuals that were calculated by ordinary GMDR, M-estimator GMDR, and L-estimator GMDR, respectively. Each cell displays three boxplots of GMDR methods. The red boxplot represents the high-risk group with the sum of the residuals larger than 0, whereas the blue boxplot represents the low-risk group with the sum of the residuals smaller than 0. The high/low classification pattern of L-estimator GMDR is the same as that of M-estimator GMDR. However, the pattern of ordinary GMDR differs from that of robust GMDRs. For example, in the first cell, when the numbers of minor alleles of rs4915657 and rs7500316 are (0,0), respectively, ordinary GMDR classified this cell as a low-risk group, whereas the two robust GMDRs classified it as a high-risk group. Note that the sum of residuals for ordinary GMDR becomes negative, which is due to the negative outlying observations in the tail part. Similarly, when the numbers of minor alleles in rs4915657 and rs7500316 are (0,1), (1,1), and (2,0), the cells showed the same tendency. The different results between the robust GMDRs and ordinary GMDR could be explained by the negative outlying residuals in the tail part. Because ordinary GMDR uses the original outlying residuals as they are, the sum of the residuals is highly affected by these outlying residuals. Thus, these outlying observations can cause errors in high/low classification. This investigation demonstrates the usefulness of robust GMDRs.

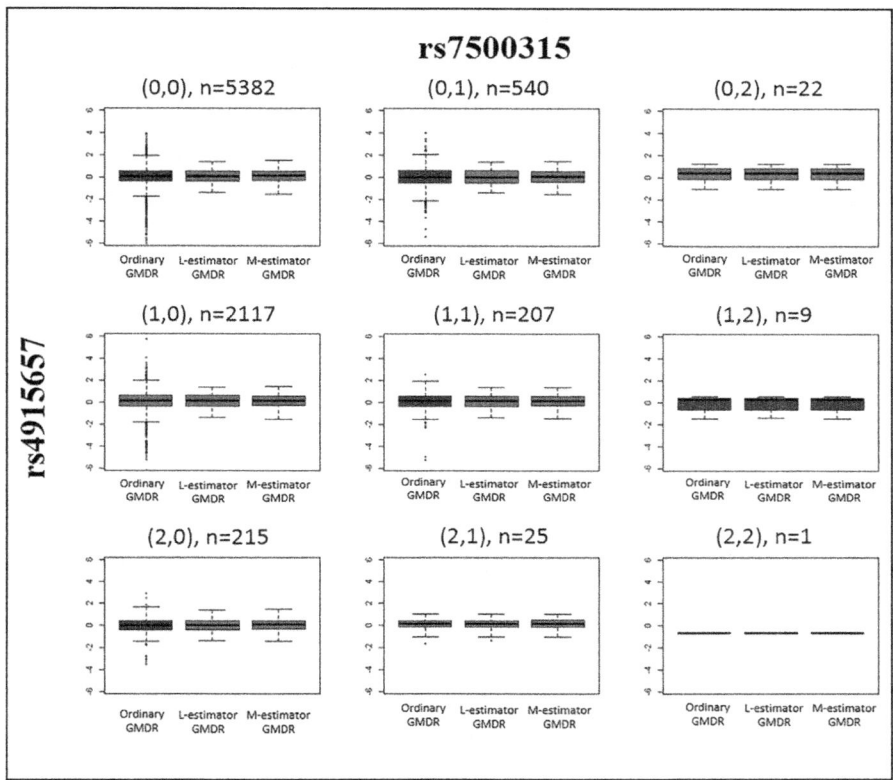

Figure 7: Boxplots of residuals calculated by ordinary GMDR, M-estimator GMDR and L-estimator GMDR.

Each cell displays three boxplots of the GMDR methods divided by the combination of rs4915657 and rs7500316. The red and blue boxplots represents the high-risk and low-risk groups with the sum of the residuals larger than and smaller than 0, respectively.

Discussion

When searching for hidden heritability, it is important to use an efficient method to establish the effects of GG interactions. To this effect, GMDR is a powerful method. However, using a simple example, it is demonstrated that the power of the GMDR method may decline in the presence of outlying observations. This problem was addressed by proposing the robust GMDR: L-estimator GMDR and M-estimator GMDR.

Simulation studies indicated that outlying observations caused the power of ordinary GMDR to decline critically. In contrast, the L-estimator and M-estimator GMDR methods were found to perform reasonably well even when outlying observations existed. When the samples are generated by assuming a Normal distribution, the L-estimator GMDR appears to outperform the other methods when the proportion of outlying observations is large. Otherwise, M-estimator GMDR and L-estimator GMDR perform similarly. If the samples are generated by assuming a Gamma distribution and there are outlying observations, L-estimator performs best. The results of the simulation study indicate that the power of GMDR with a correct distributional assumption is shown to be much larger than with an incorrect distributional assumption. Therefore, if we know the distribution of y_i, a parametric approach, such as maximum likelihood estimation, would be recommended.

In our robust GMDR analysis, we used BA as an evaluation measure for the prediction ability. BA uses less information than continuous evaluation measures when used for quantitative traits. However, as the GMDR approach is based on transforming quantitative traits into binary high/low responses, incorporating the continuous scale information directly into the robust GMDR framework is complicated. On the other hand, QMDR does not use binary responses; instead, it uses t-statistics to measure for quantitative traits [7]. Although the GMDR and QMDR methods use different frameworks, the concept of robustness used in GMDR can be used in QMDR. Thus, in future studies, we will extend our robust scheme to t-statistics with the classical statistical method.

APPENDIX

Without loss of generality, we could consider all kinds of 2nd order genetic models with a normal distribution by using the following formula (15).

$$y = \sum_{j=0}^{2} \sum_{i=0}^{2} \beta_{ij} I(SNP_1 = i, SNP_2 = j) + \varepsilon$$

(15)

Then, we could consider the variance of y to be represented by the following equations.

$$\text{var}(y) = \sigma^2 = \sigma_{envir}^2 + \sigma_{genetic}^2 = E(\text{var}(y|SNP_1, SNP_2)) + \text{var}(E(y|SNP_1, SNP_2))$$

(16)

$$= E(\sigma_{envir}^2) + \text{var}(\sum_{j=0}^{2} \sum_{i=0}^{2} \beta_{ij} I(SNP_1 = i, SNP_2 = j))$$

With this equation, we could calculate the narrow sense heritability h^2 as:

$$h^2 = \frac{\sigma^2_{genetic}}{\sigma^2} = \frac{\text{var}(\sum_{j=0}^{2} \sum_{i=0}^{2} \beta_{ij} I(SNP_1 = i, SNP_2 = j))}{\sigma^2} \tag{17}$$

AUTHOR CONTRIBUTIONS

Conceived and designed the experiments: TP YK. Performed the experiments: YK. Analyzed the data: YK. Contributed reagents/materials/analysis tools: YK TP. Wrote the paper: YK TP.

REFERENCES

1. Oh S, Lee J, Kwon MS, Weir B, Ha K, Park T (2012) A novel method to identify high order gene-gene interactions in genome-wide association studies: Gene-based MDR,BMC Bioinformatics, 13(Suppl 9):S5 doi: 10.1186/1471-2105-13-S9-S5. pmid:22901090

2. Gilbert-Diamond D, Moore JH (2011) Analysis of Gene-Gene Interactions. Current Protocols in Human Genetics. 70:1.14.1–1.14.12. doi: 10.1002/0471142905.hg0114s70

3. Freitas AA (2001) Understanding the crucial role of attribute interaction in data mining. Artif. Intel.Rev. 16:177–99.

4. Ritchie MD, Hahn LW, Roodi N, Bailey LR, Dupont WD, Parl FF et al. (2001) Multifactor-dimensionality reduction reveals high-order, Am J Hum Genet. Jul;69(1):138–47. Epub 2001 Jun 11 interactions among estrogen-metabolism genes in sporadic breast cancer. pmid:11404819 doi: 10.1086/321276

5. Lou XY, Chen GB, Yan L, Ma JZ, Zhu J, Elston RC et al. (2007) A Generalized Combinatorial Approach for Detecting Gene-by-Gene and Gene-by-Environment Interactionswith Application to Nicotine Dependence,The American Journal of Human Genetics, Volume 80, Issue 6, 1125–37, 1 June 2007 pmid:17503330 doi: 10.1086/518312

6. Cattaert T, Calle ML, Dudek SM, John JM, Lishoutet FV, Urrea V et al. (2011) Model-based multifactor dimensionality reduction for detecting epistasis in case-control data in the presence of noise.Ann Hum Genet. 2011 Jan;75(1):78–89. doi: 10.1111/j.1469-1809.2010.00604.x. pmid:21158747

7. Gui J, Moore JH, Williams SM, Andrews P, Hillege HL, van der Harst P et al. (2013) A Simple and Computationally Efficient Approach to Multifactor Dimensionality Reduction Analysis of Gene-Gene Interactions for Quantitative Traits. PLoS ONE 8(6): e66545. doi: 10.1371/journal.pone.0066545. pmid:23805232

8. Bennett SN, Caporaso N, Fitzpatrick AL, Agrawal A, Barnes K, Boyd HA et al. (2011) Phenotype Harmonization and Cross-Study Collaboration in GWAS Consortia: The GENEVA Experience, Genetic Epidemiology 35: 159–73. doi: 10.1002/gepi.20564. pmid:21284036

9. Most PJ, Vaez A, Prins BP, Munoz ML, Snieder H, Alizadeh BZ et al. (2014) QCGWAS: A flexible R package for automated quality control of genome-wide association results, Bioinformatics, 30 (8): 1185–86. doi: 10.1093/bioinformatics/btt745

10. Yang J, Loos RJ, Powell JE, Medland SE, Speliotes EK, Chasman DI et al. (2012) FTO genotype is associated with phenotypic variability of body mass index, Nature, 267–73.

11. Cho YS, Go MJ, Kim YJ, Heo JY, Oh JH, Ban HJ et al. (2009) A large-scale genome-wide association study of Asian populations uncovers genetic factors influencing eight quantitative traits. Nature Genetics. May;41(5):527–34 doi: 10.1038/ng.357. pmid:19396169

12. Gayoso-Diz P, Otero-Gonzalez A, Rodriguez-Alvarez MX, Gude F, Cadarso-Suarezet C, García F et al. (2011) Insulin resistance index (HOMA-IR) levels in a general adult population: curves percentile by gender and age. The EPIRCE study. Diabetes Res Clin Pract. Oct;94(1):146–55 doi: 10.1016/j.diabres.2011.07.015. pmid:21824674

13. Cho S, Kim K, Kim YJ, Lee JK, Cho YS, Lee JY et al. (2010) Joint Identification of Multiple Genetic Variants via Elastic-Net Variable Selection in a Genome-Wide Association Analysis. Ann Hum Genet. Sep 1;74(5):416–28 doi: 10.1111/j.1469-1809.2010.00597.x. pmid:20642809

14. Velez DR, White BC, Motsinger AA, Bush WS, Ritchie MD, Williams SM et al. (2007) A balanced accuracy function for epistasis modeling in imbalanced datasets using multifactor dimensionality reduction. Genet. Epidemiol., 31: 306–15. pmid:17323372 doi: 10.1002/gepi.20211

15. Koenker R, Zhao Q (1994) L-estimatton for linear heteroscedastic models, Journal of Nonparametric Statistics, 3:3–4, 223–35 doi: 10.1080/10485259408832584

16. Bickel PJ, Lehmann EL (1975) Descriptive Statistics for Nonparametric Models 11: Location, Ann. Statist., 3, 1045–69 doi: 10.1214/aos/1176343240

17. Owen M (2010) Tukey's Biweight Correlation and the Breakdown. Master's thesis. Pomona College.

18. Fan J, Lv J (2008) Sure independence screening for ultrahigh dimensional feature space, JRSSB, 70(5), 849–911. doi: 10.1111/j.1467-9868.2008.00674.x

19. Namkung J, Kim K, Yi S, Chung W, Kwon MS, Park T (2009) New evaluation measures for multifactor dimensionality reduction classifiers in gene–gene interaction analysis, Bioinformatics, 25 (3), 338–45. doi: 10.1093/bioinformatics/btn629. pmid:19164302

20. Yan L, Figueroa DJ, Austin CP, Liu Y, Bugianesi RM, Slaughter RS et al. (2004) Expression of voltage-gated potassium channels in human and rhesus pancreatic islets, Diabetes 53, 597–607. pmid:14988243 doi: 10.2337/diabetes.53.3.597

21. Yaras N, Ugur M, Ozdemir S, Gurdal H, Purali N, Lacampagne A et al. (2005) Effects of Diabetes on Ryanodine Receptor Ca Release Channel (RyR2) and Ca2+ Homeostasis in Rat Heart, Diabetes, 54(11) 3082–88 pmid:16249429 doi: 10.2337/diabetes.54.11.3082

22. Wang H, Chu W, Wang X, Zhang Z, Elbein SC (2005) Evaluation of sequence variants in the pre-B cell leukemia transcription factor 1 gene: a positional and functional candidate for type 2 diabetes and impaired insulin secretion, Mol Genet Metab. 86(3):384–91. Epub 2005 Sep 2 pmid:16140554 doi: 10.1016/j.ymgme.2005.07.008

Chapter 5

DYNAMIC PATHWAY ANALYSIS OF GENES ASSOCIATED WITH BLOOD PRESSURE USING WHOLE GENOME SEQUENCE DATA

Pingzhao Hu[1,4] and Andrew D Paterson[1,2,3]

[1] The Centre for Applied Genomics, The Hospital for Sick Children, 686 Bay Street, Toronto, ON, M5G 0A4, Canada

[2] Program in Genetics and Genome Biology, The Hospital for Sick Children, 686 Bay Street, Toronto, ON, M5G 0A4, Canada

[3] Dalla Lana School of Public Health, University of Toronto, Health Sciences Building, 155 College St, Toronto, ON, M5T 3M7, Canada

[4] Department of Biochemistry and Medical Genetics and George and Fay Yee Centre for Healthcare Innovation, University of Manitoba, 745 Bannatyne Avenue, Winnipeg, MB, R3E 0W3, Canada

ABSTRACT

Groups of genes assigned to a pathway, also called a module, have similar functions. Finding such modules, and the topology of the changes of the modules over time, is a fundamental problem in understanding the mechanisms of complex diseases. Here we investigated an approach that categorized variants into rare or common and used a hierarchical model to jointly estimate the group effects of the variants in a pathway for identifying enriched pathways over time using whole genome sequencing data and blood pressure data. Our results suggest that the method can identify potentially biologically meaningful genes in modules associated with blood pressure over time.

Background

It has long been recognized that genetic analysis of longitudinal phenotypic data is important for understanding the genetic architecture and biological variations of complex diseases. The analysis can help identify the stage of disease development at which specific genetic variants play a role. However, the statistical methods to analyze longitudinal genetic data are limited. A commonly used approach is to analyze the longitudinal genetic traits by averaging multiple response measurements obtained at different time points from the same individual. This approach may miss a lot of useful information related to the variability of repeated genetic traits, although it is simple and computationally less expensive. Linear mixed models have also been used for repeated measures data [1].

Recently, there has been a shift to testing rare variants, mostly using next-generation sequence technologies, for association with complex diseases. We explored dynamic pathway-based analysis of genes associated with blood pressure over time using whole genome sequencing data. We first performed gene-based association analysis at each of the 3 time points by stratifying the variants into rare and common. Then we performed pathway enrichment analysis separately at each time point. Finally, we built pathway crosstalk network maps using the enriched pathways to identify potential subnetworks associated with blood pressure over time.

METHODS

Data Description

For genotype data, we analyzed sequencing data of the 142 unrelated individuals on chromosome 3, which includes 1,215,120 variants. For phenotype data, we analyzed the simulated phenotypes of replicate 1. We analyzed 2 quantitative traits: systolic blood pressure (SBP) and Q1. SBP was measured at 3 time points (T1, T2, and T3), and was close to normally distributed (data not shown) after treatment effect adjustment (see below). There are 31 functional loci (genes) on chromosome 3 that influence the simulated SBP. Q1 was simulated as a normally distributed phenotype but not influenced by any of the genotyped single-nucleotide polymorphisms. It also has no correlation with SBP measured at T1, T2, and T3. The Pearson correlation of SBP at the 3 time points with Q1 based on the 142 unrelated individuals is −0.09 (p value = 0.27), −0.02 (p value = 0.78), and −0.006 (p value = 0.94), respectively. Q1 was generated primarily to facilitate assessment of type 1 error.

Adjust Treatment Effects

It has been shown that the association between measured blood pressure and underlying genotype is potentially confounded by antihypertensive treatment [2]. Following Cui et al [3], we adjusted SBP of subjects receiving antihypertensive medications by adding a constant value of 10 mm Hg at the 3 study exams (n = 22, 51, and 73; 15.5%, 35.9%, and 51.4%, respectively). Such a strategy had higher power than the alternatives [2].

Analyze Common and Rare Genetic Variants Using Hierarchical Models

We applied an extended hierarchical generalized linear model [4] to simultaneously analyze rare and common variants at the gene level. The model can be summarized as follows: Assume that the observed values of a given quantitative trait (SBP or Q1) are denoted as $Y = (y_1, \cdots, y_n)$ and the predictor variables, that are variants, can be categorized into 2 groups: rare (minor allele frequency <1%) and common variants (minor allele frequency $\geq$1%). The number of variants in the rare and common groups are J_1 and J_2, respectively. The extended hierarchical generalized linear model to fit the rare and common variants in a given gene can be expressed as a multiplicative form for the linear predictor η of individual i:

$$\eta_i = \beta_0 + \sum_{k=1}^{2} g_k \sum_{j \in G_k}^{J_k} \beta_j z_{ij}$$

where z_{ij} is the predictor of main-effect for individual i at genetic variant j in group G_k, equaling to the number of minor alleles for an additive coding and $Z_i = (z_{i1}, \ldots, z_{iJ})$, where $J = \sum_{k=1}^{2} J_k$ is the total number of variants. g_k represents the group effect for J_k variants in group G_k. β is a vector of all the coefficients and the intercept β.

The mean of Y is related to the linear predictor η via a link function h:

$$E(y_i|Z_i) = h^{-1}(Z_i\beta)$$

The data distribution is expressed as $p(y|Z\beta, \theta) = \prod_{i=1}^{n} p(y_i|Z_i\beta, \theta)$, where θ is a dispersion parameter, and the distribution $p(y_i|Z_i\beta, \theta)$ takes normal distribution. Because there are many highly correlated variants in a given gene in next-generation sequencing studies, a hierarchical framework is constructed for priors of the distributions of coefficients (g and β) in the model. The method was implemented in R package BhGLM.

We assigned the genetic variants to a gene if they were in the gene or within 10 kilobases (kb) of either side of the gene. We performed 2 analyses to evaluate the association between genotype and SBP at each study exam separately. First, we divided the variants within a gene into rare ($k = 1$) and common variants ($k = 2$). Separately we analyzed all the genetic variants in a gene, irrespective of allele frequency. Our main objective was to estimate gene effects g_k and to test the hypothesis $g_k = 0$, $k = 1$ (rare variants) and $k = 2$ (common variants) for the first analysis and $k = 1$ (rare and common variants) for the second analysis. We corrected for multiple testing using the Benjamini and Hochberg method [5].

Dynamic Pathway Analysis

We mapped the approximately 1200 genes on chromosome 3 to the c2 curated pathways (version 3) from the Broad Institute, which includes 2934 gene sets collected from 186 Kyoto Encyclopedia of Genes and Genomes, 430 Reactome, 217 BioCarta pathways, 880 canonical pathways, 825 biological process, and 396 molecular function gene ontology terms. We kept only the pathways with at least 5 genes in our data set, which left 531 pathways for analysis.

There are different ways to test for genes associated with an excess of SBP in the same pathway. We used the "gene set enrichment test" implemented in the limma R package [6]. The approach uses the Wilcoxon signed rank test to compute a p value to test the hypothesis that a given gene set tends to be more highly ranked than would be expected by chance. The ranking is based on a t-like test statistic, and here we used the z statistics from the hierarchical model described in above Section (Analyze common and rare genetic variants using hierarchical models). The test is essentially a streamlined version of the gene set enrichment analysis approach introduced by Subramanian et al [7].

We performed dynamic pathway crosstalk analysis between each pair of time points using the enriched pathways with a nominal p value of <0.05. Two pathways were considered to crosstalk if they shared at least 1 functional locus (gene). This ensures that each of pathway and its crosstalk has biological meaningfulness. We built pathway crosstalk subnetworks using Cytoscape

RESULTS

Given a false discovery rate (FDR) of 0.05 at the gene-level analysis, we identified 116, 57, 2, and 0 significant genes for SBP measured at T1, T2, T3, and Q1, respectively, using rare variants. However, there were no significant genes for SBP measured at the 3 time points and Q1 using common variants. Of those significant genes from the rare variant analysis, 4, 1, 0, and 0 were true positives (Table 1) for SBP measured at T1, T2, T3, and Q1, respectively. We observed that these 4 genes had significant positive associations with SBP (Table 1). For the gene-level analysis irrespective of allele frequency we identified many significant genes (468, 415, 306, and 214 for SBP measured at T1, T2, T3, and Q1, respectively) (Table 2). However, the vast majority of them were false positives (see false-positive rate analysis below), implying that, irrespective of allele frequency, the analysis strategy had a grossly inflated type I error, possibly as a result of linkage disequilibrium between variants. We calculated the false-positive rate and false-negative rare for T1, T2, T3, and Q1 using the 2 analysis approaches. We defined the positive as those genes that had an adjusted p value smaller than 0.05, and the negative as those genes that had an adjusted p value larger than or equal to 0.05. As shown in Table 2, irrespective of allele frequency, the analysis approach had many false positives compared with the approach that stratified by allele frequency. The genes based on common variants used in the first analysis had no power to detect the genetic association between genotypes and SBP.

Table 1: Significant association of causal genes with rare variants with SBP at T1 and T2 based on chromosome 3 gene-based tests

Time period and gene		Rare variants				Common variants			
		Estimate	SE	z Value	Adjust-ed p value	Esti-mate	SE	z Value	Adjust-ed p value
T1	*ABTB1*	0.71	0.24	3.01	0.0321	−0.15	0.21	−0.70	0.99
	SCAP	0.87	0.22	4.00	0.0058	−0.02	0.08	−0.27	0.99
	PROK2	3.30	0.90	3.66	0.011	−0.19	0.21	−0.93	0.99
	MUC13	0.65	0.22	2.96	0.035	0.00	0.08	0.02	1.0
T2	*SCAP*	1.06	0.27	3.89	0.0076	−0.08	0.08	−1.00	0.95

Note: There were no significant associations for causal genes including only either rare or common variants at T3 (or for Q1).

Table 2: False-positive rate (FPR) and false-negative rate (FNR) of gene-based analyses

Time period or trait	Gene-based stratified by allele frequency				Gene based	
	Rare variants		Common variants		Rare and common variants	
	FPR (%)	FNR (%)	FPR (%)	FNR (%)	FPR (%)	FNR (%)
T1	9.6	86.7	0.0	100.0	39.2	53.3
T2	4.8	96.7	0.0	100.0	34.9	63.3
T3	0.2	100.0	0.0	100.0	25.6	66.7
Q1	0.0	100.0	0.0	100.0	17.8	76.7

To further evaluate the whole-spectrum of power of the approaches to identify causal genes, we drew the receiver operating characteristic curves for each type of strategy at each time point and estimated its area under the curve. We found that each analysis strategy has different power to identify causal genes at different time points. Overall, the analysis based on rare variants had the largest power at T3, which also had larger power to detect disease-causing genes than common variants at T2.

Using the z statistic obtained from modeling rare variants in a gene, we did not find significant pathways associated with SBP at FDR of 0.05 level based on gene set enrichment analysis. However, given a nominal p value cutoff of 0.05, we identified the same 3 enriched pathways for SBP measured at 3 time points but not for Q1 (Table 3). Each of these 3 pathways included 1 «functional» gene (*FLNB*), which had 286 rare variants with 1 functional variant (chr3: 58109162, explained 0.00273 of the variance for SBP).

Table 3: Enriched pathways found in T1, T2, and T3

Pathway names	No. of genes	No. of genes on chr. 3	No. of function-al loci	*p* Value of T1	*p* Value of T2	*p* Value of T3
Actin filament-based process	114	10	1	0.021	0.016	0.018
Actin cytoskel-eton organization and biogenesis	104	10	1	0.021	0.016	0.018
Cytoskeleton organization and biogenesis	205	12	1	0.030	0.013	0.022

To identify pathway crosstalk, we built 2 pathway subnetworks (Figure 1) for the pathways with nominal p value smaller than 0.05. Two pathways crosstalk if at 2 time points they included at least 1 common true gene. As shown in Figure 1, we found that there was a subnetwork formed by the «actin filament based process,» the «actin cytoskeleton organization and biogenesis,» and the «cytoskeleton organization and biogenesis and organelle organization and biogenesis» pathways that were consistently enriched across adjacent time periods (T1 $\rightarrow$ T2 and T2 $\rightarrow$ T3).

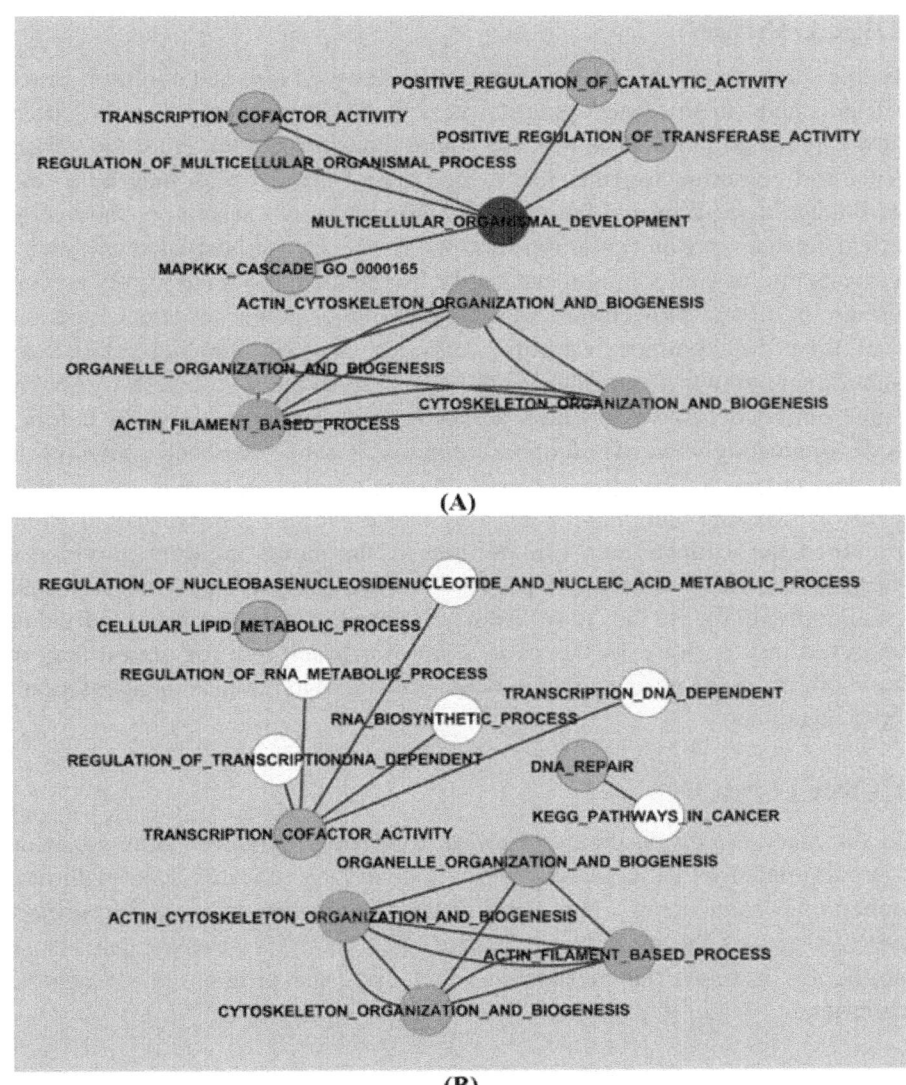

(A)

(B)

Figure 1: Dynamic pathway crosstalk: (A) T1 and T2; (B) T2 and T3. Pathways with blue, red and yellow are enriched at T1, T2, and T3, respectively. Pathways with green are enriched at both T1 and T2 (A) and at both T2 and T3 (B). A single line between 2 pathways indicates that each of the 2 pathways is enriched in only 1 of the 2 time points; a double-line between 2 pathways indicates that both pathways are enriched in both T1 and T2 (A) and in both T2 and T3 (B).

DISCUSSION

In this study, we evaluated the associations between rare and common genetic variants and the simulated quantitative trait (SBP) measured at 3 time points at the gene and pathway levels. We found that joint modeling all the variants (rare and common) together had a high type I error, which may be a result of linkage disequilibrium between common and rare variants, or the average effect between rare and common variants. However, a strategy that categorized variants into rare or common and used a hierarchical model to jointly estimate the group effects showed rare variants had higher power to detect functional loci than did common variants. Although we did not find statistically significant pathways associated with SBP (FDR of the 0.05 level), we showed some enriched pathways shared across time at a nominal p value cutoff of 0.05. Interestingly, we also found a subnetwork with 3 enriched pathways that showed crosstalk between each pair of time points, suggesting the dynamic pathway crosstalk may have a key role in the pathogenesis of SBP. It should be noted the «›functional» loci defined in simulation answers provided by Genetic Analysis Workshop 18 (GAW18) organizers were polymorphic based on all individuals, but they may be not polymorphic in the unrelated individuals analyzed in this study. In this case, some functional loci (or genes) may not have effects in the unrelated data, which may lead to the bias in calculation of false negatives.

CONCLUSIONS

In summary, we proposed a framework to identify dynamic pathways which have the potential in regulating SBP via analyzing repeated traits with next-generating sequencing. This can generate insights into the progressive mechanisms of the underlying disease. This analysis strategy can also be applied to examine the mechanisms that drive the progression of complex diseases.

ACKNOWLEDGEMENTS

This work was supported by a grant from Genome Canada through the Ontario Genomics Institute. ADP holds a Canada Research Chair in Genetics of Complex Diseases.

The GAW18 whole genome sequence data were provided by the T2D-GENES Consortium, which is supported by NIH grants U01 DK085524, U01 DK085584, U01 DK085501, U01 DK085526, and U01 DK085545. The other genetic and phenotypic data for GAW18 were provided by the San Antonio Family Heart Study and San Antonio Family Diabetes/Gallbladder Study, which are supported by NIH grants P01 HL045222, R01 DK047482, and R01 DK053889. The Genetic Analysis Workshop is supported by NIH grant R01 GM031575.

This article has been published as part of *BMC Proceedings* Volume 8 Supplement 1, 2014: Genetic Analysis Workshop 18. The full contents of the supplement are available online at Publication charges for this supplement were funded by the Texas Biomedical Research Institute.

AUTHORS' CONTRIBUTIONS

PH designed the study, performed the data analysis, and drafted the manuscript. ADP participated in designing the study. ADP supervised the study. All authors helped revise the manuscript. All authors read and approved the final manuscript.

REFERENCES

1.   Paterson AD, Waggott D, Boright AP, Hosseini SM, Shen E, Sylvestre MP, Wong I, Bharaj B, Cleary PA, Lachin JM, et al.: A genome-wide association study identifies a novel major locus for glycemic control in type 1 diabetes, as measured by both A $_{1C}$ and glucose.Diabetes 2010, 59:539–549.

2.   Tobin MD, Sheehan NA, Scurrah KJ, Burton PR: Adjusting for treatment effects in studies of quantitative traits: antihypertensive therapy and systolic blood pressure. Stat Med 2005, 24:2911–2935.

3.   Cui JS, Hopper JL, Harrap SB: Antihypertensive treatments obscure familial contributions to blood pressure variation. Hypertension2003, 41:207–210.

4.   Yi N, Liu N, Zhi D, Li J: Hierarchical generalized linear models for multiple groups of rare and common variants: jointly estimating group and individual-variant effects. PLoS Genet 2011, 7:e1002382.

5.   Benjamini Y, Hochberg Y: Controlling the false discovery rate: a practical and powerful approach to multiple testing. J R Stat Soc Series B Stat Methodol 1995, 85:289–300.

6.   Smyth GK: Linear models and empirical Bayes methods for assessing differential expression in microarray experiments. Stat Appl Genet Mol Biol 2004., 3:: Article 3

7.   Subramanian A, Tamayo P, Mootha VK, Mukherjee S, Ebert BL, Gillette MA, Paulovich A, Pomeroy SL, Golub TR, Lander ES, Mesirov JP:Gene set enrichment analysis: a knowledge-based approach for

interpreting genome-wide expression profiles. Proc Natl Acad Sci USA 2005, 102:15545–15550.

Chapter 6

NETWORK ANALYSIS OF GENE ESSENTIALITY IN FUNCTIONAL GENOMICS EXPERIMENTS

Peng Jiang[1] , Hongfang Wang[2] , Wei Li[1] , Chongzhi Zang[1] , Bo Li[1] , Yinling J. Wong[2] , Cliff Meyer[1] , Jun S. Liu[3] , Jon C. Aster[2] and X. Shirley Liu[1,4]

[1]Department of Biostatistics and Computational Biology, Dana-Farber Cancer Institute, Harvard T.H. Chan School of Public Health, Boston, MA 02215, USA

[2] Department of Pathology, Brigham and Women's Hospital, Boston, MA 02115, USA

[3] Department of Statistics, Harvard University, Cambridge 200092, USA

[4] School of Life Science and Technology, Tongji University, Shanghai, MA 02138, China

ABSTRACT

Many genomic techniques have been developed to study gene essentiality genome-wide, such as CRISPR and shRNA screens. Our analyses of public CRISPR screens suggest protein interaction networks, when integrated with gene expression or histone marks, are highly predictive of gene essentiality. Meanwhile, the quality of CRISPR and shRNA screen results can be significantly enhanced through network neighbor information. We also found network neighbor information to be very informative on prioritizing ChIP-seq target genes and survival indicator genes from tumor profiling. Thus, our study provides a general method for gene essentiality analysis in functional genomic experiments.

Background

Essential genes are those genes critical for cell viability under certain contexts. Recent years have seen the rapid development of functional genomics techniques for studying gene essentiality genome-wide. For example, large-scale shRNA screens have been used to search for essential genes in diverse cell lines [1]. If a specific transcription factor drives the cell viability under certain condition, ChIP-seq technique can be used to profile the regulatory targets to further find essential genes [2]. Many computational methods have also been developed to predict context specific gene essentiality through integration of gene expression, molecular alterations, and biological pathways [3].

Recently, the CRISPR (clustered regularly interspaced short palindromic repeats) screen emerged as an exciting new approach to profile gene essentiality at genome scale [4–11]. In the CRISPR system, single-guide RNAs (sgRNA) direct Cas9 nucleases to induce double-strand breaks (DSB) at targeted genomic regions [12, 13]. When the error-prone non-homologous end-joining mechanism repairs the DSBs, insertions and deletions occur with high frequency, which produce a non-functional protein. Catalytically inactive Cas9 fused with a transcriptional activator or repressor has also been used to modulate gene expression at targeted loci [8, 9, 14–17]. Combined with lentiviral delivery method, CRISPR systems enable genome-scale functional screening in a cost-effective manner [4–11]. In CRISPR screens, sgRNAs targeting candidate genes are synthesized, and viral integration enables readout through next-generation sequencing [18]. The relative abundances of each integrated sgRNA between different conditions are compared and the importance of sgRNA target gene is inferred according to its sgRNAs' effect on cell growth.

The progress of CRISPR screen technology enabled systematic and reliable determination of gene essentiality under diverse conditions. The high quality gene essentiality profiles from CRISPR could enable a better comparison among essentiality prediction methods and better identification of distinct features of the essential genes. Such features not only facilitate a better understanding of the CRISPR screen data, but also can help prioritize the leads from CRISPR screens. From the analysis of yeast protein interactions, it is well known that highly connected proteins in a network (degree hubs) are more likely to be essential for viability [19–21]. Thus, we hypothesize that the gene essentiality outcome in CRISPR screens might depend on the gene connectivity in biological networks. Protein interaction networks have been integrated to improve the quality of RNAi screen results, which are very noisy due to off-target effect and low knockdown efficiency [22–25]. These previous works on RNAi screen indicate that the CRISPR screen result quality may also be improved by integration with protein interaction networks.

In this study, we took a network perspective and developed a method called NEST (Network Essentiality Scoring Tool) to systematically analyze the recent genome wide CRISPR screen data. We found that gene essentiality determined by CRISPR screen largely depends on the expression level of interacting genes in the biological network. Moreover, the quality of CRISPR and shRNA screen data can be further improved by NEST after considering the gene neighborhood screen outcome. Besides applications on CRISPR and shRNA screens, NEST is also generally applicable on many other types of genomics data analysis, such as ChIP-seq target gene prioritization and survival gene identification from tumor profiling data.

RESULTS AND DISCUSSION

NEST Predicts Gene Essentiality in CRISPR Screen

We first collected recently published CRISPR loss-of-function screen data [4, 5, 8], and selected three cell lines (K562, HL60, and A375) with publicly available gene expression data [26–28]. The significant CRISPR screen gene hits are called with software MAGeCK [29]. In CRISPR screens for growth phenotype, most significant genes are negatively selected, which means these genes are essential in the corresponding experimental condition (Additional file 1: Figure S1). To identify distinct features of gene essentiality in CRISPR screens, we developed a network-based method called NEST (Network Essentiality Scoring Tool), and found the following metric to give reliable performance.

For each gene, NEST calculates neighbor expression measure as the sum of normalized expression of its neighbor genes connected in the protein interaction network, weighted by the interaction confidence (Fig. 1a). The calculation of NEST score can also be formulated as product between connectivity matrix, which is composed of interaction weights between protein pairs, and gene expression vector. Each gene's relative expression in one cell is normalized against its average expression across all cell lines, and the protein interaction network information is from STRING [30] (Additional file 1: Figure S2). For essential genes selected by CRISPR screen, we defined the gold standard set as the genes hits called by MAGeCK with FDR threshold 0.05 [29]. For each measure, such as NEST score or network degree, all genes were ranked by their values in descending order. Receiver operating characteristic (ROC) curve was used to test the performance of predicting the CRISPR screen gold standard set (Fig. 1b).

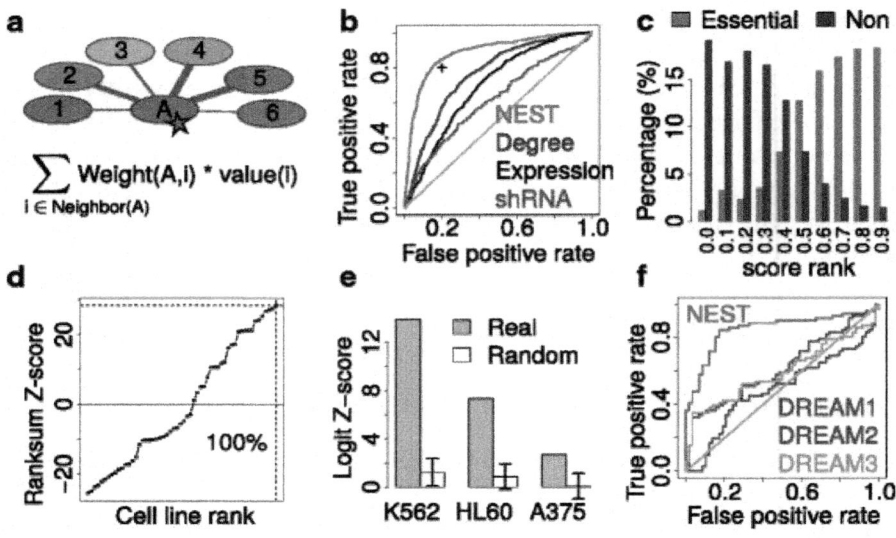

Figure. 1: Prediction of CRISPR screen outcome. **a** NEST calculates the neighbor expression of a gene as the sum of expression values of its neighbor genes connected in the network, weighted by the interaction weight. **b** Receiver operating characteristic (ROC) curve is used to test the performance of predicting gene essentiality determined by K562 CRISPRi screen. The performance of NEST score, network degree, gene expression, and shRNA screen are shown. The black point represents false positive rate 0.2 and true positive rate 0.8. **c** The NEST scores are converted to rank percentiles from 0 to 1, and shown for essential genes and non-essential genes determined in K562 CRISPRi screen. **d** For each Roadmap expression profile, we calculated the prediction power of NEST score on gene essentiality in K562 screen by Wilcoxon rank-sum test. The rank-sum Z-scores for all cell lines are ranked and the K562 profile has the largest value. **e** The STRING network was randomized 1,000 times, and the NEST scores were calculated for random networks. We used multivariate logistic regression to test the association of NEST score with gene essentiality after controlling the effects of network degree and gene expression (Table 1). The Logit Z-scores are shown for real and random networks. **f** In DREAM gene essentiality prediction challenge, A375 cell line also has CRISPR screen data available. Using essential genes selected in CRISPR screen as gold standard, the prediction performance is compared between NEST (red) and the top three winners in DREAM

For gene essentiality prediction in K562 CRISPRi screen, NEST achieved a false positive rate of 0.2 and a true positive rate of 0.8, with an area under the ROC curve (AUC) of 0.89. The AUC of NEST score is consistently better than network degree, gene expression, and shRNA screen data from the Achilles project [1] (Delong P value <1e-10 for all comparisons). Similar performance differences were also observed in CRISPR screen in HL60 and

A375 (Additional file 1: Figure S3a). To visualize the CRISPR prediction performance in an intuitive way, we plotted the rank percentile of NEST scores for essential genes and non-essential genes in CRISPR screen (Fig. 1c and Additional file 1: Figure S3B). The NEST ranks are significantly higher for essential genes than non-essential genes (Wilcoxon rank-sum P value <1e-10 for cell lines). Besides STRING network, we also used other large-scale networks for CRISPR outcome prediction. However, we did not find any performance improvement using either other network or merged network among several data sources (Additional file 1: Figure S4).

The results above suggest that if a gene's network neighbors are over-expressed in some conditions, the gene itself becomes more essential. We also found that genes with high NEST scores are tightly clustered in protein interaction network. The STRING network genes were grouped into 2,271 dense complexes using SPICi [31]. Gene with high NEST scores tend to stay in fewer number of STRING clusters than clusters with gene names shuffled (Additional file 1: Figure S5). Thus, a high NEST score may indicate the gene to be member of an active protein complex.

To test the prediction specificity, we applied NEST for gene expression profiles of 56 cell lines profiled by Roadmap project [26]. Measured by rank-sum test Z-scores, K562 CRISPRi screen data achieved the highest association with NEST score in K562 cell than all other cell lines (Fig. 1d). Similarly, HL60 and A375 CRISPR screen data also achieved higher associations with NEST scores in the same cell line (Additional file 1: Figure S3C). Housekeeping genes, such as ribosome members are often selected as essential genes in CRISPR screens [5, 8]. Thus, we further tested that the high prediction power of NEST scores was not purely derived from the same set of housekeeping genes. The prediction performance of NEST remains high after removal of housekeeping genes annotated previously [32] (Additional file 1: Figure S6). Notably, the majority of essential genes selected in CRISPR screen do not overlap between K562, HL60, and A375 (Additional file 1: Table S1). Thus, our NEST score is an orthogonal feature of CRISPR selected gene essentiality other than the universal housekeeping genes shared across conditions.

Since gene network degree and gene expression are also predicative of gene essentiality (Fig. 1b), we then tested whether the prediction performance of NEST is simply an additive effect of network degree and gene expression (Table 1). Using the gene essentiality in CRISPR screen as responsive variable, we did a multivariate logistic regression among all three covariates (NEST score, gene network degree, and expression). While all covariates are predictive of gene essentiality jointly, NEST has the largest statistical significance defined as Logit Z-score (Table 1). Moreover, the logistic regression fitted value, combining all covariates together, did not improve the CRISPR prediction performance comparing to NEST score alone (Additional file 1: Figure S7). As a further control, we randomized the STRING network but preserved the network degree for each gene [33]. The Logit Z-scores for the NEST score in random networks are significantly smaller than in real data (Fig. 1e, empirical P value <0.001 for K562 and HL60, and P value=0.003 for A375).

Table 1: Confounding factors for NEST prediction performance

Covariate	Coefficient	Standard error	Z-score	P value
NEST	0.02329	0.001748	13.32	1.72e-40
Degree	0.00415	0.000846	4.91	9.33e-07
Expression	0.12937	0.054223	2.39	1.70e-02
A. K562				
NEST	0.03494	0.00505	6.91	4.73e-12
Degree	0.00608	0.00146	4.16	3.13e-05
Expression	0.33873	0.16595	2.04	4.12e-02
B. HL60				
NEST	0.07296	0.02483	2.94	0.00329
Degree	0.00792	0.00357	2.22	0.02647
Expression	1.12266	0.48343	2.32	0.02022
C. A375				

The prediction power of NEST score on gene essentiality decided by CRISPR screen is tested through logistic regression with gene network degree and gene expression as covariates. The Logit Z-score is defined as Coef/Stderr. The P value is calculated by Ward test. The result is shown for (A) K562, (B) HL60, and (C) A375

There have been many previous methods developed for gene essentiality prediction. Since CRISPR screen measures the gene essentiality, any previous methods can be predictive for CRISPR outcome. In a recent DREAM challenge, contenders were asked to develop algorithms to predict cell specific gene essentiality [3]. Among cell lines included in the DREAM challenge, A375 has CRISPR screen data available. We compared the CRISPR outcome prediction performance between our method and the top three methods from the DREAM challenge, and found NEST to consistently outperform all DREAM winners (Fig. 1f). Besides the methods in DREAM, we also compared the performance of NEST with other methods using gene expression and network to predict gene essentiality [34, 35]. NEST significantly outperformed all other methods (Additional file 1: Figure S8 and Additional file 1: Methods).

Besides gene expression, we also used H3K27ac histone mark data to compute NEST scores and tested the gene essentiality prediction performance. Previously, we developed a method to calculate the regulatory potential (RP) scores of a histone modification on each gene promoter from the ChIP-seq profile [36, 37]. Based on our previous method, gene level RP scores in K562 cell were computed using the Roadmap H3K27ac ChIP-seq profile [26]. For each gene, NEST computed neighbor H3K27ac score as the sum of H3K27ac RP scores of its neighbor genes connected in the protein interaction network, weighted by the interaction confidence (Fig. 1a). H3K27ac NEST scores could also reliably predict the gene essentiality in K562 CRISPRi screen (Additional file 1: Figure S3), suggesting the applicability of NEST analysis on histone modification data.

NEST Enhances the Quality of CRISPR Screen Results

Since early CRISPR screens might have inefficient sgRNA selection and few sgRNA per gene, these screens may not give very strong hits. Encouraged by the prediction performance, we checked whether the network neighbor information could enhance the quality of CRISPR screen results. To measure the quality of a screen data, we need to know the expected outcome. In a K562 CRISPRi screen, the authors performed a genome-scale selection for genes that modulate sensitivity to Cholera/Diphtheria toxin [8]. For genes that work with the toxin, their knock out will protect the cell against the toxin and induce a positive gene selection in screen For genes that are targeted by toxin, their knock out will sensitize the cell for toxin effect and induce a negative gene selection The positively selected genes, which played a protective role against toxin, were enriched in KEGG pathways 'Vibrio Cholerae Infection' and 'Glycosphingolipid Biosynthesis' [8]. The negatively selected genes, which played a sensitizing role for toxin, were enriched in 'Ribosome' and 'Proteasome' pathways [8]. We used these enriched pathway genes as gold standard and tested how well network interaction could improve the CRISPR screen result (Fig. 2).

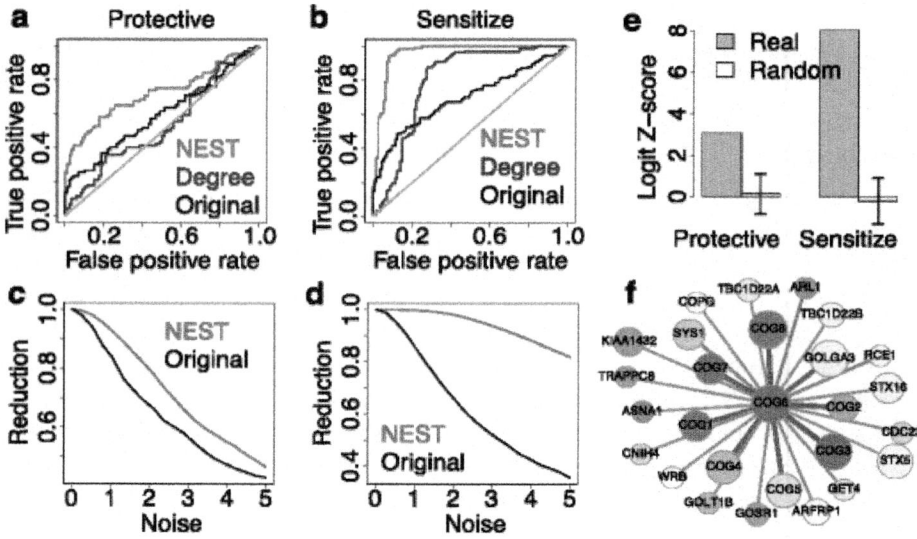

Figure. 2: Enhancement of CRISPR screen result. **a** For K562 CRISPRi screen under Cholera/Diptheria toxin selection, the gold standard of toxin protective genes comes from KEGG pathways 'Vibrio Cholera Infection' and 'Glycosphingolipid biosynthesis'. For each gene, NEST computes the neighbor CRISPR score as the sum of CRISPR screen fold change scores of neighbor genes connected in the network. The prediction performance is compared between NEST and original CRISPR scores. **b** The gold standard of toxin sensitizing genes comes from KEGG pathways 'Ribosome' and 'Proteasome'. The prediction performances of NEST and original CRISPR values are compared. **c** The original CRISPR values were randomized by Gaussian white noise. The standard deviation of all original CRISPR values was used as base level. At each noise level relative to the base level, the area under ROC curve (AUC) of prediction is compared with the initial AUC for toxin protective genes in K562 Cholera toxin screen. The reduction ratios were plotted for NEST and original CRISPR scores. **d** The reduction ratios were plotted for toxin sensitizing genes. **e** The STRING network was randomized and the NEST scores were calculated for 1,000 random networks. We used multivariate logistic regression to test the association of NEST scores with gold standards, after controlling the effects from network degree and original CRISPR score. The Logit Z-scores are shown for real and random networks. **f** As an example of high NEST score gene, *COG6* is a component of Golgi complex, and connected with several other components in Golgi complex. The thickness of each edge represents the interaction weight. The color of each gene represents the CRISPR screen fold change score; red color indicates toxin protective and blue color indicates toxin sensitizing

For each gene, NEST calculated a neighbor CRISPR score by adding up the CRISPR fold change scores among neighbor genes connected in the STRING network, weighted by the interaction confidence. This NEST score is significantly more predictive on the gold standard outcome than the original CRISPR scores for both protective and sensitizing genes (Fig. 2a, b, Delong test P value=0.010 for protective genes, P value=9.92e-14 for sensitizing genes). Moreover, when we put different levels of Gaussian white noise into CRISPR screen scores, the prediction performance of NEST score diminishes slower than original CRISPR scores (Fig. 2c, d). As a control, if we calculated the NEST scores from randomized network, the prediction power became significantly worse (Fig. 2e, P value <0.001 for both protective and sensitizing genes). Thus, through the connectivity of protein interaction network, NEST can enhance the quality of CRISPR screen result.

As an example of gene with high NEST score, *COG6* is a member of Golgi complex and its NEST score is significantly larger than expected (permutation test P value <0.001). *COG6* is connected with many other members of Golgi complex (Fig. 2f), and most of them have positive CRISPR screen fold change scores. Since they are connected with each other in network, they mutually boosted each other's NEST scores. Our result is consistent with the knowledge that cholera toxin needs to enter host cells and travel through the trans-Golgi network to take effect [38].

The above results suggest that if a gene's network neighbors are under CRISPR screen selection, the gene itself is more likely to be under CRISPR screen selection in the same direction. Besides CRISPR screen, we applied NEST on the Achilles shRNA screen data [1]. NEST can also significantly enhance the quality of shRNA screen result (Additional file 1: Figure S9). Thus, in general, the quality of functional genomic screen result can be improved by considering the gene network neighbor information.

Previously, there were methods developed to improve the quality of RNAi screen results from integration with protein interaction networks [25]. For CRISPR enhancement, we compared our method NEST against NePhe, which was a leading method on RNAi screen network analysis [24]. Using K562 toxin screen as the gold standard, we found that NePhe and NEST show similar performance as measured by ROC curves (Additional file 1: Figure S10AB). However, while NePhe used 14 GB memory and 6.2 h running time, NEST only used 8.3 MB memory and 10.8 s (Additional file 1: Figure S10C). Thus, NEST maintains reliable screen enhancement performance of previous method with much better computational efficiency.

NEST Prioritizes ChIP-Seq Essential Targets

Besides functional genomic CRISPR/shRNA screen, many other genomic experimental techniques can be used to search for essential genes. For example, if a specific transcription factor (TF) drives the cell viability under certain condition, ChIP-seq technique can be used to profile its regulatory targets to further find essential genes [2]. The previous analyses demonstrate that NEST can identify the essential genes in a CRISPR screen. We further explored whether NEST can help prioritize key target genes in a ChIP-seq experiment. ChIP-seq often finds tens of thousands in vivo binding sites for a TF. Since target genes can be regulated by TF binding through long range DNA looping, often thousands of genes near the TF binding sites can be putative targets, and it is hard to prioritize the functional target genes directly from a ChIP-seq experiment. We therefore investigated using network neighbor information to prioritize the functional TF target genes.

Our previous studies of NOTCH1 ChIP-seq and gene expression profiles in the T-lymphoblastic leukemia (TLL) cell line CUTLL1 identified 1,012 differential NOTCH1 binding sites between the *NOTCH* on and off conditions [2]. Based on the ChIP-seq peaks, we calculated a regulatory potential (RP) score for each gene [36, 37, 39], a distance-weighted sum of binding sites measuring the overall regulatory impact of differential NOTCH1 binding on target genes. We set the KEGG NOTCH signaling pathway members as the gold stand, and tested the prediction performance of expression, ChIP-seq RP and NEST scores (Fig. 3). In addition to NEST scores computed from gene expression (NEST E), we also computed NEST scores from ChIP-seq (NEST C), which measures the sum of ChIP-seq RP scores of neighbor genes connected in network. While expression and ChIP-seq measures are barely better than random, NEST scores can predict the annotated KEGG NOTCH signaling pathway members at AUC 0.90 and 0.95 (Fig. 3). It suggests that if a gene's network neighbors are enriched in the binding target of a TF, the gene itself is more likely to be regulated by the same TF.

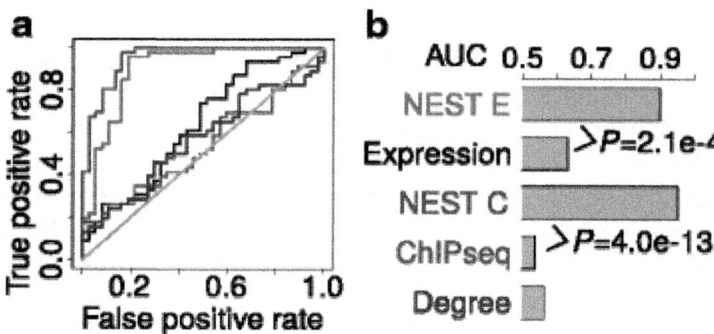

Figure. 3: Prediction of NOTCH signaling pathway members. **a** The differential gene expression between *NOTCH* on and off conditions is used to calculate the NEST E score. The NOTCH1 ChIP-seq regulatory potential score for each gene is used to calculate the NEST C score. The KEGG Notch Signaling pathway members are

used as gold standard and the prediction performances of all measures are shown by ROC curves. **b** The area under ROC curve (AUC) is shown for each measure. The comparison between AUCs is done by Delong test

NEST Predicts Cancer Patient Survival

Encouraged by the above analyses in cell lines, we checked whether NEST could facilitate the analysis of tumor profiling data. There have been previous studies integrating biological networks with cancer (or disease) biology data to understand the mechanisms of pathogenesis [35, 40–43]. Inspired by these studies, we examined the TCGA tumor profiling data to see whether NEST score computed from gene expression can better predict patient survival than gene expression. For example, over-activation of oncogene *EGFR* is a key feature of Glioblastoma (GBM) [44]. In TCGA GBM profiles [45], while*EGFR* over-expression does not correlate with worse survival (Fig. 4a, Cox-PH *P* value=0.109), higher *EGFR* NEST score is significantly associated with worse survival (Fig. 4b, Cox-PH *P* value=0.001).

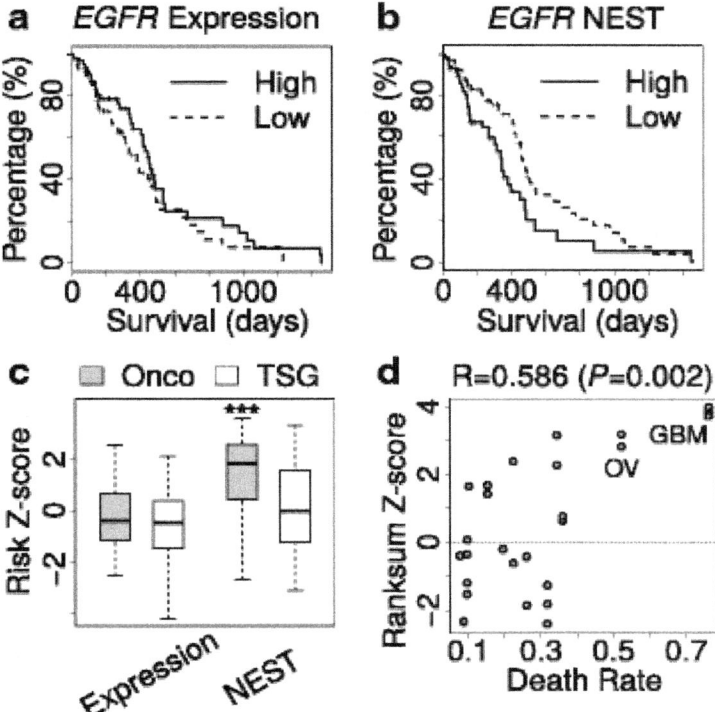

Figure. 4: Prediction of patient survival. **a** All TCGA Glioblastoma (GBM) patients are ranked by *EGFR* expression; the top half patients are assigned as high group and

the lower half are assigned as low. Kaplan Meier (KM) survival plot is shown for two groups. **b** The survival analysis is done in the same way as A for *EGFR* NEST scores. **c** For each gene, we calculated a death risk Z-score by Cox-PH model from either gene expression or NEST score. We compared the Z-scores for oncogenes (Onco) and tumor suppressor genes (TSG) based on the annotation from Vogelstein et al. The bottom and top of the boxes are the 25th and 75th percentiles (interquartile range). Whiskers on the top and bottom represent the maximum and minimum data points within the range represented by 1.5 times the interquartile range. The distribution of Z-scores is compared by Wilcoxon rank-sum test and three stars represent *P* value <0.001. **d** For each TCGA cohort, the difference of risk Z-scores computed from NEST was tested by Wilcoxon rank-sum test. The rank-sum Z-scores are plotted against the death rate of each cancer type, with Spearman's rank correlation as title

To systematically evaluate the survival prediction performance, we hypothesized that a good gene-wise survival predictor should show significant higher death risk for oncogenes than for tumor suppressors. We tested this hypothesis on all the annotated oncogenes and tumor suppressors [46] using the TCGA GBM data (Fig. 4c). While gene expression showed no significant difference on survival risk Z-scores, NEST gave significantly higher survival risk for oncogenes than tumor suppressors (Fig. 4c). This observation was corroborated in another independent GBM cohort [47] (Additional file 1: Figure S11), suggesting NEST score to be a much better indicator of GBM survival than gene expression alone. To examine the survival prediction performance of NEST in other cancer types, we used the Wilcoxon rank-sum test to measure the difference of survival risk Z-scores between oncogenes and tumor suppressors. A positive rank-sum Z-score indicates oncogenes with higher survival risk than tumor suppressors, and a negative Z-score indicates the opposite. For low death rate cancers, the Cox-PH survival regression may not get accurate risk estimation for each gene. In contrast, cancer types with high death rate, such as GBM and ovarian cancer (OV), seemed to give positive rank-sum Z-scores that separate oncogenes from tumor suppressors (Fig. 4d). These results suggest that if a gene's neighbors are over expressed in tumors, the gene itself is more likely to be an oncogene with associated survival risk.

We conducted pathway analysis on all the genes whose NEST scores are associated with GBM survival (FDR <= 0.05), and found 'cytokine cytokine-receptor interaction' to be the most enriched KEGG pathway (Additional file 1: Table S2). It was known that cytokines played a pivotal role in the pathogenesis of GBM [48], so we plotted the outcome-associated cytokine genes using CytoScape [49] (Fig. 5). Many of them are known oncogenes in GBM, such as *EGFR* and *CSF1R* [46], and several also have known targeted drugs from Drug Bank [50]. For example, the inhibitors of *EGFR*, *CSF1R*, and *CXCR4* were shown to reduce the invasiveness of Glioma cells or block

GBM progression [51–53]. Besides the known druggable genes, many other genes in our prediction could serve as promising targets. For example, NEST predicted *KITLG* as indicator of poor GBM survival which is consistent with the finding that downregulation of *KITLG* inhibits angiogenesis and Glioma growth [54]. Thus, our predictions could sketch a general landscape to investigate therapeutic possibilities for GBM and other cancers.

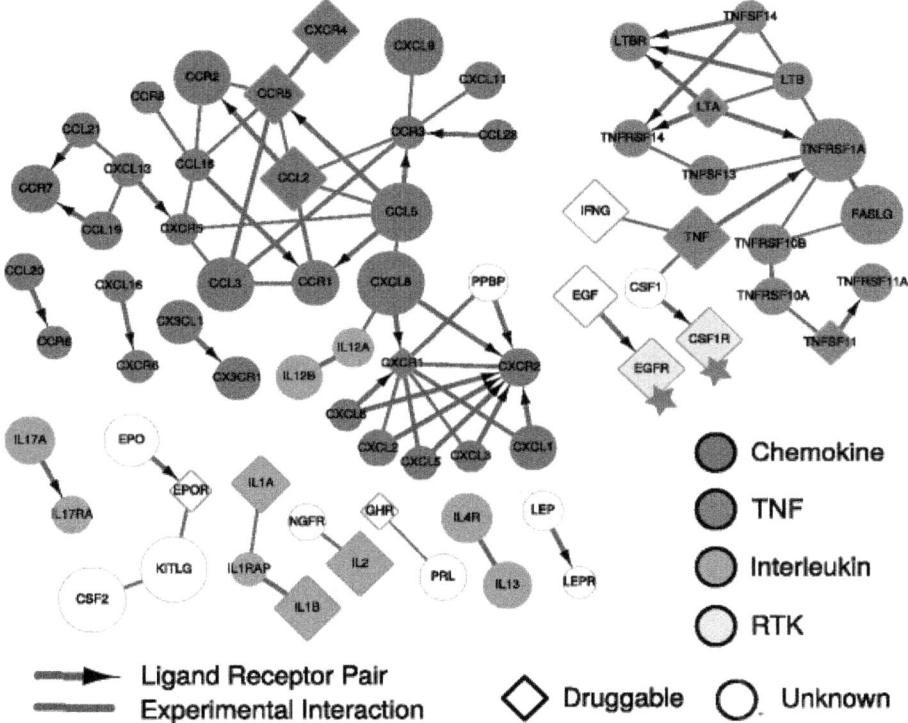

Figure. 5: Cytokine receptor interaction network for GBM survival. For all members in KEGG pathway Cytokine Cytokine-Receptor Interaction, we selected the genes indicating death risk with FDR threshold 0.05. The pathway members are colored by their gene family categories, including Chemokine, Tumor Necrosis Factor (TNF), Interleukin, and Receptor Tyrosine Kinase (RTK). A diamond shape indicates the gene to be drug targetable in Drug Bank. The node size is proportional to the NEST score averaged among all GBM patients. Stars are used to label known oncogenes annotated by Vogelstein et al. The directed edge indicates a cytokine-to-receptor relation in KEGG, and undirected edge indicates an experimental protein interaction curated by STRING. The thickness of each edge indicates the interaction confidence

Conclusion

To identify distinct features of gene essentiality in CRISPR screens, we developed a network-based method called NEST (Network Essentiality Scoring Tool). We found that essential genes selected in CRISPR screens showed characteristic higher expression level of neighbor genes connected in protein interaction network. Our analysis of Cholera toxin screen in K562 cell also suggests that the quality of CRISPR screen result can be enhanced through the neighbor CRISPR selection score. For a ChIP-seq experiment, NEST can also reliably identify the key TF target genes. Last but not least, NEST score can better predict patient survival than gene expression alone from TCGA tumor profiles. Historically, protein interaction networks were widely used to infer discrete labels such as gene functions, phenotypes [55–57], or gene categories [58]. Our study is different from these previous works in that continuous expression or screen change fold values are integrated with the protein networks. Despite these differences, all of these studies indicate that network information can greatly help biological inference.

NEST significantly outperformed previous methods on gene essentiality prediction and functional screen result enhancement, including all winning methods in the DREAM challenge (Fig. 1f). According to the rule of DREAM challenge, all DREAM methods can gene expression as well as any other features they could utilize. However, NEST outperformed all top DREAM methods. One possible reason is that the gene essentiality gold standard of DREAM is the Achilles shRNA screen data, which is poorly correlated with CRISPR screen (Fig. 1b and Additional file 1: Figure S3A). Because we used CRISPR data as gold standard, those top DREAM methods, optimized to fit Achilles shRNA screen, may not have satisfactory performance.

Several limitations should be noted for our study. NEST computed gene activity is based on network interaction partners, which could have either an activating or a repressive effect. Meanwhile, for compensating interaction such as synthetic lethality, the activation of interaction partners indicates gene loss of function. For example, *PLK1*, an interaction partner of *TP53* in STRING network, was consistently upregulated in cancer cells with inactivated *TP53* compared with those with wild type [59]. We currently summed all neighbor values without distinguishing between activating, repressive, or synthetic lethal relations. Thus, further categorization of network interaction types will be critical for better gene prioritization. Another limitation of our study is that current data on protein interaction network only covered a subset of well-studied genes [60]. Because of the dependence on interaction knowledge, our method may not reliably infer the activity for under-studied genes. As a third limitation, we only tested NEST on gene loss-of-function CRISPR screens. However, for CRISPRa gain of function screen [8, 9], it remains to see whether network-based analysis can bring any predictive power and result enhancement.

In summary, we derived a network-based method, NEST, to interpret and enhance the outcome of genome-wide CRISPR screens, and NEST showed significantly better performance than previous related methods. We recommend researchers using NEST to calculate neighbor CRISPR values from their CRISPR screen result. Moreover, the candidate essential genes in a cell condition might be prioritized before running a large-scale screen to reduce the total number of genes under the screen, which might improve the results and practicality of in vivo CRISPR screens. Besides CRISPR analysis, our method can also identify key targets from ChIP-seq experiments, and find clinical outcome associated genes from tumor profiling data. Thus, we foresee NEST as generally applicable to many applications related with gene essentiality prioritization.

MATERIALS AND METHODS

Availability

The web application and source code of NEST are freely available under the GNU Public License v3 at . The source code and testing data of NEST are additionally deposited at .

Data Collection

For CRISPR screen data, we searched published studies with data publicly available and sgRNA coverage on genome scale for human cell lines until 1 June 2015. There are three studies fulfilling our criterion. In K562 cell, growth phenotype and toxin selection phenotype are screened with CRISPRi technology [8]. In HL60 and A375 cell lines, growth phenotype is screened on genome scale with CRISPR technology [4, 5]. Significant gene hits are called from these datasets by MAGeCK 0.5 with default parameters and FDR threshold 0.05 [29]. For gene essentiality prediction in each cell line, only negatively selected gene hits were considered as gold standard, because most significant gene hits are negatively selected in collected CRISPR experiments (Additional file 1: Figure S1). For gold standard control set, we extracted the same number of genes ranked by MAGeCK on bottom.

For K562, the gene expression profile was downloaded from the Roadmap project [26]. For HL60, the gene expression profile by exon-array was downloaded from the ENCODE project [27] and converted to gene level values by JETTA [61]. For A375, the gene expression profile was downloaded from the CCLE project [28]. For each gene, we normalized the expression value by subtracting the mean across all samples in each cohort. Compared to absolute expression level, the normalized expression value can achieve a better CRISPR prediction performance of NEST (Additional file 1: Figure S12). The TCGA tumor gene expression data was downloaded from TCGA Data Portal on 27 July 2014. Only cohorts that are not embargoed are used. For each cancer cohort, the expression values of all normal control samples were averaged as background, and the difference of gene expression between tumor sample and normal background was analyzed. For NOTCH signaling pathway analysis, the NOTCH off condition is defined as gamma secretase inhibitors (GSI) treatment 3 days, and NOTCH on condition is defined as GSI wash 4 h [62]. The differential expression value between on/off conditions was analyzed [62]. The NOTCH1 ChIP-seq data are generated in our previous work, and the dynamic binding peaks between NOTCH on/off conditions were used [2].

For H3K27ac ChIP-seq profiles, we downloaded data from the Roadmap project [26]. Among all cell lines with CRISPR data collected, K562 is the only one having H3K27ac profile available. Previously, we developed a BETA method to calculate the regulatory potential (RP) on gene promoters from the ChIP-seq profile of a transcription factor or histone modification [36,37]. We used the implementation in RABIT package with default parameters to calculate the H3K27ac RP scores [39]. For each gene, the RP scores were normalized, by subtracting the mean across all cell lines profiled.

Network Randomization and Permutation Test

We used stub rewiring method to randomize unweighted STRING network, which preserves gene degree [33]. The edges from each gene are first detached from its partners, and then randomly connected with each other. Since we do not allow self-interaction and duplicated edges, the connection process may fail to finish. In this case, we restart the rewiring process until 98 % edges are reconnected.

Based on random networks, we derived a permutation test to access whether the NEST score of each gene is significantly larger (or smaller) than expected. For each random network, we calculated the NEST values as random NEST. For each gene, we computed the Z-score as (real NEST − average random NEST)/(Stderr of random NEST). If the Z-score is positive, we computed the P value as the fraction of random NEST scores that are larger than or equal to the real NEST score. If the Z-score is negative, we computed the P values as the fraction of random NEST scores that are smaller than or equal to the real NEST score.

Survival Analysis

We used Cox-PH model to analyze the effect of gene expression or NEST scores on survival. For GBM, there are several factors that affect the survival and we included them as covariates in survival regression, including age, gender, G-CIMP status, and treatment status [45]. So the final survival effect was corrected with the effects of these confounding factors. For TCGA pan-cancer analysis, we only included cancer types with more than 50 patients and 5 % death rate. In the Cox-PH regression, we only included age, gender, and stage (if available) to enable uniform comparison among different cancer types.

ACKNOWLEDGEMENTS

The authors would like to thank Michael Love, Eric Severson, Han Xu, Yiwei Chen, and Jing Mi for helpful discussions. The project was supported by the U01 CA180980 grant from NIH and the Claudia Adams Barr Award in Innovative Basic Cancer Research from the Dana-Farber Cancer Institute.

AUTHORS' CONTRIBUTIONS

PJ, HW, WL, CZ, BL, CM, YW, JSL, JCA, and XSL designed the study and interpreted the results. PJ developed the algorithm and performed the analyses. PJ and XSL wrote the paper. All authors read and approved the final manuscript.

REFERENCES

1.   Cowley GS, Weir BA, Vazquez F, Tamayo P, Scott JA, Rusin S, et al. Parallel genome-scale loss of function screens in 216 cancer cell lines for the identification of context-specific genetic dependencies. Sci Data. 2014;1:140035.

2.   Wang H, Zang C, Taing L, Arnett KL, Wong YJ, Pear WS, et al. NOTCH1-RBPJ complexes drive target gene expression through dynamic interactions with superenhancers. Proc Natl Acad Sci U S A. 2014;111:705–10.

3.   Broad-DREAM Gene Essentiality Prediction Challenge. Available at:

4. Shalem O, Sanjana NE, Hartenian E, Shi X, Scott DA, Mikkelsen TS, et al. Genome-scale CRISPR-Cas9 knockout screening in human cells. Science. 2014;343:84–7.

5. Wang T, Wei JJ, Sabatini DM, Lander ES. Genetic screens in human cells using the CRISPR-Cas9 system. Science. 2014;343:80–4.

6. Koike-Yusa H, Li Y, Tan EP, Velasco-Herrera Mdel C, Yusa K. Genome-wide recessive genetic screening in mammalian cells with a lentiviral CRISPR-guide RNA library. Nat Biotechnol. 2014;32:267–73.

7. Zhou Y, Zhu S, Cai C, Yuan P, Li C, Huang Y, et al. High-throughput screening of a CRISPR/Cas9 library for functional genomics in human cells. Nature. 2014;509:487–91.

8. Gilbert LA, Horlbeck MA, Adamson B, Villalta JE, Chen Y, Whitehead EH, et al. Genome-scale CRISPR-mediated control of gene repression and activation. Cell. 2014;159:647–61.

9. Konermann S, Brigham MD, Trevino AE, Joung J, Abudayyeh OO, Barcena C, et al. Genome-scale transcriptional activation by an engineered CRISPR-Cas9 complex. Nature. 2015;517:583–8.

10. Chen S, Sanjana NE, Zheng K, Shalem O, Lee K, Shi X, et al. Genome-wide CRISPR screen in a mouse model of tumor growth and metastasis. Cell. 2015;160:1246–60.

11. Shi J, Wang E, Milazzo JP, Wang Z, Kinney JB, Vakoc CR. Discovery of cancer drug targets by CRISPR-Cas9 screening of protein domains. Nat Biotechnol. 2015;33:661–7.

12. Jinek M, Chylinski K, Fonfara I, Hauer M, Doudna JA, Charpentier E. A programmable dual-RNA-guided DNA endonuclease in adaptive bacterial immunity. Science. 2012;337:816–21.

13. Gasiunas G, Barrangou R, Horvath P, Siksnys V. Cas9-crRNA ribonucleoprotein complex mediates specific DNA cleavage for adaptive immunity in bacteria. Proc Natl Acad Sci U S A. 2012;109:E2579–86.

14. Maeder ML, Linder SJ, Cascio VM, Fu Y, Ho QH, Joung JK. CRISPR RNA-guided activation of endogenous human genes. Nat Methods. 2013;10:977–9.

15. Qi LS, Larson MH, Gilbert LA, Doudna JA, Weissman JS, Arkin AP, et al. Repurposing CRISPR as an RNA-guided platform for sequence-specific control of gene expression. Cell. 2013;152:1173–83.

16. Gilbert LA, Larson MH, Morsut L, Liu Z, Brar GA, Torres SE, et al. CRISPR-mediated modular RNA-guided regulation of transcription in eukaryotes. Cell. 2013;154:442–51.

17. Zalatan JG, Lee ME, Almeida R, Gilbert LA, Whitehead EH, La Russa M, et al. Engineering complex synthetic transcriptional programs with CRISPR RNA scaffolds. Cell. 2015;160:339–50.

18. Shalem O, Sanjana NE, Zhang F. High-throughput functional genomics using CRISPR-Cas9. Nat Rev Genet. 2015;16:299–311.

19. Jeong H, Mason SP, Barabasi AL, Oltvai ZN. Lethality and centrality in protein networks. Nature. 2001;411:41–2.

20. Han JD, Bertin N, Hao T, Goldberg DS, Berriz GF, Zhang LV, et al. Evidence for dynamically organized modularity in the yeast protein-protein interaction network. Nature. 2004;430:88–93.

21. Hahn MW, Kern AD. Comparative genomics of centrality and essentiality in three eukaryotic protein-interaction networks. Mol Biol Evol. 2005;22:803–6.

22. Kaplow IM, Singh R, Friedman A, Bakal C, Perrimon N, Berger B. RNAiCut: automated detection of significant genes from functional genomic screens. Nat Methods. 2009;6:476–7.

23. Tu Z, Argmann C, Wong KK, Mitnaul LJ, Edwards S, Sach IC, et al. Integrating siRNA and protein-protein interaction data to identify an expanded insulin signaling network. Genome Res. 2009;19:1057–67.

24. Wang L, Tu Z, Sun F. A network-based integrative approach to prioritize reliable hits from multiple genome-wide RNAi screens in Drosophila. BMC Genomics. 2009;10:220.

25. Ma X, Chen T, Sun F. Integrative approaches for predicting protein function and prioritizing genes for complex phenotypes using protein interaction networks. Brief Bioinform. 2014;15:685–98.

26. Roadmap Epigenomics C, Kundaje A, Meuleman W, Ernst J, Bilenky M, Yen A, et al. Integrative analysis of 111 reference human epigenomes. Nature. 2015;518:317–30.

27. Consortium EP, Bernstein BE, Birney E, Dunham I, Green ED, Gunter C, et al. An integrated encyclopedia of DNA elements in the human genome. Nature. 2012;489:57–74.

28. Barretina J, Caponigro G, Stransky N, Venkatesan K, Margolin AA, Kim S, et al. The Cancer Cell Line Encyclopedia enables predictive modelling of anticancer drug sensitivity. Nature. 2012;483:603–7.

29. Li W, Xu H, Xiao T, Cong L, Love MI, Zhang F, et al. MAGeCK enables robust identification of essential genes from genome-scale CRISPR/Cas9 knockout screens. Genome Biol. 2014;15:554.

30. Franceschini A, Szklarczyk D, Frankild S, Kuhn M, Simonovic M, Roth A, et al. STRING v9.1: protein-protein interaction networks, with increased coverage and integration. Nucleic Acids Res. 2013;41:D808–15.

31. Jiang P, Singh M. SPICi: a fast clustering algorithm for large biological networks. Bioinformatics. 2010;26:1105–11.

32. Eisenberg E, Levanon EY. Human housekeeping genes are compact. Trends Genet. 2003;19:362–5.

33. Milo R, Shen-Orr S, Itzkovitz S, Kashtan N, Chklovskii D, Alon U. Network motifs: simple building blocks of complex networks. Science. 2002;298:824–7.

34. Chuang HY, Lee E, Liu YT, Lee D, Ideker T. Network-based classification of breast cancer metastasis. Mol Syst Biol. 2007;3:140.

35. Hofree M, Shen JP, Carter H, Gross A, Ideker T. Network-based stratification of tumor mutations. Nat Methods. 2013;10:1108–15.

36. Tang Q, Chen Y, Meyer C, Geistlinger T, Lupien M, Wang Q, et al. A comprehensive view of nuclear receptor cancer cistromes. Cancer Res. 2011;71:6940–7.

37. Wang S, Sun H, Ma J, Zang C, Wang C, Wang J, et al. Target analysis by integration of transcriptome and ChIP-seq data with BETA. Nat Protoc. 2013;8:2502–15.

38. Wernick NL, Chinnapen DJ, Cho JA, Lencer WI. Cholera toxin: an intracellular journey into the cytosol by way of the endoplasmic reticulum. Toxins (Basel). 2010;2:310–25.

39. Jiang P, Freedman ML, Liu JS, Liu XS. Inference of transcriptional regulation in cancers. Proc Natl Acad Sci U S A. 2015;112:7731–6.

40. Goh KI, Cusick ME, Valle D, Childs B, Vidal M, Barabasi AL. The human disease network. Proc Natl Acad Sci U S A. 2007;104:8685–90.

41. Menche J, Sharma A, Kitsak M, Ghiassian SD, Vidal M, Loscalzo J, et al. Disease networks. Uncovering disease-disease relationships through the incomplete interactome. Science. 2015;347:1257601.

42. Greene CS, Krishnan A, Wong AK, Ricciotti E, Zelaya RA, Himmelstein DS, et al. Understanding multicellular function and disease with human tissue-specific networks. Nat Genet. 2015;47:569–76.

43. Wang X, Wei X, Thijssen B, Das J, Lipkin SM, Yu H. Three-dimensional reconstruction of protein networks provides insight into human genetic disease. Nat Biotechnol. 2012;30:159–64.

44. Hatanpaa KJ, Burma S, Zhao D, Habib AA. Epidermal growth factor receptor in glioma: signal transduction, neuropathology, imaging, and radioresistance. Neoplasia. 2010;12:675–84.

45. Brennan CW, Verhaak RG, McKenna A, Campos B, Noushmehr H, Salama SR, et al. The somatic genomic landscape of glioblastoma. Cell. 2013;155:462–77.

46. Vogelstein B, Papadopoulos N, Velculescu VE, Zhou S, Diaz Jr LA, Kinzler KW. Cancer genome landscapes. Science. 2013;339:1546–58.

47. Gravendeel LA, Kouwenhoven MC, Gevaert O, de Rooi JJ, Stubbs AP, Duijm JE, et al. Intrinsic gene expression profiles of gliomas are a better predictor of survival than histology. Cancer Res. 2009;69:9065–72.

48. Zhu VF, Yang J, Lebrun DG, Li M. Understanding the role of cytokines in Glioblastoma Multiforme pathogenesis. Cancer Lett. 2012;316:139–50.

49. Shannon P, Markiel A, Ozier O, Baliga NS, Wang JT, Ramage D, et al. Cytoscape: a software environment for integrated models of biomolecular interaction networks. Genome Res. 2003;13:2498–504.

50. Law V, Knox C, Djoumbou Y, Jewison T, Guo AC, Liu Y, et al. DrugBank 4.0: shedding new light on drug metabolism. Nucleic Acids Res. 2014;42:D1091–7.

51. Taylor TE, Furnari FB, Cavenee WK. Targeting EGFR for treatment of glioblastoma: molecular basis to overcome resistance. Curr Cancer Drug Targets. 2012;12:197–209.

52. Pyonteck SM, Akkari L, Schuhmacher AJ, Bowman RL, Sevenich L, Quail DF, et al. CSF-1R inhibition alters macrophage polarization and blocks glioma progression. Nat Med. 2013;19:1264–72.

53. Ehtesham M, Winston JA, Kabos P, Thompson RC. CXCR4 expression mediates glioma cell invasiveness. Oncogene. 2006;25:2801–6.

54. Sun L, Hui AM, Su Q, Vortmeyer A, Kotliarov Y, Pastorino S, et al. Neuronal and glioma-derived stem cell factor induces angiogenesis within the brain. Cancer Cell. 2006;9:287–300.

55. Schwikowski B, Uetz P, Fields S. A network of protein-protein interactions in yeast. Nat Biotechnol. 2000;18:1257–61.

56. Hishigaki H, Nakai K, Ono T, Tanigami A, Takagi T. Assessment of prediction accuracy of protein function from protein--protein interaction data. Yeast. 2001;18:523–31.

57. Sharan R, Ulitsky I, Shamir R. Network-based prediction of protein function. Mol Syst Biol. 2007;3:88.

58. Warde-Farley D, Donaldson SL, Comes O, Zuberi K, Badrawi R, Chao P, et al. The GeneMANIA prediction server: biological network integration for gene prioritization and predicting gene function. Nucleic Acids Res. 2010;38:W214–20.

59. Sur S, Pagliarini R, Bunz F, Rago C, Diaz Jr LA, Kinzler KW, et al. A panel of isogenic human cancer cells suggests a therapeutic approach for cancers with inactivated p53. Proc Natl Acad Sci U S A. 2009;106:3964–9.

60. Rolland T, Tasan M, Charloteaux B, Pevzner SJ, Zhong Q, Sahni N, et al. A proteome-scale map of the human interactome network. Cell. 2014;159:1212–26.

61. Seok J, Xu W, Gao H, Davis RW, Xiao W. JETTA: junction and exon toolkits for transcriptome analysis. Bioinformatics. 2012;28:1274–5.

62. Wang H, Zou J, Zhao B, Johannsen E, Ashworth T, Wong H, et al. Genome-wide analysis reveals conserved and divergent features of Notch1/RBPJ binding in human and murine T-lymphoblastic leukemia cells. Proc Natl Acad Sci U S A. 2011;108:14908–13.

Chapter 7

CASTOR BEAN ORGANELLE GENOME SEQUENCING AND WORLDWIDE GENETIC DIVERSITY ANALYSIS

Maximo Rivarola[1]., Jeffrey T. Foster[2], Agnes P. Chan[3], Amber L. Williams[4], Danny W. Rice[5], Xinyue Liu[1], Admasu Melake-Berhan[3], Heather Huot Creasy[1], Daniela Puiu[3], M. J. Rosovitz[3], Hoda M. Khouri[1], Stephen M. Beckstrom-Sternberg[2,6], Gerard J. Allan[4], Paul Keim[2], Jacques Ravel[1,7], Pablo D. Rabinowicz[1,3,8]

[1] Institute for Genome Sciences, University of Maryland School of Medicine, Baltimore, Maryland, United States of America

[2] Center for Microbial Genetics and Genomics, Northern Arizona University, Flagstaff, Arizona, United States of America

[3] J. Craig Venter Institute, Rockville, Maryland, United States of America

[4] Department of Biological Sciences, Environmental Genetics and Genomics Laboratory, Northern Arizona University, Flagstaff, Arizona, United States of America

[5] Department of Biology, Indiana University, Bloomington, Indiana, United States of America

[6] Pathogen Genomics Division, Translational Genomics Research Institute, Phoenix, Arizona, United States of America

[7] Department of Microbiology and Immunology, University of Maryland School of Medicine, Baltimore, Maryland, United States of America

[8] Department of Biochemistry and Molecular Biology, University of Maryland School of Medicine, Baltimore, Maryland, United States of America

ABSTRACT

Castor bean is an important oil-producing plant in the Euphorbiaceae family. Its high-quality oil contains up to 90% of the unusual fatty acid ricinoleate, which has many industrial and medical applications. Castor bean seeds also contain ricin, a highly toxic Type 2 ribosome-inactivating protein, which has gained relevance in recent years due to biosafety concerns. In order to gain knowledge on global genetic diversity in castor bean and to ultimately help the development of breeding and forensic tools, we carried out an extensive chloroplast sequence diversity analysis. Taking advantage of the recently published genome sequence of castor bean, we assembled the chloroplast and mitochondrion genomes extracting selected reads from the available whole genome shotgun reads. Using the chloroplast reference genome we used the methylation filtration technique to readily obtain draft genome sequences of 7 geographically and genetically diverse castor bean accessions. These sequence data were used to identify single nucleotide polymorphism markers and phylogenetic analysis resulted in the identification of two major clades that were not apparent in previous population genetic studies using genetic markers derived from nuclear DNA. Two distinct sub-clades could be defined within each major clade and large-scale genotyping of castor bean populations worldwide confirmed previously observed low levels of genetic diversity and showed a broad geographic distribution of each sub-clade.

INTRODUCTION

Castor bean (*Ricinus communis*) is an oilseed crop in the Euphorbiaceae family, which includes 245 genera comprising 6,300 species [1]. Other important members of this family include tropical crops such as cassava (*Manihot esculenta*) and rubber tree (*Hevea brasiliensis*), as well as ornamental poinsettias (*Euphorbia pulcherrima*), the invasive weed leafy spurge (*Euphorbia esula*), and the parasitic *Rafflesia* that has the largest flower in the plant kingdom[2].

Castor bean is cultivated in tropical and subtropical areas of the world for oil production as well as an ornamental. The world annual castor oil production in 2004–5 was around 0.5 million tons[3] and the U.S. is among the world largest importers of castor oil and its derivatives [4], [5]. Castor oil contains 90% of the unusual fatty acid ricinoleate (12-hydroxy-oleate), whose hydroxy group has unique chemical and physical properties that make castor oil a vital industrial raw material for numerous products [6], including high-quality lubricants, paints, coatings, plastics, soaps, medications for skin affections, and cosmetics. Castor oil can also be used as a lubricity additive to replace sulfur-based lubricant components in petroleum diesel helping to reduce sulfur emissions [7]. Moreover, Brazil, a highly advanced country in the use of ethanol as motor fuel, is starting to develop the industry of castor oil as a biodiesel component[8].

Castor bean seeds also contain ricin, a highly toxic water-soluble protein. Ricin is a Type 2 ribosome-inactivating enzyme [9] that can be extracted from seeds through a relatively simple process, raising concerns that it could be used as a bioweapon. This, added to its importance as an oil crop, poses a need to gain knowledge of the castor bean plant and its genetic variation in order to help the development of breeding and forensic tools. To this end, organelle genomic sequences can be used to develop robust markers for genetic and phylogenetic studies due to their high level of conservation.

The chloroplast genome contains genes that code for structural and functional components of the organelle. In land plants, chloroplast genomes are contained in a circular molecule of DNA ranging between 115 and 165 kb in size [10]. It contains two copies of a duplicated region arranged as inverted repeats (IR), separated by a large single-copy (LSC) sequence and a small single-copy (SSC) sequence [11], [12].

Complete chloroplast genomic sequences can be obtained using a shotgun strategy [13]. This approach involves DNA purification from isolated organelles [14], mechanical shearing of the DNA, and cloning of random fragments for sequencing [15], [16]. Cloned restriction fragments from purified chloroplast DNA have also been used to complete chloroplast genome sequencing [17], [18]. If a reference chloroplast genome from a related species is available, primers can be designed to amplify multiple PCR fragments to cover and assemble the whole chloroplast genome [19], [20]. Recently, the rolling circle amplification technique [21] has been used to amplify chloroplast DNA and generate the necessary quantities of DNA for standard Sanger capillary sequencing [22], [23] or next-generation sequencing [24]. In addition, chloroplast genomes are often assembled as byproducts of plant genome sequencing projects. Complete chloroplast genome sequences have been obtained using genomic bacterial artificial chromosome (BAC) libraries. In these cases, the libraries are screened by hybridization or PCR amplification of a conserved chloroplast sequence to identify large-insert BAC clones that contain chloroplast DNA. A few BAC clones that cover the complete chloroplast genome are then sequenced by a shotgun strategy [25].

Here we report the compete castor bean chloroplast and mitochondrion genome sequences generated from a whole genome shotgun (WGS) sequencing project of the cultivar (cv.) Hale[26] along with draft chloroplast genome sequences obtained by methylation filtration (MF) [27]from additional castor bean cultivars. MF is used for selectively cloning and sequencing hypomethylated (*i.e.* low-copy) sequences from plant genomes, reducing the recovery of repetitive elements that are usually methylated [28], [29], [30]. As the chloroplast DNA is not methylated [31], [32], MF libraries are typically constructed using nuclear DNA to reduce the recovery of mitochondrion and, mainly, chloroplast DNA clones, which are preferentially selected in MF libraries [33]. Therefore, we used a MF approach to sequence the chloroplast genomes of seven additional castor bean accessions in order to identify polymorphisms for phylogenetic and population genetic studies.

Castor bean has low genetic diversity based on AFLP and SSR studies [34], and recent single nucleotide polymorphism (SNP) analysis in a worldwide collection of accessions identified five main groups of castor bean showing a mixture of genotypes in most geographical regions studied [35]. The chloroplast SNPs analysis described here, contributes to the characterization of the genetic diversity and global castor bean population structure.

RESULTS AND DISCUSSION

Assembly of the Castor Bean Cv. Hale Organelle Genomes

Organelle genomic sequences were extracted from the castor bean cultivar Hale sequence database [26]. The chloroplast genome was assembled using the Celera Assembler [36] using short (3.5–9 kbp) and long (40 kbp) paired-end Sanger reads that showed high similarity to other dicot chloroplast genomes. The 40 kbp insert size fosmid reads allowed us to resolve the proper assembly and orientation of the large inverted repeat, common in angiosperm chloroplasts. This assembly was used to identify additional chloroplast sequences from a larger pool of castor bean sequences that were used to close the gaps in the draft assembly. The resulting closed circular molecule contained 163,161 bp with a near-perfect inverted repeat of 27,347 bp.

Annotation of the finished chloroplast molecule predicted 129 genes, including 38 tRNA genes, 8 ribosomal RNA genes, and 83 protein-coding genes (Figure 1). Each of the 27 kbp inverted repeats contains 17 genes. Fourteen genes contain one intron and 3 genes contain 2 introns. Cassava is the closest relative of castor bean whose chloroplast genome has been sequenced and annotated [37]. Sequence alignment of these two genomes showed a 93% sequence identity over approximately 150 of the 161 kbp of the castor bean chloroplast genome. The gene content of both chloroplasts is identical, and the presence of the conserved *atpF* group II intron in castor bean, which is absent in cassava [37], was confirmed. At the amino acid level, an average of 91% identity was observed between the castor bean and cassava protein-coding genes.

The large (LSC, 89,652 bp) and small (SSC, 18,817 bp) single copy regions are separated by two inverted repeats (IRa and IRb, 27,347 bp each), which are colored in grey. The outer most whorl (4) contains genes transcribed clockwise, and the genes on whorl 3 are transcribed counter-clockwise. The inner-most whorl (1) shows the alignment of a nucleotide blast of *R. communis* against all cassava (*M. esculenta*) tRNAs and rRNAs from the plastid genome. Whorl 2 shows the alignment of a nucleotide to protein blast (BLASTX) of *R. communis* against all proteins of the cassava plastid genome (*M. esculenta*). * Split genes or genes with introns. ** Trans-spliced gene*rps12*.

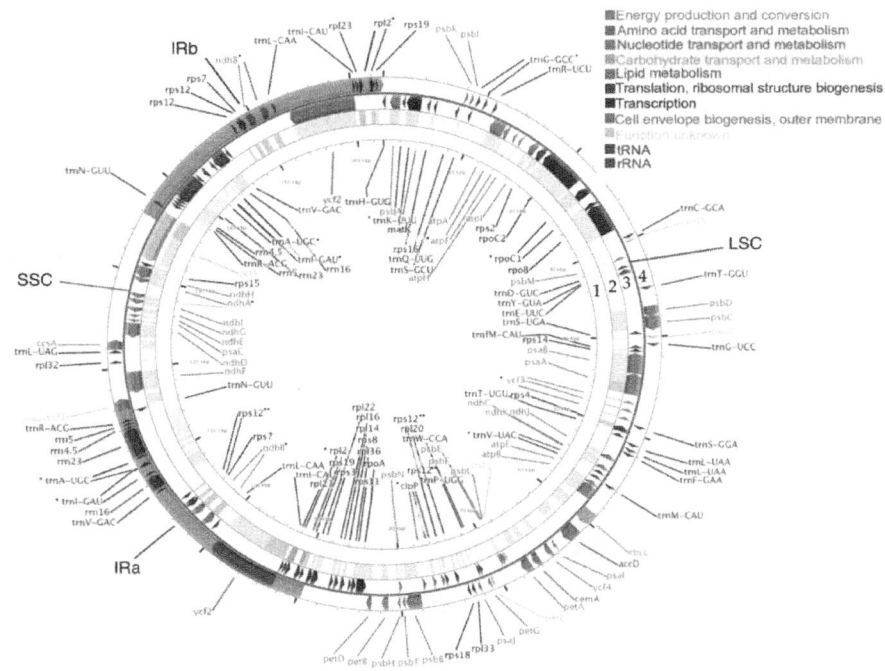

Figure 1: Map of the castor bean chloroplast.

Sequence reads belonging to the mitochondrion genome were extracted using a custom script and assembled using the CAP3 assembler [38]. The assembled genome represents a consensus of the multiple possible genomic and subgenomic configurations that can exist in vivo [39]. The high confidence consensus mitochondrial genome spans 502,773 bp. Due to its larger size and lower abundance, the mitochondrial genome is less suitable for marker development in multiple cultivars than the chloroplast counterpart. Therefore, we did not further utilize the mitochondrion genome for genetic diversity analysis.

The mitochondrial genomes typically code for a few RNA genes as well as some ribosomal proteins and subunits of the oxidative phosphorylation complexes. The castor bean mitochondrial genome contains 37 protein-coding and 3 ribosomal RNA genes, commonly found in angiosperms (Figure S1). Horizontal gene transfer to the nuclear genome often results in variability regarding presence or absence of genes in the mitochondrial genome. In castor bean, the genes coding for ribosomal proteins L2 and S14, often found in plant mitochondrial genomes [40], have been transferred to the nuclear genome, while the gene coding for ribosomal protein L10, which has been recently found in the nuclear or mitochondrial genomes in different angiosperms [41], [42], has been retained in the castor bean mitochondrial genome.

Sequencing the Chloroplast Genomes of Additional Castor Bean Accessions

Based on diversity groupings previously determined by AFLP analysis [34], we selected seven castor bean accessions representing genetically and geographically diverse varieties that originated in different parts of the world (Ethiopia, India, U.S. Virgin Islands, Puerto Rico, El Salvador, Greece, and Mexico) for chloroplast genome sequencing and subsequent SNP discovery.

In order to rapidly obtain chloroplast sequence information we constructed a MF library for each accession and sequenced both ends of several thousand clones per library (Table 1). To estimate the enrichment in chloroplast sequences obtained by MF of leaf genomic DNA from castor bean we also prepared standard shotgun genomic libraries for the accessions from Ethiopia and El Salvador. While the shotgun sequences produced 10% and 9% chloroplast sequences in these two accessions, respectively, the proportion of plastid sequences went up to 29% and 24%, respectively, (nearly a 3-fold enrichment) in the MF libraries. On average, the proportion of chloroplast sequences in the seven accessions' MF libraries was 35% with a maximum of 58% in the Greek accession (Figure 2).

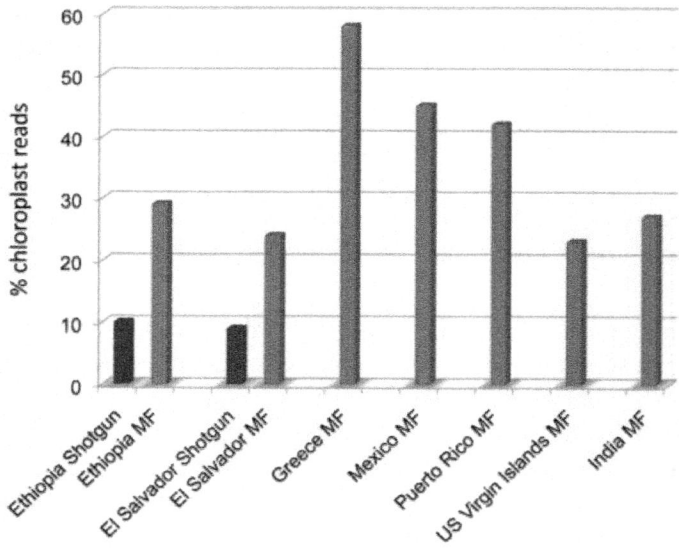

Figure 2: Chloroplast read enrichment in methylation filtration libraries.

Percentage of chloroplast reads in each shotgun (blue) or MF (green) library. Available shotgun reads for cultivars from Ethiopia and El Salvador are shown to highlight the enrichment in chloroplast sequences in MF libraries.

Table 1: MF and WGS sequencing of 7 additional castor bean accessions

Origin	Castor bean accession	Number of shotgun reads	% chloroplast shotgun reads	Number of MF reads	% chloroplast MF reads	Number of contigs
Ethiopia	PI 193851	9,216	10	2,304	29	2
El Salvador	PI 197048	192	9	7,104	24	21
Greece	PI 280219	N/A	N/A	4,032	58	3
Mexico	PI 255238	N/A	N/A	6,144	45	3
Puerto Rico	PI 209132	N/A	N/A	5,376	42	8
US Virgin Islands	PI 209326	N/A	N/A	6,912	23	15
India	PI 173946	N/A	N/A	5,376	27	38

doi:10.1371/journal.pone.0021743.t001

The MF (or the combination of MF and shotgun) reads were assembled for each accession using the AMOS-Cmp assembler, which uses a reference sequence to guide the assembly[43]. We used the closed chloroplast genome that we obtained from the cv. Hale as reference and generated a draft assembly for the chloroplast genome of each cultivar. In this way we were able to order and orient all contigs in each genome, producing a single scaffold for each one. In general, we obtained fairly complete assemblies that would allow SNP identification. Thus, we did not attempt to do any manual work to finish the sequences. Nevertheless, alignment of all the reads in each genome to the reference showed that most of the gaps in the assemblies could potentially be covered by existing sequence reads after manual analysis (Figure 3). In order to obtain genotypic information for some SNPs that fell within gaps in some of the genomes, we performed additional assemblies of the same sequences with the Celera assembler, which uses a different algorithm [36]. In this way we managed to close some gaps, increasing the chances of identifying SNPs with genotypic information in all seven cultivars.

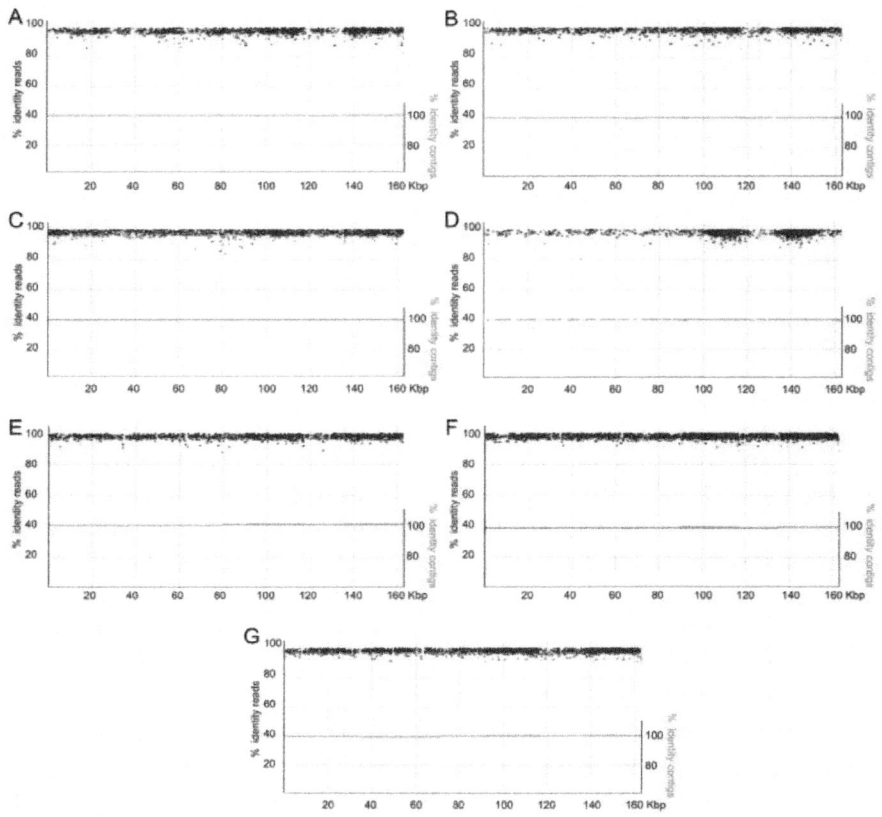

Figure 3: Draft assemblies of the chloroplast genomes of 7 castor bean accessions.

Assembled contigs corresponding to the chloroplast genomes of the different accessions are shown in orange, mapped to the reference v. Hale chloroplast genome (x-axis). Individual reads are shown in blue, aligned to the same reference genome. Left y-axis corresponds to the percent identity of each read relative to the reference cv. Hale sequence. Right y-axis shows the percent identity of contig relative to the Hale sequence. Each panel corresponds to a different cultivar: A) US Virgin Islands, B) El Salvador, C) Puerto Rico, D) India, E) Ethiopia, F) Greece, and G) Mexico.

Chloroplast DNA sequences are useful for phylogenetics and population genetics [44], [45], as well as cultivar identification and forensic analyses [46]. Our results show that MF is a cost-effective way to obtain sequence information of plant organelle genomes given that it does not require time-consuming plastid preparations, library screenings, or amplifications. With this approach, we were able to identify SNPs that allowed us to conduct a worldwide analysis of the castor bean accessions as described below.

SNP Identification and Phylogenetic Analysis

We searched for candidate SNPs in pair-wise alignments between each of the seven castor bean chloroplast genome assemblies and that of cv. Hale. Using custom bioinformatics analyses and stringent criteria (see Methods), 112 SNPs were identified (Table S1). One of them showed a triallelic nature so it was discarded and 29 were not covered in all seven chloroplast genome assemblies. Of these 29, SNP#111 was covered in all genome assemblies except that of the Indian cv. However, because the Greek cv. showed a rare allele for this SNP and was specific to this cultivar (see below) we genotyped this SNP using a TaqMan assay in the Indian cv. to verify that the Greek cv. allele was not present in any of the other genomes. As a result we selected a set of 83 SNPs for which there is high quality sequence data for all cultivars (including SNP#111, Figure 4A) and we used this SNP set to construct a chloroplast-based phylogeny (Figure 4B). Two main groups were evident among the seven castor bean cultivars and we called them clades A and B, which are separated by 69 of the 83 high-quality SNPs.

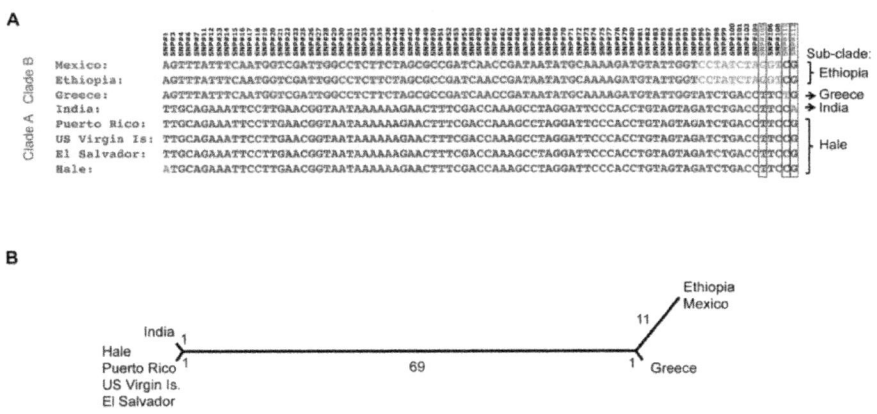

Figure 4: SNP identification and phylogeny obtained from genomic sequencing of 8 castor bean accessions.

A) 83 high quality SNPs identified in the chloroplast genomes of the 8 sequenced chloroplast accessions. The two major clades are shown in red and blue and SNPs that differentiate different sub-clades are shown in orange or green. B) Phylogeny of the 8 castor bean accessions based on the identified SNPs. Two major clades, "A" and "B", are shown separated by 69 SNPs. Members of each sub-clade are indicated in each branch of the phylogeny.

Members of clade A are highly similar, with only SNP#112 unique to the Indian accession (designating the India sub-clade). The accessions from Puerto Rico, Virgin Islands, and El Salvador were monomorphic for all SNPs and differed from cv. Hale only at SNP#1 (Figure 4A). However, because no assay could be designed for this SNP, these four cultivars are hereafter referred to as the Hale sub-clade. For clade B, the Ethiopian and Mexican accessions did not show polymorphisms for any of the SNPs analyzed and we refer to these accessions as the Ethiopia sub-clade as defined by SNP#105. SNP#111 allowed the identification of the Greek cv. genotype (Greece sub-clade). Thus, the eight accessions with chloroplast sequences were divisible into four sub-clades (Hale, India, Greece, and Ethiopia). We designed TaqMan assays specific to these sub-clades in order to distinguish individuals belonging to each sub-clade among 894 castor bean samples from accessions collected around the world. Overall, genotyping of the worldwide population with these assays showed that limited chloroplast genetic diversity exists throughout the world in castor bean, which is consistent with the low genetic variation observed for nuclear genomic sequences [34], [35].

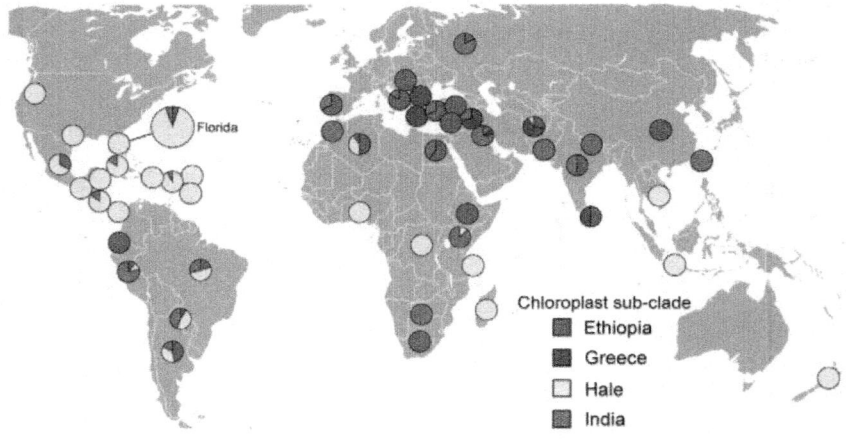

Figure 5: Worldwide distribution of chloroplast genotypes of castor bean from a collection of 894 accessions.

Nevertheless, we could determine that each sub-clade was broadly spread geographically (Figure 5). Within clade A, the Hale sub-clade genotype was distributed throughout the world, dominating in collections from North and Central America, the Caribbean, and parts of South America, Africa, and Asia (Table 2). Members of the India sub-clade, within clade A as well, were also found widespread on every continent sampled. Members of the clade B were found less frequently than those of clade A, with the Ethiopia sub-clade being more widely distributed while the Greece sub-clade was mostly limited to the Middle East and Sri Lanka. The geographic origin of the different accessions could not always be correlated with its genotype. For example, the Mexican and Ethiopian cvs. showed no polymorphisms for any of the SNPs we tested, although the majority of samples from Mexico fell in the Hale sub-clade. A similar scenario was found in clade A, within which there is very little variation but included the Indian cv. together with mostly American varieties. Distribution of chloroplast genotypes corresponding to the sub-clades shown in Figure 4, based on the origin of each accession. The pie chart corresponding to Florida is expanded to reflect the larger number of samples ($n=272$) that came from that state, relative to other parts of the world.

Table 2: Country or state of origin and the chloroplast sub-clade for 894 *Ricinus communis* samples collected worldwide.

Origin	Number of samples per sub-clade			
	Ethiopia	Greece	Hale	India
Afghanistan	3	7	1	
Algeria	5		4	1
Argentina	25		19	9
Bahamas			8	
Benin			8	
Botswana				11
Brazil	9		22	13
Cambodia			8	
China	5			
Costa Rica			5	1
Cuba			17	3
Ecuador	4			
Egypt	3			2
El Salvador			7	
Ethiopia	4			
Florida	13		244	15
Greece		3		
Guatemala			8	
Hale			5	
Hungary		4		
India	52		3	49

Indonesia			8	
Iran	5	3	1	24
Israel	5			
Jamaica			11	1
Jordan	5			
Kenya			1	7
Madagascar			8	
Mexico	3		6	
Morocco				5
Nepal				8
New Zealand			1	
Oregon			3	
Pakistan				6
Panama			8	
Paraguay	1		5	5
Peru	3		3	23
Portugal	4			2
Puerto Rico			8	
Russia	1			5
Serbia	5			1
South Africa	4			
Sri Lanka	1	1		
Syria		8		3
Taiwan				2
Texas			1	
Turkey	40			18
Uruguay			11	
US Virgin Islands			8	
Yugoslavia	5			
Zaire			5	
Zanzibar			2	
Total	205	26	449	214

doi:10.1371/journal.pone.0021743.t002

The two main phylogenetic groups obtained based on chloroplast DNA sequence variation argue for two ancient and differentiated populations that merged to generate the current global populations. In fact, two robust ancestral populations appear to have been involved in the creation of modern hybrid *Ricinus* populations —perhaps even different subspecies. This division into two groups was much less apparent from previous analysis of nuclear DNA sequence variation data [34], [35], likely because of extensive recombination. This has not happened in the cytoplasmic DNAs due to the lack of recombination. It is somewhat fortuitous that the two main groups still exist since cytoplasmic DNA would tend to rapidly go to fixation as one type or the other, especially in small populations.

We compared the nuclear SNP data where five main groups where encountered worldwide [35]with the four chloroplast sub-clades. A strong relationship occurred between groupings based on nuclear and chloroplast SNPs, particularly for the Hale cultivar genotype. Because 451 of the samples that we analyzed in this study had also been analyzed with nuclear SNPs [35], we could compare the grouping results from both types of SNP markers for exactly the same samples. For this subset of samples, 96.1% (171 of 178) of the Hale chloroplast sub-clade genotypes were also grouped together when analyzed with nuclear SNPs. This correspondence was not complete for comparisons between other groups and their most likely population assignment: India 68% (34 of 50), Ethiopia 57% (59 of 86), and Greece 8% (11 of 137).

Global patterns of genetic variation in castor bean have likely been influenced by agricultural and horticultural practices. The pattern of introduction of *R. communis* from multiple chloroplast groupings in many countries is consistent with genetic patterns from nuclear SNPs [35]. In some areas however, such as North America, this is not evident from chloroplast DNA probably because of the limited diversity found in the chloroplast genome. We caution that this low level of genetic diversity may be related to the fact that our samples were limited primarily to those available from the U.S. Department of Agriculture (USDA) germplasm collection, which may not represent the totality of the world's geographic and genetic diversity for castor bean.

MATERIALS AND METHODS

Plant Material

Castor bean accessions were obtained from USDA's National Plant Germplasm System, Agricultural Research Center in Griffin, Georgia and some of the accessions for the worldwide population analysis were also obtained from commercial sources or collected in the wild in the state of Florida, USA. The accessions selected for chloroplast genome sequencing were: Ethiopian cv.: PI193851, Greek cv.: PI280219, Mexican cv.: PI255238, Puerto Rican cv.: PI209132, Indian cv.: PI173946, Salvadorian cv.: PI197048, and cv. from U.S. Virgin Islands: PI209326. Plants were grown in normal greenhouse conditions and leaf samples were collected and dried before DNA extraction as described [34]. Total DNA was prepared using DNeasy kits (Qiagen) following the manufacturer's instructions.

Organelle Genome Sequencing, Assembly and Annotation

Assembly of the castor bean cv. Hale chloroplast genome was carried out by selecting chloroplast reads from the database of whole genome shotgun Sanger reads generated for genome assembly [26]. This database includes paired-end sequence reads from multiple plasmid and fosmid libraries. From this database, we extracted a set of approximately 7,000 3–4 kbp-insert paired-end plasmid reads and selected 765 reads that could be aligned at high stringency against available dicot chloroplast genome sequences from GenBank using sequence similarity searches. These reads were assembled using the Celera Assembler [36]. Subsequently, additional chloroplast reads from fosmid clones were added to the assembly to allow building and orienting the 27 kbp inverted repeats. We then performed standard finishing work to close sequence gaps and resolve ambiguities [47].

Because the sequence and structure of angiosperm mitochondrial genomes are poorly conserved and include unique sequences, it is relatively difficult to identify all reads corresponding to the mitochondrial genome when mixed with a nuclear genome WGS project. To address this difficulty, we developed a method for extending initially assembled reads by iteratively comparing all reads to the ends of the extending assemblies. By examining the read depths of the resulting assemblies, mitochondrial sequences could be separated from those of nuclear and chloroplast origin. The resulting reads were assembled using CAP3 [38].

The chloroplast genomes of the seven additional castor bean cultivars were sequenced using MF libraries. Briefly, castor bean total DNA was purified from leaves and was randomly sheared by nebulization, end-repaired with consecutive BAL31 nuclease and T4 DNA polymerase treatments, and 1.5 to 3 kb fragments were eluted from a 1% low-melting-point agarose gel after electrophoresis. After ligation to BstXI adaptors, DNA was purified by three rounds of gel electrophoresis to remove excess adaptors, and the fragments were ligated into the vector pHOS2 (a modified pBR322 vector) linearized with BstXI. The pHOS2 plasmid contains two BstXI cloning sites immediately flanked by sequencing primer binding sites. The ligation reactions were introduced by electroporation into *E. coli* strain GC10 (*McrBC⁻*) for regular shotgun libraries or strain DH5α (*McrBC⁺*) for MF libraries.

Draft assemblies of the genomes of the seven castor bean cultivars sequenced were conducted using AMOS-Cmp [43] using the closed cv. Hale genome sequence as reference. Additional assemblies of the same sequence data were performed with the Celera assembler [36] in order to cover some of the gaps that contained candidate SNPs.

Chloroplast genome annotation was carried out using the software tool DOGMA (Dual Organellar GenoMe Annotator) [48] followed by manual inspection of intron-containing genes. Comparative analysis of castor bean and cassava chloroplast genomes was carried out using CGView [49]. The mitochondrion genome was annotated using comparative analysis to identify mitochondrial genes, taking into account exons and introns. Intergenic regions were further annotated as possible using sequence similarity searches against GenBank.

The finished and annotated castor bean chloroplast and mitochondrion genome sequences from the cv. Hale can be found under GenBank accession numbers JF937588 and HQ874649, respectively. The draft sequence assemblies of the 7 castor bean accessions chloroplast genomes can be found under GenBank accession numbers JF940518 (Ethiopia), JF940520 (India), JF940515 (U.S. Virgin Islands), JF940521 (Puerto Rico), JF940517 (El Salvador), JF940519 (Greece), and JF940516 (Mexico).

SNP Discovery

In order to call candidate SNPs in pair-wise comparisons of the seven castor bean chloroplast genomes versus the Hale cv. genome we applied the following criteria: reads must be longer than 500 bp, the polymorphic base should be covered by at least 3 reads and should not be located in the first or last 30 bp of any read, the Phred quality score of the polymorphic base and mean scores of the 5 bases upstream and 5 bases downstream should be greater than 30, and the polymorphic base should be covered in all cultivars meeting the previous criteria.

Genotyping Assay Development and Screening

We developed three Real-Time TaqMan (Applied Biosystems) minor groove binding (MGB) assays for the four castor bean sub-clades. Primers and probes were designed using Primer Express for TaqMan MGB assays for Allelic

Discrimination version 3.0 (Applied Biosystems). Each assay had dual-probes targeting the specific SNP (Table 3). Although a single SNP separates the Hale cultivar from the cultivars from Puerto Rico, U.S. Virgin Islands, and El Salvador, sequence composition prevented the development of a reliable assay, so these closely related cultivars were grouped together.

Table 3: Sequence for TaqMan primers and probes for defining castor bean chloroplast sub-clades

SNP	Sub-clade	Forward primer sequence	Reverse primer sequence	Sub-clade specific probe*	Alternate probe*
105	Ethiopia	GACATTCCGTCTTCTGAAACCAA	CTAACTATAGTGCAAAGTCGCATCTCTT	TTTCTATATGCCGATTATGG	TTTCTATATGCCTATTATG
111	Greece	GATTGATTGGCTGATGTTTCAAAA	AAAGAAACGTCTGTATTCAGCTACAAAG	ATACCCAAAGTTCC	ATACCCAAAGCTCCCA
112	India	TGTCAGGCTATTGTTCTCCTGTTC	GGGAGTCCATCATGTAATCAAAAGA	CTAAAAGTAATGAAGTAAGAC	AAGTAATGGGAGTAAGACATC

*SNP-state is underlined in each probe.
doi:10.1371/journal.pone.0021743.t003

We ran the genotyping assays on an ABI Prism 7900HT Sequence Detection System (Applied Biosystems). Each 10 µl PCR mixture contained 1× AB TaqMan Universal Master mix, 0.9 µM each primer, and 0.25 µM each probe. We added 0.2 U of Platinum Taq (Invitrogen) per reaction to increase the efficiency of the amplification. Each assay was run at standard conditions: 2 min inactivation at 50°C, a 10 min hot start at 95°C, followed by 40 cycles of 15 sec denaturation and 1 min annealing at 60°C. Amplification and allelic discrimination plots were visualized using SDS version 2.3 software. We screened 894 samples from 49 countries and 3 US states against all three SNP assays.

ACKNOWLEDGMENTS

We thank the J. Craig Venter Institute's Joint Technology Center and the Informatics group for carrying out sequencing and bioinformatics support, respectively.

AUTHOR CONTRIBUTIONS

Conceived and designed the experiments: JR PK PDR. Performed the experiments: MR JTF APC ALW. Analyzed the data: MR JTF APC DWR SMB-S GJA PDR JR PK. Contributed reagents/materials/analysis tools: AM-B HHC DP MJR HMK XL. Wrote the paper: PDR MR JTF. Edited the manuscript: PK GJA JR DWR

REFERENCES

1.   Wurdack KJ, Hoffmann P, Chase MW (2005) Molecular phylogenetic analysis of uniovulate Euphorbiaceae (euphorbiaceae sensu stricto)

Using plastid rbcl and trnl-f dna sequences. American Journal of Botany 92: 1397–1420.

2.   Davis CC, Latvis M, Nickrent DL, Wurdack KJ, Baum DA (2007) Floral gigantism in Rafflesiaceae. Science 315: 1812.

3.   Gunstone F (2005) Fatty acid production for human consumption. Inform 16: 736–737.

4.   Roetheli J, Glaser L, Brigham R (1991) Castor: Assessing the feasibility of US production. Washington, DC: USDA-CSRS, Office of Agricultural Materials.

5.   DeVries J, Toenniessen G (2001) Securing the Harvest. Biotechnology, Breeding and Seed Systems for African Crops. New York: CABI Publishers.

6.   Brigham R (1993) Castor: Return of an old crop. In: Simon JJaJE, editor. New crops. New York, USA: Wiley & Sons.

7.   Goodrum JW, Geller DP (2005) Influence of fatty acid methyl esters from hydroxylated vegetable oils on diesel fuel lubricity. Bioresour Technol 96: 851–855.

8.   da Silva Nde L, Maciel MR, Batistella CB, Filho RM (2006) Optimization of biodiesel production from castor oil. Appl Biochem Biotechnol 129–132: 405–414.

9.   Endo Y, Mitsui K, Motizuki M, Tsurugi K (1987) The mechanism of action of ricin and related toxic lectins on eukaryotic ribosomes. The site and the characteristics of the modification in 28 S ribosomal RNA caused by the toxins. J Biol Chem 262: 5908–5912.

10.   Palmer M (1991) Plastid chromosomes: structure and evolution. In: Vasil IKaB L, editor. The Molecular Biology of Plastids. San Diego: Academic Press. pp. 5–53.

11.   Sugiura M, Hirose T, Sugita M (1998) Evolution and mechanism of translation in chloroplasts. Annu Rev Genet 32: 437–459.

12.   Sugiura M (1992) The chloroplast genome. Plant Mol Biol 19: 149–168.

13.   Anderson S (1981) Shotgun DNA sequencing using cloned DNase I-generated fragments. Nucleic Acids Res 9: 3015–3027.

14.   Sandbrink JM, Vellekoop P, Vanham R, Vanbrederode J (1989) A method for evolutionary studies on RFLP of chloroplast DNA, applicable to a range of plant species. Biochem Syst Ecol 17: 45–49.

15.   Steane DA (2005) Complete nucleotide sequence of the chloroplast

genome from the Tasmanian blue gum, Eucalyptus globulus (Myrtaceae). DNA Res 12: 215–220.

16. Chumley TW, Palmer JD, Mower JP, Fourcade HM, Calie PJ, et al. (2006) The complete chloroplast genome sequence of Pelargonium×hortorum: organization and evolution of the largest and most highly rearranged chloroplast genome of land plants. Mol Biol Evol 23: 2175–2190.

17. Maier RM, Neckermann K, Igloi GL, Kossel H (1995) Complete sequence of the maize chloroplast genome: gene content, hotspots of divergence and fine tuning of genetic information by transcript editing. J Mol Biol 251: 614–628.

18. Kim KJ, Lee HL (2004) Complete chloroplast genome sequences from Korean ginseng (Panax schinseng Nees) and comparative analysis of sequence evolution among 17 vascular plants. DNA Res 11: 247–261.

19. Shahid Masood M, Nishikawa T, Fukuoka S, Njenga PK, Tsudzuki T, et al. (2004) The complete nucleotide sequence of wild rice (Oryza nivara) chloroplast genome: first genome wide comparative sequence analysis of wild and cultivated rice. Gene 340: 133–139.

20. Chung HJ, Jung JD, Park HW, Kim JH, Cha HW, et al. (2006) The complete chloroplast genome sequences of Solanum tuberosum and comparative analysis with Solanaceae species identified the presence of a 241-bp deletion in cultivated potato chloroplast DNA sequence. Plant Cell Rep 25: 1369–1379.

21. Dean FB, Nelson JR, Giesler TL, Lasken RS (2001) Rapid amplification of plasmid and phage DNA using Phi 29 DNA polymerase and multiply-primed rolling circle amplification. Genome Res 11: 1095–1099.

22. Lee SB, Kaittanis C, Jansen RK, Hostetler JB, Tallon LJ, et al. (2006) The complete chloroplast genome sequence of Gossypium hirsutum: organization and phylogenetic relationships to other angiosperms. BMC Genomics 7: 61.

23. Bausher MG, Singh ND, Lee SB, Jansen RK, Daniell H (2006) The complete chloroplast genome sequence of Citrus sinensis (L.) Osbeck var 'Ridge Pineapple': organization and phylogenetic relationships to other angiosperms. BMC Plant Biol 6: 21.

24. Moore MJ, Dhingra A, Soltis PS, Shaw R, Farmerie WG, et al. (2006) Rapid and accurate pyrosequencing of angiosperm plastid genomes. BMC Plant Biol 6: 17.

25. Saski C, Lee SB, Daniell H, Wood TC, Tomkins J, et al. (2005) Complete chloroplast genome sequence of Gycine max and comparative analyses with other legume genomes. Plant Mol Biol 59: 309–322.

26. Chan AP, Crabtree J, Zhao Q, Lorenzi H, Orvis J, et al. (2010) Draft genome sequence of the oilseed species Ricinus communis. Nat Biotechnol 28: 951–956.

27. Rabinowicz PD, Schutz K, Dedhia N, Yordan C, Parnell LD, et al. (1999) Differential methylation of genes and retrotransposons facilitates shotgun sequencing of the maize genome. Nat Genet 23: 305–308.

28. Whitelaw CA, Barbazuk WB, Pertea G, Chan AP, Cheung F, et al. (2003) Enrichment of gene-coding sequences in maize by genome filtration. Science 302: 2118–2120.

29. Palmer LE, Rabinowicz PD, O'Shaughnessy A, Balija V, Nascimento L, et al. (2003) Maize genome sequencing by Methylation Filtration. Science. in press.

30. Bedell JA, Budiman MA, Nunberg A, Citek RW, Robbins D, et al. (2005) Sorghum genome sequencing by methylation filtration. PLoS Biol 3: e13.

31. Fojtova M, Kovarik A, Matyasek R (2001) Cytosine methylation of plastid genome in higher plants. Fact or artefact? Plant Sci 160: 585–593.

32. McCullough AJ, Kangasjarvi J, Gengenbach BG, Jones RJ (1992) Plastid DNA in Developing Maize Endosperm : Genome Structure, Methylation, and Transcript Accumulation Patterns. Plant Physiol 100: 958–964.

33. Rabinowicz PD (2007) Plant genomic sequencing using gene-enriched libraries. Chem Rev 107: 3377–3390.

34. Allan G, Williams A, Rabinowicz PD, Chan AP, Ravel J, et al. (2008) Worldwide genotyping of castor bean germplasm (*Ricinus communis* L.) using AFLPs and SSRs. Genet Resour Crop Evol 55: 365–378.

35. Foster JT, Allan GJ, Chan AP, Rabinowicz PD, Ravel J, et al. (2010) Single nucleotide polymorphisms for assessing genetic diversity in castor bean (Ricinus communis). BMC Plant Biol 10: 13.

36. Myers EW, Sutton GG, Delcher AL, Dew IM, Fasulo DP, et al. (2000) A whole-genome assembly of Drosophila. Science 287: 2196–2204.

37. Daniell H, Wurdack KJ, Kanagaraj A, Lee SB, Saski C, et al. (2008) The complete nucleotide sequence of the cassava (Manihot esculenta) chloroplast genome and the evolution of atpF in Malpighiales: RNA editing and multiple losses of a group II intron. Theor Appl Genet 116: 723–737.

38. Huang X, Madan A (1999) CAP3: A DNA sequence assembly program. Genome Res 9: 868–877.

39. Woloszynska M (2010) Heteroplasmy and stoichiometric complexity of

plant mitochondrial genomes–though this be madness, yet there›s method in›t. J Exp Bot 61: 657–671.

40. Adams KL, Qiu YL, Stoutemyer M, Palmer JD (2002) Punctuated evolution of mitochondrial gene content: high and variable rates of mitochondrial gene loss and transfer to the nucleus during angiosperm evolution. Proc Natl Acad Sci U S A 99: 9905–9912.

41. Mower JP, Bonen L (2009) Ribosomal protein L10 is encoded in the mitochondrial genome of many land plants and green algae. BMC Evol Biol 9: 265.

42. Kubo N, Arimura S (2010) Discovery of the rpl10 gene in diverse plant mitochondrial genomes and its probable replacement by the nuclear gene for chloroplast RPL10 in two lineages of angiosperms. DNA Res 17: 1–9.

43. Pop M, Phillippy A, Delcher AL, Salzberg SL (2004) Comparative genome assembly. Brief Bioinform 5: 237–248.

44. Newton AC, Allnutt TR, Gillies AC, Lowe AJ, Ennos RA (1999) Molecular phylogeography, intraspecific variation and the conservation of tree species. Trends Ecol Evol 14: 140–145.

45. Provan J, Powell W, Hollingsworth PM (2001) Chloroplast microsatellites: new tools for studies in plant ecology and evolution. Trends Ecol Evol 16: 142–147.

46. Sucher NJ, Carles MC (2008) Genome-based approaches to the authentication of medicinal plants. Planta Med 74: 603–623.

47. Birren B, Green ED, Klapholz S, Myers RM, Roskams J (1997) Genome Analysis. A Laboratory Manual. Analyzing DNA. Cold Spring Harbor, New York: Cold Spring Harbor Laboratory Press.

48. Wyman SK, Jansen RK, Boore JL (2004) Automatic annotation of organellar genomes with DOGMA. Bioinformatics 20: 3252–3255.

49. Grant JR, Stothard P (2008) The CGView Server: a comparative genomics tool for circular genomes. Nucleic Acids Res 36: W181–184.

Chapter 8

ANALYSIS OF EAST ASIA GENETIC SUBSTRUCTURE USING GENOME-WIDE SNP ARRAYS

Chao Tian[1] , Roman Kosoy[1] , Annette Lee[2] , Michael Ransom[1] , John W. Belmont[3] , Peter K. Gregersen[2] , Michael F. Seldin[1]

[1] Rowe Program in Human Genetics, Departments of Biochemistry and Medicine, University of California Davis, Davis, California, United States of America

[2] The Robert S. Boas Center for Genomics and Human Genetics, Feinstein Institute for Medical Research, North Shore LIJ Health System, Manhasset, New York, United States of America

[3] Department of Molecular and Human Genetics, Baylor College of Medicine, Houston, Texas, United States of America

ABSTRACT

Accounting for population genetic substructure is important in reducing type 1 errors in genetic studies of complex disease. As efforts to understand complex genetic disease are expanded to different continental populations the understanding of genetic substructure within these continents will be useful in design and execution of association tests. In this study, population differentiation (Fst) and Principal Components Analyses (PCA) are examined using >200 K genotypes from multiple populations of East Asian ancestry. The population groups included those from the Human Genome Diversity Panel [Cambodian, Yi, Daur, Mongolian, Lahu, Dai, Hezhen, Miaozu, Naxi, Oroqen, She, Tu, Tujia, Naxi, Xibo, and Yakut], HapMap [Han Chinese (CHB) and Japanese (JPT)], and East Asian or East Asian American subjects of Vietnamese, Korean, Filipino and Chinese ancestry. Paired Fst (Wei and Cockerham) showed close relationships between CHB and several large East Asian population groups (CHB/Korean, 0.0019; CHB/JPT, 00651; CHB/ Vietnamese, 0.0065) with larger separation with Filipino (CHB/Filipino, 0.014). Low levels of differentiation were also observed between Dai and Vietnamese (0.0045) and between Vietnamese and Cambodian (0.0062).

Similarly, small Fst's were observed among different presumed Han Chinese populations originating in different regions of mainland of China and Taiwan (Fst's <0.0025 with CHB). For PCA, the first two PC's showed a pattern of relationships that closely followed the geographic distribution of the different East Asian populations. PCA showed substructure both between different East Asian groups and within the Han Chinese population. These studies have also identified a subset of East Asian substructure ancestry informative markers (EASTASAIMS) that may be useful for future complex genetic disease association studies in reducing type 1 errors and in identifying homogeneous groups that may increase the power of such studies.

INTRODUCTION

Analysis of population genetic substructure has been enhanced by the ability to perform large genome array studies. The differences and patterns of variation within continental populations are useful for several reasons including recapitulating population migration and origins of ethnic groups, forensic identification, and for defining and applying an understanding of allele frequency variation to genetic association studies. Recent studies by several groups including our own have examined European population substructure [1]–[4]. Importantly, these studies have shown that discerning and accounting for differences in substructure can improve error rates in association studies [5]. With the availability of East Asian (EAS) SNP genotypes, we undertook the current study to perform similar studies for this sub-continental region that contains the largest contribution to the world's population. East Asian population genetic structure is particularly important since multiple genetic studies of complex disease are currently underway including studies of autoimmune diseases in Korean, Chinese and Japanese populations[6]–[11]. An understanding of the relationship among these different populations and ascertaining ancestry informative markers (AIMs) that can discern East Asian substructure will undoubtedly facilitate accurate interpretation of such studies[5].

This study combines high density SNP array genotypes from studies of EAS population groups within the Human Genome Diversity Panel (HGDP) [12], [13] with those of several additional population groups of EAS ancestry. The use of high density SNP genotypes containing over 200 K common autosomal genotypes allows a more comprehensive analyses than those previously performed using limited number of autosomal genotypes. It also complements studies of mitochondrial and Y chromosome haplogroups as well as classical markers that provide important information with respect to part of the history of particular EAS ethnic groups[14]–[20]. Our study expands on previous analyses using HGDP population groups [13] by examining additional parameters of population structure/diversity and by including many additional samples including those from several of the most populous EAS groups (Korean, Filipino and Vietnamese) and Chinese American participants of diverse origin. We apply the genotypic information to identify a set of SNPs that may be useful in the design and execution of association studies.

RESULTS

Population Differentiation Between East Asian Populations

To examine similarities and differences in population differentiation among EAS populations paired Fst values were determined between 19 EAS population groups that were typed with genome-wide SNP arrays (see **Methods**). The studies included samples derived from HapMap[21], [22], HGDP[13], samples collected in Korea and East Asian American participants (see**Methods**). The Fst values were obtained using three random non-overlapping sets of 3500 SNPs distributed over the autosomal genome (minimum of 50 kb distance between SNPs). This approach was taken to limit potential bias from SNPs in close linkage disequilibrium and to measure of variability of Fst. The small differences in these independent samplings (mean SD=0.0015; median SD=0.0013) indicate that this approach resulted in good estimations of paired Fst values. Relatively large Fst values were evident between many of the relatively small ethnic groups within China (Table 1 and see Figure 1 for geographical information). In particular, those population groups derived from Mongolia or near by provinces including Oroqen, Hezhen, and Daur show relatively large differences with Han Chinese. Similarly, two of the ethnic groups in the southeastern region of China, Lahu and Dai, also showed large paired Fst values with Han Chinese. With respect to population groups derived from very populous groups, the data indicate that Japanese and Korean were very closely related, as were Korean and Han Chinese but that these groups are much further from the south-east Asian populations (Filipino and Vietnamese). The Han Chinese and Japanese groups showed larger separation than either with Korean, although the paired Fst values were still small relative to Chinese/Filipino Fst. The Fst values also showed a close relationship between the Dai ethnic group in China and the Vietnamese population sample. Each of the groups had large paired Fst values with the Yakut from Siberia with the exception of the Mongolian, Hezhen and Oroqen ethnic groups that derive from north-eastern China or Mongolia. The relative size of the Fst values also generally corresponded to the geographical separation of the EAS population groups (depicted in Figure 1).

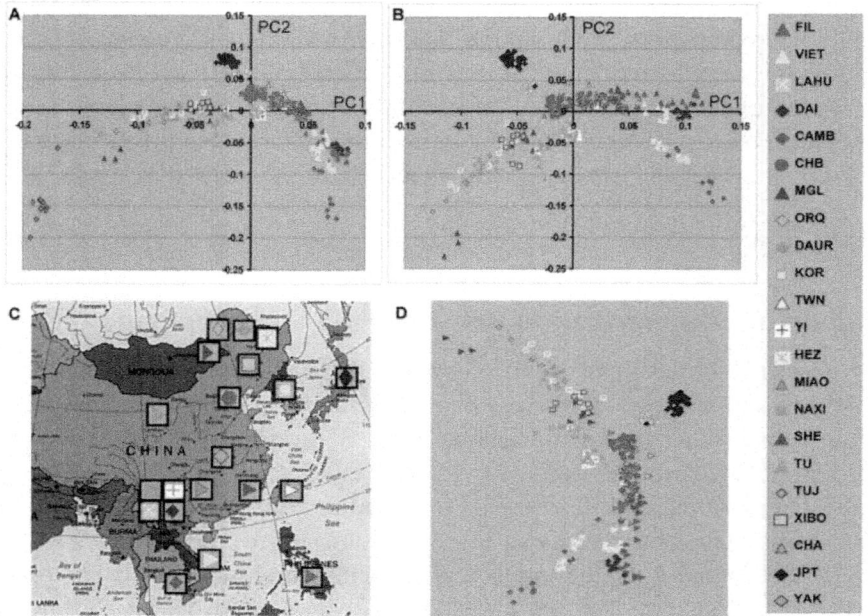

Figure 1: Principal component analyses of substructure in a diverse set of subjects of East Asian descent.

Graphic representation of the first two PCs based on analysis with >200 K SNPs are shown. Color code shows subgroup of subjects for each population group. The subjects included Filipino (FIL), Vietnamese (VIET), Lahu, Dai, Cambodian (CAMB), Han Chinese (CHB), Mongola (MGL), Oroqen (ORQ), Daur, Korean (KOR), Chinese Americans from Taiwan (TWN),Yi, Hezhen (HEZ), Miaozu (MIAO), Naxi, She, Tu, Tujia (TUJ), Xibo, Chinese Americans (CHA), Japanese (JPT), and Yakut (YAK). A, Analyses including the Yakut population group. B, Analysis without Yakut is shown. C, Approximate geographic origin of population group is depicted on a map of East Asia (downloaded from University of Texas Library website). The positions of the HGDP population groups are based on the collection site information[12] and the other population groups were placed based on self-identified country or region of origin. [Note: Yakut are not shown on the map since this population is from Siberia and is a considerable distance north of the depicted region.] D, Shows rotated results of PC1 and PC2 to assist illustration of geographic correspondence of ethnic group locations.

Table 1: Paired Fst values and Standard Deviations for EAS Population Groups[a]

	CHB	KOR	FIL	VIET	CAMB	YAK	YI	DUAR	MGL	LAHU	DAI	HEZH	MIAO	NAXI	OROQ	SHE	TU	TUJ	XIBO
JPT[b]	0.0085+/ −0.0003																		
KOR	0.0019+/ −0.0002	0.0028+/ −0.0003																	
FIL	0.0140+/ −0.0007	0.0204+/ −0.0004	0.0182+/ −0.0015																
VIET	0.0065+/ −0.0009	0.0146+/ −0.0014	0.0106+/ −0.0002	0.0112+/ −0.0007															
CAMB	0.0136+/ −0.0011	0.0210+/ −0.0009	0.0191+/ −0.0021	0.0158+/ −0.0006	0.0062+/ −0.0011														
YAK	0.0289+/ −0.0009	0.0297+/ −0.0020	0.0279+/ −0.0010	0.0430+/ −0.0002	0.0376+/ −0.0008	0.0377+/ −0.0023													
YI	0.0059+/ −0.0008	0.0127+/ −0.0002	0.0083+/ −0.0006	0.0207+/ −0.0015	0.0126+/ −0.0012	0.0156+/ −0.0021	0.0296+/ −0.0016												
DUAR	0.0088+/ −0.0023	0.0103+/ −0.0016	0.0072+/ −0.0023	0.0262+/ −0.0017	0.0185+/ −0.0028	0.0227+/ −0.0017	0.0163+/ −0.0010	0.0143+/ −0.0021											
MGL	0.0049+/ −0.0010	0.0082+/ −0.0014	0.0052+/ −0.0008	0.0212+/ −0.0009	0.0139+/ −0.0008	0.0161+/ −0.0020	0.0141+/ −0.0013	0.0085+/ −0.0006	0.0081+/ −0.0096										
LAHU	0.0223+/ −0.0016	0.0295+/ −0.0015	0.0271+/ −0.0021	0.0306+/ −0.0010	0.0219+/ −0.0029	0.0037+/ −0.0016	0.0511+/ −0.0016	0.0261+/ −0.0016	0.0346+/ −0.0010	0.0280+/ −0.0010									
DAI	0.0110+/ −0.0004	0.0195+/ −0.0007	0.0171+/ −0.0023	0.0119+/ −0.0020	0.0045+/ −0.0008	0.0076+/ −0.0030	0.0039+/ −0.0015	0.0152+/ −0.0018	0.0226+/ −0.0006	0.0179+/ −0.0016	0.0229+/ −0.0016								
HEZH	0.0084+/ −0.0011	0.0104+/ −0.0016	0.0069+/ −0.0010	0.0264+/ −0.0009	0.0191+/ −0.0061	0.0230+/ −0.0009	0.0195+/ −0.0032	0.0130+/ −0.0010	0.0036+/ −0.0007	0.0042+/ −0.0005	0.0343+/ −0.0037	0.0252+/ −0.0016							
MIAO	0.0059+/ −0.0010	0.0141+/ −0.0003	0.0100+/ −0.0001	0.0152+/ −0.0010	0.0068+/ −0.0008	0.0129+/ −0.0005	0.0355+/ −0.0019	0.0105+/ −0.0008	0.0166+/ −0.0017	0.0128+/ −0.0017	0.0240+/ −0.0032	0.0097+/ −0.0014	0.0166+/ −0.0004						
NAXI	0.0100+/ −0.0014	0.0177+/ −0.0021	0.0130+/ −0.0010	0.0255+/ −0.0025	0.0172+/ −0.0004	0.0209+/ −0.0008	0.0355+/ −0.0023	0.0078+/ −0.0013	0.0183+/ −0.0030	0.0133+/ −0.0033	0.0310+/ −0.0007	0.0214+/ −0.0016	0.0185+/ −0.0010	0.0157+/ −0.0019					
OROQ	0.0164+/ −0.0011	0.0175+/ −0.0003	0.0166+/ −0.0007	0.0334+/ −0.0027	0.0273+/ −0.0038	0.0308+/ −0.0021	0.0177+/ −0.0006	0.0213+/ −0.0014	0.0055+/ −0.0014	0.0095+/ −0.0008	0.0415+/ −0.0010	0.0320+/ −0.0028	0.0066+/ −0.0013	0.0247+/ −0.0028	0.0246+/ −0.0032				
SHE	0.0090+/ −0.0006	0.0168+/ −0.0012	0.0123+/ −0.0014	0.0187+/ −0.0006	0.0110+/ −0.0018	0.0181+/ −0.0025	0.0418+/ −0.0008	0.0159+/ −0.0006	0.0202+/ −0.0025	0.0160+/ −0.0005	0.0303+/ −0.0010	0.0148+/ −0.0005	0.0230+/ −0.0021	0.0117+/ −0.0011	0.0211+/ −0.0020	0.0293+/ −0.0034			
TU	0.0035+/ −0.0008	0.0093+/ −0.0010	0.0055+/ −0.0003	0.0191+/ −0.0009	0.0120+/ −0.0010	0.0145+/ −0.0014	0.0243+/ −0.0011	0.0063+/ −0.0014	0.0078+/ −0.0010	0.0045+/ −0.0000	0.0261+/ −0.0002	0.0163+/ −0.0003	0.0101+/ −0.0021	0.0086+/ −0.0013	0.0098+/ −0.0007	0.0173+/ −0.0006	0.0145+/ −0.0015		
TUJ	0.0003+/ −0.0002	0.0083+/ −0.0006	0.0038+/ −0.0011	0.0120+/ −0.0013	0.0037+/ −0.0025	0.0122+/ −0.0007	0.0318+/ −0.0007	0.0054+/ −0.0012	0.0113+/ −0.0007	0.0065+/ −0.0038	0.0195+/ −0.0027	0.0076+/ −0.0019	0.0119+/ −0.0015	0.0034+/ −0.0014	0.0047+/ −0.0007	0.0195+/ −0.0026	0.0076+/ −0.0011	0.0032+/ −0.0005	
XIBO	0.0034+/ −0.0014	0.0076+/ −0.0012	0.0033+/ −0.0011	0.0197+/ −0.0004	0.0120+/ −0.0026	0.0167+/ −0.0004	0.0185+/ −0.0005	0.0096+/ −0.0007	0.0045+/ −0.0012	0.0017+/ −0.0003	0.0286+/ −0.0025	0.0182+/ −0.0026	0.0052+/ −0.0032	0.0214+/ −0.0028	0.0139+/ −0.0011	0.0099+/ −0.0006	0.0148+/ −0.0009	0.0035+/	0.0055+/ −0.0009

[a]Paired Fst values are the mean+/−S.D determined from three nonoverlapping sets of 3500 SNPs using the Weir and Cockerham algorithm (see **Methods**).
[b]The EAS population groups included Japanese from Tokyo (JPT) and Chinese from Beijing (CHB) both from HapMap data, Korean (KOR), Filipino (FIL), Vietnamese (VIET) from new typing studies and the following HGDP groups: Cambodian (CAMB), Yakut (YAK), Yi, Daur, Mongolian (MGL), Lahu, Dai, , Hezhen (HEZH), , Miaozu (MIAO), Naxi, Oroqen (OROQ), She, Tu, Tujia (TU), and Xibo. The geographic positions of these populations are provided in Figure 1 with the exception of the Yakut that derive from northern Siberia.
doi:10.1371/journal.pone.0003862.t001

Fis values were also determined for each of the population sample and did not indicate a strong inbreeding component for any of the tested sample groups (**Supplemental Table S1**).

The different Chinese subjects derived from different regions of origin were also examined. For each of the Chinese American groups with self reported origin from North China, South China and Taiwan the paired Fst values with the Han Chinese from Beijing was small (<0.0025) (**Supplemental Table S2**).

Principal Component Analyses Using >200 K SNPs Show Substructure Relationships

To further explore the relationship among EAS population groups and examine population substructure PCA was performed using the genotype results from a set of >200 K SNPs. Analyses were done with and without the inclusion of the Yakut population thought to originate in central Asia, since PCA results are influenced by the inclusion or exclusion of different population groups and we were interested in the relationship between EAS and central Asian populations. The first two principal components in these analyses display the largest genotype variation (**Table 2**) and are graphically depicted in **Figure 1.** Inclusion of the Yakut group showed a possible cline in PC1/PC2 that extends from the current Siberian location of the Yakut to the northern East Asian population groups (**Figure 1A**). Interestingly, the position of the different population groups shows a remarkable correspondence with the geographic origin of each group. This is more clearly suggested when the Yakut population is excluded (**Figure 1B**) and is best illustrated by comparing these geographic locations with rotated PCA results (**Figure 1C and D**). Additional, PCA analyses including the central Asian Uygur and Hazara population groups were also performed but these did not show a clear relationship with the EAS (Supplemental **Figure S1**).

Table 2: Evaluation of Principal Components Analyses in East Asian Populations using 200 K SNPs

PC	All EAS Population Groups[a]			Five Population Groups		
	% Eigen[b]	SHT[c]	K-W Test[d]	% Eigen	SHT	K-W Test
1	17.9%	0.969+/−0.030	2.39E-41	18.0%	0.985+/−0.001	2.66E-26
2	12.0%	0.950+/−0.017	1.01E-40	12.0%	0.951+/−0.010	4.43E-24
3	10.6%	0.798+/−0.109	9.81E-39	9.2%	0.774+/−0.045	1.74E-14
4	10.0%	0.690+/−0.198	2.56E-36	8.8%	0.301+/−0.127	9.32E-01
5	8.9%	0.738+/−0.139	1.43E-31	8.8%	0.011+/−0.013	2.54E-01
6	8.4%	0.481+/−0.055	1.05E-28	8.7%	0.051+/−0.041	4.71E-02
7	8.2%	0.177+/−0.028	6.39E-23	8.7%	0.038+/−0.041	1.79E-01
8	8.0%	0.129+/−0.162	4.97E-10	8.6%	0.069+/−0.032	1.89E-01
9	8.0%	0.033+/−0.016	6.09E-05	8.6%	0.016+/−0.011	5.67E-01
10	7.9%	0.006+/−0.004	4.88E-02	8.6%	0.005+/−0.007	2.00E-01

PC	CHB, KOR, JPT			"Chinese" Groups Alone		
	% Eigen	SHT	K-W Test	% Eigen	SHT	K-W Test
1	15.2%	0.982+/−0.002	6.75E-22	11.5%	0.685+/−0.049	5.39E-06
2	9.8%	0.616+/−0.081	3.26E-08	10.0%	0.059+/−0.060	6.02E-01
3	9.5%	0.003+/−0.003	1.55E-01	10.0%	0.120+/−0.019	3.07E-01
4	9.5%	0.036+/−0.032	1.87E-01	9.9%	0.098+/−0.065	6.90E-01
5	9.5%	0.038+/−0.032	4.98E-01	9.9%	0.014+/−0.018	2.68E-01
6	9.3%	0.053+/−0.045	7.27E-01	9.8%	0.051+/−0.067	1.83E-01
7	9.3%	0.037+/−0.008	2.50E-02	9.8%	0.069+/−0.084	3.77E-01
8	9.3%	0.024+/−0.013	1.18E-01	9.7%	0.113+/−0.073	9.36E-01
9	9.3%	0.035+/−0.041	1.26E-01	9.7%	0.040+/−0.063	4.70E-01
10	9.3%	0.014+/−0.010	4.65E-02	9.7%	0.018+/−0.018	3.05E-01

[a]EAS population groups included each of the populations indicated in Figure 1.
[b]The % eigenvalue (Eigen) is the percentage of the total variance in the first ten PCs.
[c]The Spearman-Brown split half reliability test (SHT)[39] r^2 is the mean+/−SD from the adjusted correlations between: 1) every other chromosomes; 2) half chromosomes (first half each chromosome and second half each chromosome); and 3) first half genome and second half genome (see **Methods**). These correlations were determined after PCA of each individual set.
[d]The Kruskal-Wallis test [23], a nonparametric alternative to the ANOVA was used to examine the statistical significance of the difference in PC scores among subject groups pre-assigned based on self-identification.
doi:10.1371/journal.pone.0003862.t002

The PCA results for PC1 and PC2 are generally consistent with the relative paired Fst values with respect to the distance separation among the different population groups. For example the position of the Korean group approximately midway between the HapMap CHB and JPT groups both graphically (**Figure 1**) and as discussed above for paired Fst values. It is also consistent with the closer relationship between the Dai ethnic group and the Vietnamese subjects. However, the first two PCs do not show the full relationships among the population groups. For example the Lahu ethnic group appears to be closely related to the Cambodian ethnic group (**Figure 1**), although the paired Fst value is relatively large (Table 1). Examination of additional PCs shows the large difference between the Lahu and Cambodian ethnic groups in PCs 3, 4 and 5 (**Figure 2**). Using both the Kruskal-Wallis test [23], a nonparametric alternative to the ANOVA, and a split half reliability test (see Methods) substructure was present in multiple principal components (**Table 2**). Substantial population substructure can be observed by the nonrandom grouping of population groups that extends through PC7.

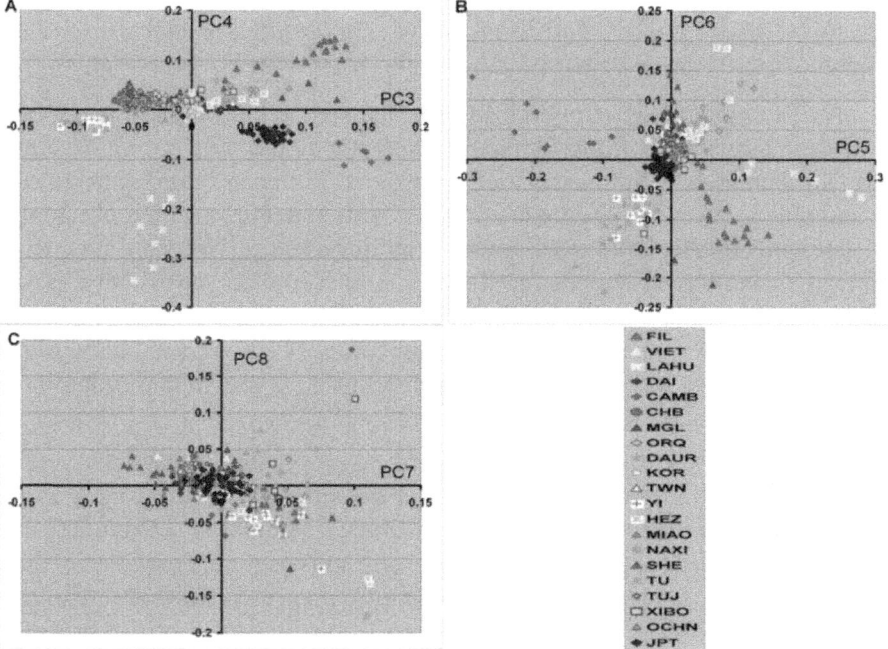

Figure 2: Graphic representation of additional principal components (PCs 3–8) in a diverse set of subjects of East Asian Descent

Color key shows groups as defined in Fig 1. A, PC3 and PC4. B, PC5 and PC6. C, PC7 and PC8.

For the entire EAS population groups studied, the majority of substructure variation defined by PCA appears to be within the first 4 PCs (**Table 2**). The eigenvalues plateau after PC4 with only small differences observed in subsequent PCs (**Figure 3a**). The proportion of the sum of the eigenvalues above this plateau provides a measure of the relative amount of substructure variation defined by each PC (**Figure 3b**). For the total EAS group, >90% of the substructure is defined in the first four PCs by this measurement. For the group of the five populations representing the most populous ethnic groups studied the first two PCs account for 90% of the variation above the plateau.

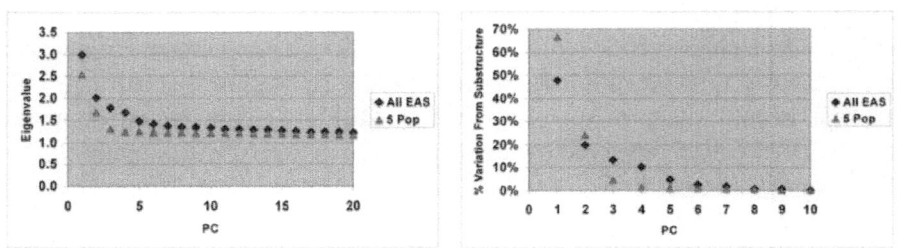

Figure 3: Eigenvalue distribution for principal components.

A, The eigenvalues for each PC are shown for both the entire group of EAS (excluding Yakut), and for the five most populous ethnic groups (Chinese, Korean, Japanese, Filipino and Vietnamese). B, The proportion of the adjusted eigenvalue for each PC for the first 10 PCs is shown. For this measurement the PC10 eigenvalue for each group was used as the baseline. [Note: the eigenvalues plateau as shown in panel A and there is no discernable substructure beyond PC10 for these analyses (Table 2)]. For each PC, the PC10 eigen value was subtracted to determine an "adjusted" eigenvalue. The % substructure variation measurement was the proportion of each adjusted eigenvalue divided by the sum of the adjusted eigenvalues (PC1 through PC10).

Similar analyses were also performed using population sets restricted to the more closely related Han Chinese, Japanese, and Korean groups, as well as a group restricted to Han Chinese and Chinese Americans (**Table 2**). These results as expected indicated substantially less substructure. However, even the subject set limited to Han Chinese and Chinese Americans showed substructure in PC1 using the split half reliability test and with the self identified groupings (ANOVA result). The relationship among the Han Chinese can be demonstrated in PCAs performed either including or excluding other EAS populations (**Figure 4**). Although there is variability in the distribution of many of the self-identified groups there was a general northwest/southeast gradient within these Chinese participants. In PC1 the North Han Chinese (HGDP from north central China[12]) were most separated from the southern Chinese participants including the Chinese American participants from Taiwan or with self-reported southern China origin.

A, Results from PCA performed together with EAS populations. B, PCA performed using only Chinese and Chinese American participants. The color coded population groups included the HapMap Han Chinese from Beijing (CHB), HGDP Han Chinese (HAN), HGDP North Han Chinese (HAN_N), Chinese American North (CHAN), Chinese American South (CHAS), Chinese American Central (CHAC), Taiwan Chinese American (TWN), Korean (KOR), and Hezhen (HEZ).

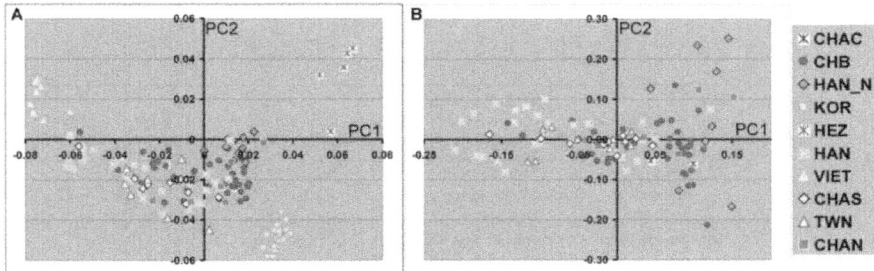

Figure 4: PCA analyses of Han Chinese and Chinese American population groups.

Informativeness of Smaller Sets of SNPS for Large East Asian Population Groups

We next examined the ability of smaller sets of SNPs to define population genetic structure in EAS populations. Random sets of 20 K, 5 K and 1 K SNPs were used to examine substructure in the combined population set and a subset of subjects from the most populous EAS groups (Han Chinese, Japanese, Korean, Filipino and Vietnamese). Correlation values (r^2) were calculated comparing these SNP subsets with the 200 K SNP set. These results, summarized in Table 3, showed that the 20 K random SNP set and 5 K random SNP set corresponded closely with the >200 K SNP set for the first 4 PCs, with decreased correlations observed for the 1 K random SNP set. The relatively poor performance of the 1 K random sets was more pronounced when more closely related population groups were considered e.g. Japanese and Korean for PC1, 20 K/200 K $r^2=0.82+/-0.12$ (mean+/-SD), 5 K/200 K $r^2=0.69+/-0.03$, and 1 K/200 K $r^2=0.28+/-0.06$. These results suggest that random sets of 5 K SNPs may be necessary for resolving and adjusting for substructure in these EAS populations (see discussion).

Table 3: Correlation of PCA Results using Random and Selected Sets with 200 K SNPs

	PC1	PC2	PC3	PC4
All EAS Groups[a]				
20 K random[b]	0.992+/−0.005	0.977+/−0.002	0.854+/−0.139	0.851+/−0.137
5 K random	0.956+/−0.005	0.888+/−0.021	0.725+/−0.086	0.705+/−0.081
1 K random	0.813+/−0.007	0.514+/−0.019	0.228+/−0.047	0.125+/−0.052
Five Population Groups (CHB, JPT, KOR, FIL, VIET)				
	PC1	PC2	PC3	PC4
20 K random	0.991+/−0.002	0.961+/−0.005	0.897+/−0.073	0.754+/−0.198
5 K random	0.961+/−0.004	0.862+/−0.019	0.616+/−0.141	0.708+/−0.02
1 K random	0.814+/−0.024	0.419+/−0.074	0.192+/−0.165	0.218+/−0.074
Test Population Group[c]				
	PC1	PC2		
3 K random	0.862+/−0.035	0.446+/−0.096		
3.0 K AIMs	0.953	0.848		
1.5 K AIMs	0.939	0.819		
750 AIMs	0.886	0.579		

[a]Includes all EAS population groups (see Methods).
[b]Summary of analyses is provided for correlations of three independent random marker sets for each random marker group. For each random group the correlation with the full array set (>200 K SNPs) and is expressed as the mean r²+/−S.D.
[c]The tester population panel consisted of 20 Chinese, 20 Japanese, 4 Korean, 3 Filipino, 1 Vietnamese, 10 Dai and 10 Cambodian. . This test group did not contain any subjects used in the selection of the EAS-AIMs. As with other comparisons the correlation with the full array set (>200 K SNPs) is expressed as the mean r²+/−S.D. The EAS-AIMs are provided in Supplemental **Table S3**.
doi:10.1371/journal.pone.0003862.t003

East Asian Substructure Ancestry Informative Markers

AIMs that discern population substructure are likely to be useful in candidate gene, chromosomal position based association studies and defining homogeneous subject sets [24]. Since the application of these methods is most applicable to large population groups we restricted our ascertainment to five populations (Han Chinese, Japanese, Korean, Vietnamese and Filipino) (See **Methods**). To access the potential usefulness of these AIMs an independent set of samples was used and compared with the same number of random SNPs. For this assessment we included Cambodian and Dai samples since we had limited samples from the Vietnamese and Filipino populations. 3 K AIMs showed close correlation between the 200 K results for the first two PCs (Table 3). A set of the best 1.5 K AIMs also showed close correlation (**Figure 5** and **Table 3**). A reduced set of 750 AIMs showed a fall-off in correlation but was still equivalent to 3 K random SNPs. None of the AIM sets correlated with PC3 or PC4 (r2<0.01, p<0.05), however, these PCs distinguished the Dai and Cambodian from the other population groups and these were not included in our AIM selections. Nevertheless, for the common EAS populations these data suggest that the EAS-AIMs (**Table S3**) will be useful for association studies in the majority of EAS and EAS-American populations.

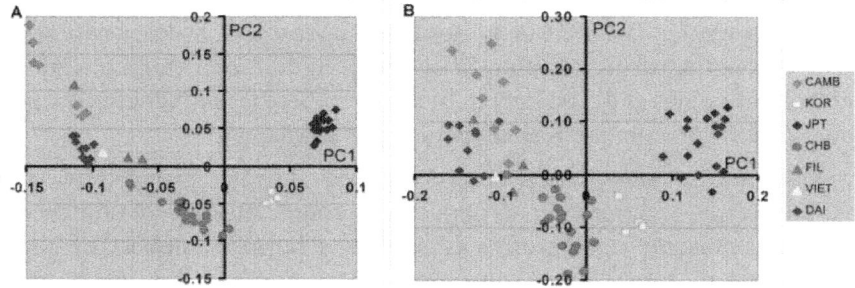

Figure 5: Ability of EAS-AIMs to discern population substructure

A, PCA analysis of tester population samples (see Table 3) using 200 K SNPs. B, PCA analysis of same tester population samples using 1500 EAS-AIMs.

DISCUSSION

The current study extends the definition of EAS population substructure and the relationships among these ethnic groups. The inclusion of participant groups from populous countries in this region with large contributions to the USA population is an important aspect of our study. These population groups complement those included within HapMap studies as well as the HGDP in showing relationships between EAS groups and demonstrating that autosomal genotypes can be used to ascertain membership to various EAS groups. These results emphasize that EAS substructure, similar that previously shown for European substructure, will likely be important for complex disease association studies in defining study participants and reducing type 1 and type 2 error rates.

Our study extends the results of PCA analyses of EAS populations including those of HGDP populations that was recently reported [13]. The graphic representation of the first two PCs showed close correspondence to the historical geographical location and/or sample collection site for most of the EAS population groups. Thus, despite admixture and perhaps uncertain migration patterns, overall the largest component of genotypic variation that is discernable by reducing high order data (all genotypes) to lower order variations (PCs) is consistent with the population geography. This finding supports hypotheses that the relationships among the EAS populations are largely explained by clines formed by demic expansion(s). We speculate that the inclusion of many different related ethnic groups has recapitulated the most common events that separated these ethnic groups. The first PC axis accounting for the largest variation has a north/south orientation. One major part of this pattern forms a line from Siberia (Yakut) to Mongolia to Eastern China (**Figure 1**). The PCA analyses also suggest that at least two separate clines originating or terminating in eastern China at one end and Cambodia and the Philippines at the other end. In addition there is another cline extending from Eastern China to the Korean peninsular and Japan.

Multiple previous studies have examined the relationship between and possible origins of different EAS population groups. Analysis of mitochondrial and Y chromosome haplogroups as well as a limited numbers of classical markers and microsatellite polymorphisms have also provided results that are generally consistent with a north/south orientation of relationships between different EAS population groups [15]–[18]. However, there are exceptions with some studies failing to show this relationship [19], [25]. Summarized by a recent review [26] there are three different postulates regarding the origins of EAS population groups: 1) South East Asian origin [14]–[18], 2) North Asian origin [27] and 3) a combination of northern and southern origin[19], [20]. However, the majority of studies have supported a South-East Asian origin for most EAS populations and include detailed analyses of the age of specific mitochondrial haplogroups, Y chromosome sequences as well as limited marker studies [26]. In contrast, hierarchical trees in the recent HGDP study [13] show branching points consistent with a Yakut derivation. Recent studies using a novel copying model statistical approach appear to suggest an initial northern and southern origin (Cambodians, Mongolians, Xibo, Yi , Tu, Daur, and Naxi receiving large contributions from central-Asian populations) that contribute to Han ancestry[28]. These studies also provide data supporting the derivation of many other EAS groups from a Han expansion (including She, Japanese, Dai, Lahu and Miao). While the current study does not strongly support any of these hypotheses, it does suggest that eastern China is central to the events shaping the population groups in this region.

Multiple additional PCs are necessary to define the overall substructure relationships for the entire group of EAS populations studied as shown in **Figure 2**. However, most of the variation is discerned in the first four PCs for the EAS populations examined and in the first two PCs for the five most populous EAS groups studied. There was no geographic correspondence of the additional PCs and it is unclear whether these additional patterns correspond to individual or multiple different events in the histories of these population groups. Overall the size of the paired Fst values, as expected, showed a strong correlation with the PC eigenvalues summed over the first four PCs (data not shown). Although Fis values do not provide evidence for inbreeding in the current populations, it is unclear whether inbreeding or other factors including bottlenecks during the history of particular EAS ethnic groups may have contributed to the relationships between these populations.

An important aspect of the current study was the identification of EAS-AIM sets. The results show that these AIMs can distinguish the major variation between the populous population groups including Han Chinese, Japanese, Korean, Vietnamese, and Filipino. Additional testing to examine correction for stratification with these population groups was not possible due to limited genotypes currently available. However, by analogy with previous studies in European population groups, these AIMs particular the 1500 EAS-AIM set should be effective in addressing population stratification. The close correspondences of the relative positions in the first two PCs in individual subjects, even within the Han Chinese group, support the potential use of these SNP AIMs. Furthermore, the SHT analysis suggests that studies within the Han Chinese population and Chinese-Americans will benefit from the use of such AIMs in candidate gene studies.

METHODS

Populations Studied

The populations including those from the HGDP, HapMap, the I-control database, a Korean sample set and East Asian Americans. For all but the East Asian American and Korean samples set, genotypes were available from online databases. These included HapMap subjects (44 CHB and 44 JPT) and HGDP subjects (10 Cambodian, 10 Dai, 24 Hazara, 9 Hezhen, 27 Japanese, 10 Miaozu, 7 Naxi, 8 Oroqen, 10 She, 10 Tu, 10 Tujia, 8 Xibo, 13 Yakut and 44 Han Chinese) from the I-ControlDB (www.illumina.com/iControlDB, Illumina, San Diego, CA). Genotypes from other HGDP subjects (10 Daur, 8 Lahu, 9 Mongola, 10 Uygur, 10 Yi,) were from the NIH Laboratory of Neurogenetics (http://neurogenetics.nia.nih.gov/paperdata/public/).

For all EAS American and Korean subjects, blood cell samples were obtained from all individuals, according to protocols and informed-consent procedures approved by institutional review boards, and were labeled with an anonymous code number linked only to demographic information.

The Korean participants were from recruited in Korea (21 subjects). The EAS American samples were individuals born in the respective EAS country and were from Vietnam (22 subjects), Philippines (17 subjects) and different regions of the Peoples Republic of China (23 subjects) and Taiwan (9 subjects). The Filipino American participants included 15 that were recruited as part of the New York Cancer Project (NYCP); a prospective longitudinal study [29] and two recruited in Houston TX. 3 Filipino, 15 Vietnamese and 32 Chinese American samples were recruited in Houston TX. An additional 7 Vietnamese and 3 Korean genotypes were from the I-ControlDB. Of the Chinese American participants (CHA), 28 also indicated their general origin from regions within China (6 north, 10 south, 3 central and 9 subjects Taiwan).

Genotyping

Genotyping was performed using a 300 K Illumina array according to the Illumina Infinium 2 assay manual (Illumina, San Diego), as previously described [30].

Data Filters

SNPs and individual samples with less than 90% complete genotyping information from any data set were excluded from analyses. SNPs that showed extreme deviation from Hardy Weinberg equilibrium (p<0.00001) in individual population groups were also excluded from analysis. These filters resulted in a total of 215 K autosomal SNPs that were used for these studies. In addition, for samples from nonHGDP origin individuals with evidence of >10% contribution from other continents were excluded from further study. This was either performed prior to Illumina array genotyping for the Filipino, Vietnamese and CHA subjects using 128 continental AIMS [31]. Samples were also filtered for possible cryptic relationships using the PLINK program [32].

Statistical Analyses

F_{st} and F_{is} was determined using Genetix software[33] that applies the Weir and Cockerham algorithm[34]. A measure of informativeness for each SNP (I_n) was determined using an algorithm previously described [35]. Hardy-Weinberg equilibrium was determined using HelixTree 5.0.2 software (Golden Helix, Bozeman, MT, USA).

Population structure was examined using STRUCTURE v2.1[36], [37] using parameters and AIMs previously described [31]. This analysis was performed to exclude individuals with evidence of substantial continental admixture from Europe, Africa or the American continent (see Data Filters).

PCA was performed using the EIGENSTRAT statistical package[38]. All analyses were performed after deleting the MHC region on chromosome 6 since regions of high linkage disequilibrium can overly influence PCA results. The Kruskal-Wallis test [23], a nonparametric alternative to the ANOVA was used to examine the statistical significance of the difference in PC scores among subject groups pre-assigned based on self-identification.

The split half reliability test can determine whether independent (non-overlapping) SNP sets provide the same or different results. The split half reliability test was adjusted by the Spearman-Brown formula [39] and was performed three times using 1) alternate chromosomes, 2) alternate half chromosomes, and 3) half genome SNP sets. These sets were chosen to eliminate any dependency in each test between the two half data sets based on linkage disequilibrium.

Selection of EAS-AIMs

Genotypes from 32 Han Chinese (CHA and CHB), 36 Japanese (JPT), 19 Korean, 21 Filipino and 14 Vietnamese were used for SNP selection. An initial set of 3000 EAS substructure AIMs (EAS-AIMs) were based on either I_n values or using SNP scores from PCA. The best performance using a testing panel was observed using a set of SNPs selected using I_n values from a combination of 1) all five population groups (top 600 SNPs), 2) Chinese and Japanese (top 1200 SNPs), and 3) Chinese and Filipino (top 1200 SNPs). The best performance of a 1500 SNP set and a 750 SNP set were observed using a combination of 500 or 250 from each of these three groups. The testing panel consisted of 20 Chinese, 20 Japanese, 4 Korean, 3 Filipino, 1 Vietnamese, 10 Dai and 10 Cambodian. None of the samples in the testing panel overlapped with the ascertainment samples. The Dai and Cambodian samples were included since there were limited numbers of samples available from the Vietnamese group. The performance of the EAS-AIMs was evaluated using correlations in PC1 and PC2 with the >200 K SNP set.

ACKNOWLEDGMENTS

We thank Stephen Johnson and Robert Lundsten for informatics support on the New York Cancer Project samples. We also thank Anthony Liew and Houman Khalili for expert assistance with genotyping. We thank the volunteers from the different populations for donating blood samples.

AUTHOR CONTRIBUTIONS

Conceived and designed the experiments: MFS. Performed the experiments: CT AL PG MFS. Analyzed the data: CT RK MFS. Contributed reagents/materials/analysis tools: CT AL MR JWB PG MFS. Wrote the paper: CT RK JWB PG MFS

REFERENCES

1.   Seldin MF, Shigeta R, Villoslada P, Selmi C, Tuomilehto J, et al. (2006) European Population Substructure: Clustering of Northern and Southern Populations. PLoS Genetics 2: 1339–1351.

2.   Bauchet M, McEvoy B, Pearson LN, Quillen EE, Sarkisian T, et al. (2007) Measuring European population stratification with microarray genotype data. Am J Hum Genet 80: 948–956.

3.   Price AL, Butler J, Patterson N, Capelli C, Pascali VL, et al. (2008) Discerning the ancestry of European Americans in genetic association studies. PLoS Genet 4: e236.

4.   Tian C, Plenge RM, Ransom M, Lee A, Villoslada P, et al. (2008) Analysis and application of European genetic substructure using 300 K SNP information. PLoS Genet 4: e4.

5.   Tian C, Gregersen PK, Seldin MF (2008) Accounting for ancestry: population substructure and genome-wide association studies. Hum Mol Genet 17: R143–150.

6.   Suzuki A, Yamada R, Chang X, Tokuhiro S, Sawada T, et al. (2003) Functional haplotypes of PADI4, encoding citrullinating enzyme peptidylarginine deiminase 4, are associated with rheumatoid arthritis. Nat Genet 34: 395–402.

7.   Tokuhiro S, Yamada R, Chang X, Suzuki A, Kochi Y, et al. (2003) An intronic SNP in a RUNX1 binding site of SLC22A4, encoding an organic cation transporter, is associated with rheumatoid arthritis. Nat Genet 35: 341–348.

8.   Lee HS, Remmers EF, Le JM, Kastner DL, Bae SC, et al. (2007) Association of STAT4 with rheumatoid arthritis in the Korean population. Mol Med 13: 455–460.

9.   Lin YJ, Wan L, Lee CC, Huang CM, Tsai Y, et al. (2007) Disease association of the interleukin-18 promoter polymorphisms in Taiwan Chinese systemic lupus erythematosus patients. Genes Immun 8: 302–307.

10. Shin HD, Sung YK, Choi CB, Lee SO, Lee HW, et al. (2007) Replication of the genetic effects of IFN regulatory factor 5 (IRF5) on systemic lupus erythematosus in a Korean population. Arthritis Res Ther 9: R32.

11. Kobayashi S, Ikari K, Kaneko H, Kochi Y, Yamamoto K, et al. (2008) Association of STAT4 with susceptibility to rheumatoid arthritis and systemic lupus erythematosus in the Japanese population. Arthritis Rheum 58: 1940–1946.

12. Cavalli-Sforza LL (2005) The Human Genome Diversity Project: past, present and future. Nat Rev Genet 6: 333–340.

13. Li JZ, Absher DM, Tang H, Southwick AM, Casto AM, et al. (2008) Worldwide human relationships inferred from genome-wide patterns of variation. Science 319: 1100–1104.

14. Ballinger SW, Schurr TG, Torroni A, Gan YY, Hodge JA, et al. (1992) Southeast Asian mitochondrial DNA analysis reveals genetic continuity of ancient mongoloid migrations. Genetics 130: 139–152.

15. Chu JY, Huang W, Kuang SQ, Wang JM, Xu JJ, et al. (1998) Genetic relationship of populations in China. Proc Natl Acad Sci U S A 95: 11763–11768.

16. Su B, Xiao J, Underhill P, Deka R, Zhang W, et al. (1999) Y-Chromosome evidence for a northward migration of modern humans into Eastern Asia during the last Ice Age. Am J Hum Genet 65: 1718–1724.

17. Yao YG, Kong QP, Bandelt HJ, Kivisild T, Zhang YP (2002) Phylogeographic differentiation of mitochondrial DNA in Han Chinese. Am J Hum Genet 70: 635–651.

18. Shi H, Dong YL, Wen B, Xiao CJ, Underhill PA, et al. (2005) Y-chromosome evidence of southern origin of the East Asian-specific haplogroup O3-M122. Am J Hum Genet 77: 408–419.

19. Karafet T, Xu L, Du R, Wang W, Feng S, et al. (2001) Paternal population history of East Asia: sources, patterns, and microevolutionary processes. Am J Hum Genet 69: 615–628.

20. Cavalli-Sforza LL, Menozzi P, Piazza A (1996) The history and geography of human genes. Princeton, , N.J.: Princeton University Press.

21. 2003) (2003) The International HapMap Project. Nature 426: 789–796.

22. Altshuler D, Brooks LD, Chakravarti A, Collins FS, Daly MJ, et al. (2005) A haplotype map of the human genome. Nature 437: 1299–1320.

23. Hollander M, Wolfe DA (1973) Nonparametric statistical inference. New York: John Wiley and Sons. pp. 115–120.

24. Seldin MF, Price AL (2008) Application of ancestry informative markers to association studies in European Americans. PLoS Genet 4: e5.

25. Ding YC, Wooding S, Harpending HC, Chi HC, Li HP, et al. (2000) Population structure and history in East Asia. Proc Natl Acad Sci U S A 97: 14003–14006.

26. Zhang F, Su B, Zhang YP, Jin L (2007) Genetic studies of human diversity in East Asia. Philos Trans R Soc Lond B Biol Sci 362: 987–995.

27. Nei M, Roychoudhury AK (1993) Evolutionary relationships of human populations on a global scale. Mol Biol Evol 10: 927–943.

28. Hellenthal G, Auton A, Falush D (2008) Inferring human colonization history using a copying model. PLoS Genet 4: e1000078.

29. Mitchell MK, Gregersen PK, Johnson S, Parsons R, Vlahov D (2004) The New York Cancer project: rationale, organization, design, and baseline characteristics. J Urban Health 81: 301–310.

30. Duerr RH, Taylor KD, Brant SR, Rioux JD, Silverberg MS, et al. (2006) A genome-wide association study identifies IL23R as an inflammatory bowel disease gene. Science 314: 1461–1463.

31. Kosoy R, Nassir R, Tian C, White PA, Butler LM, et al. (In press) Ancestry informative marker sets for determining continental origin and admixture proportions in common populations in America. Human Mutation.

32. Purcell S, Neale B, Todd-Brown K, Thomas L, Ferreira MA, et al. (2007) PLINK: a tool set for whole-genome association and population-based linkage analyses. Am J Hum Genet 81: 559–575.

33. Belkhir K, Borsa P, Chikhi L, Raufaste N, Bonhomme F (2001) GENETIX, software under WindowsTM for the genetic of populations. 4.02 ed. Montpellier, France: Laboratory Genome, Populations, Interactions CNRS UMR 5000, University of Montpellier II.

34. Weir B, Cockerham C (1984) Estimating F-statistics for the analysis of population structure. Evolution 38: 1358–1370.

35. Rosenberg NA, Li LM, Ward R, Pritchard JK (2003) Informativeness of genetic markers for inference of ancestry. Am J Hum Genet 73: 1402–1422.

36. Pritchard JK, Stephens M, Donnelly P (2000) Inference of population structure using multilocus genotype data. Genetics 155: 945–959.

37. Falush D, Stephens M, Pritchard JK (2003) Inference of population structure using multilocus genotype data: linked loci and correlated allele frequencies. Genetics 164: 1567–1587.

38. Price AL, Patterson NJ, Plenge RM, Weinblatt ME, Shadick NA, et al. (2006) Principal components analysis corrects for stratification in genome-wide association studies. Nat Genet 38: 904–909.

39. Spearman C (1910) Correlation calculated with faulty data. British Journal of Psychology 3: 271–295.

Chapter 9

ASSEMBLY OF INFLAMMATION-RELATED GENES FOR PATHWAYFOCUSED GENETIC ANALYSIS

Matthew J. Loza[1] , Charles E. McCall[2] , Liwu Li[3] , William B. Isaacs[4,5], Jianfeng Xu6, Bao-Li Chang[7]

[1] Center for Human Genomics, Department of Internal Medicine, Wake Forest University School of Medicine, Winston-Salem, North Carolina, United States of America

[2]Department of Internal Medicine, Wake Forest University School of Medicine, Winston-Salem, North Carolina, United States of America

[3]Department of Biology, Virginia Polytechnic Institute and State University, Blacksburg, Virginia, United States of America

[4]Department of Urology, Johns Hopkins University Medical Institutions, Baltimore, Maryland, United States of America

[5]Department of Oncology, Johns Hopkins University Medical Institutions, Baltimore, Maryland, United States of America

[6] Center for Human Genomics, Department of Epidemiology and Prevention, Wake Forest University School of Medicine, Winston-Salem, North Carolina, United States of America

[7] Center for Human Genomics, Department of Pediatric Medicine, Wake Forest University School of Medicine, Winston-Salem, North Carolina, United States of America

ABSTRACT

Recent identifications of associations between novel variants in inflammation-related genes and several common diseases emphasize the need for systematic evaluations of these genes in disease susceptibility. Considering that many genes are involved in the complex inflammation responses and many genetic variants in these genes have the potential to alter the functions and expression of these genes, we assembled a list of key inflammation-related genes to facilitate the identification of genetic associations of diseases with an inflammation-related etiology. We first reviewed various phases of inflammation responses, including the development of immune cells, sensing of danger, influx of cells to sites of insult, activation and functional responses of immune and non-immune cells, and resolution of the immune response. Assisted by the Ingenuity Pathway Analysis, we then identified 17 functional sub-pathways that are involved in one or multiple phases. This organization would greatly increase the chance of detecting gene-gene interactions by hierarchical clustering of genes with their functional closeness in a pathway. Finally, as an example application, we have developed tagging single nucleotide polymorphism (tSNP) arrays for populations of European and African descent to capture all the common variants of these key inflammation-related genes. Assays of these tSNPs have been designed and assembled into two Affymetrix ParAllele customized chips, one each for European (12,011 SNPs) and African (21,542 SNPs) populations. These tSNPs have greater coverage for these inflammation-related genes compared to the existing genome-wide arrays, particularly in the African population. These tSNP arrays can facilitate systematic evaluation of inflammation pathways in disease susceptibility. For additional applications, other genotyping platforms could also be employed. For existing genome-wide association data, this list of key inflammation-related genes and associated subpathways can facilitate comprehensive inflammation pathway- focused association analyses.

INTRODUCTION

Inflammation is an essential component of immune-mediated protection against pathogens and tissue damage. Immune responses are also responsible for the unfavorable rejection of tissue/organ transplants, hypersensitivity reactions (e.g., atopy, anaphylaxis, contact hypersensitivity, delayed-type hypersensitivity), and septic shock. Aberrant or unchecked immune responses may lead to a state of chronic inflammation [1]–[3]. This may occur when the immune response: 1) is activated in the absence of 'danger' signals; 2) fails to fully turn-off (resolve) after elimination of the danger; and 3) fails to completely clear the danger stimulus. Factors that may influence the initiation, activity, and resolution of immune responses include health (physical and emotional), age, diet, medications, and genetic predisposition.

Inflammation may also be a contributing factor for some diseases. The role of chronic inflammation in a wide variety of diseases is well-appreciated, including rheumatoid arthritis and other autoimmune disorders [4], cardiovascular disease [5]–[7], gastrointestinal disorders [8],[9], and a number of cancers [10]–[14]. Perhaps the best evidence for the importance of chronic inflammation in disease is the efficacy of NSAIDs in reducing the risk or severity of these disorders [15]. There is mounting evidence that dietary factors that may influence inflammation, such as the balance of omega-3 vs. omega-6 polyunsaturated fatty acids (PUFAs), have an impact on disease risk and progression [16]. Genetic studies also provide evidence that variant alleles of genes associated with inflammatory pathways impact the risk of disease initiation, progression, and severity (see **Table 1**). The role of inflammation as a mediator of disease is currently receiving extensive attention, resulting in the National Institute of Allergy and Immunologic Diseases (NIAID) plans for an NIH Roadmap initiative with the overarching theme: "Inflammation as a Common Mechanism of Disease" (http://nihroadmap.nih.gov/inflammation/index.asp).

Table 1: Confirmed associations of genetic variants in inflammation-associated genes and disease

Disease	Gene	Encoded protein	Variant	Odds ratio [a]	p-value	Confirmation method
Age-related macular degeneration [44]	CFH	Complement factor H	rs1061170	3.40 [b]	$<1\times10^{-5}$	Case-control/meta-analysis
Atopic asthma [32]	IL4R	IL-4 receptor alpha	rs1801275	1.79	3×10^{-9}	Meta-analysis of 7 study populations
Atopic asthma [21]	TNF	TNF-alpha	-308 G/A	1.46	1×10^{-4}	Meta-analysis of 15 study populations
Crohn's disease [30]	CARD15	Nod2	1007fsinsC	4.3	7×10^{-28}	Meta-analysis of 16 study populations
Breast cancer [33]	CASP8	Caspase-8	rs1045485	0.90	0.016	Analysis of 3 study populations (6351 cases/5708 controls)
Breast cancer [33]	TGFB1	TGF-beta 1	rs1982073	1.08	0.0088	Analysis of 3 study populations (6863 cases/5587 controls)
Breast cancer [34]	TNF	TNF-alpha	rs361525	1.18	0.008	Analysis of two independent study populations
Graves' disease [24]	PTPN22	Lymphoid-specific phosphatase	C1858T	1.61	$<1\times10^{-5}$	Meta-analysis of 3 study populations
Inflammatory bowel disease [31]	IL23R	IL-23 receptor beta	rs11209026	0.26	5×10^{-9}	Genome-wide screen (raw p-value)
				0.45	8×10^{-4}	Case-control replication
				~0.5 [c]	1.3×10^{-10}	Family-based TDT replication
Psoriatic arthritis [23]	TNF	TNF-alpha	-238 G/A	2.29	2×10^{-4}	Meta-analysis of 8 study populations
Rheumatoid arthritis [24]	PTPN22	Lymphoid-specific phosphatase	C1858T (R620W)	1.68	$<1\times10^{-8}$	Meta-analysis of 12 study populations
Systemic lupus erythamatosus [25]	IRF5	Interferon response factor 5	rs2004640	1.47	4.2×10^{-21}	Case-control/meta-analysis+replication in family-based
Systemic lupus erythamatosus [24]	PTPN22	Lymphoid-specific phosphatase	C1858T (R620W)	1.49	$<1\times10^{-5}$	Meta-analysis of 5 study populations
Systemic lupus erythamatosus [22]	TNF	TNF-alpha	-308 G/A	2.1	<0.001	Meta-analysis of 10 study populations of European descent
Type 1 diabetes [29]	CTLA4	CTLA-4	rs3087243	1.18 [c]	5.6×10^{-8}	Family-based TDT
Type 1 diabetes [28]	CTLA4	CTLA-4	rs3087243	1.17 [c]	6×10^{-4}	Family-based TDT
				1.21	1.3×10^{-7}	Case-control
Type 1 diabetes [27]	CTLA4	CTLA-4	rs3087243	1.20	3.7×10^{-10}	Case-control
Type 1 diabetes [26]	IFIH1	Mda-5, Helicard	rs1990760	0.86	1.42×10^{-10}	Genome-wide, validated in case-control+family-based
Type 1 diabetes [24]	PTPN22	Lymphoid-specific phosphatase	C1858T (R620W)	1.85	$<1\times10^{-5}$	Meta-analysis of 6 study populations

[a]Odds ratio for allele test (multiplicative model), unless otherwise indicated.
[b]Odds ratio for dominant model
[c]Risk ratio from family-based transmission disequilibrium test (TDT).
doi:10.1371/journal.pone.0001035.t001

Numerous genetic linkage and case-control association studies have implicated genetic variations in genes important in immunity and inflammation and inflammatory diseases. Single missense heritable mutations can be the sole or major determinant for inflammatory diseases, such as Familial Cold Autoinflammatory Syndrome (missense mutations in exon 3 of *Cias1* account for all cases) [17]–[19] and Familial Mediterranean Fever (*MEFV*, five founder missense mutations, when homozygous, account for 74% of cases) [20]. For many complex inflammatory diseases, polymorphisms in inflammatory genes are more likely to act as modifiers for disease susceptibility rather than sole determinants. Recent meta-analyses report modest associations between single nucleotide polymorphisms (SNPs) in *TNF* (encoding TNF-α) and increased risk for asthma [21], system lupus erythamatosus (SLE) [22], and psoriatic arthritis [23]. A variant of *PTPN22* (encoding a lymphoid-specific protein tyrosine phosphatase) is modestly associated with multiple autoimmune diseases (rheumatoid arthritis, SLE, type 1 diabetes, and Graves' disease) [24]. Associations of *IRF5* (interferon regulator factor 5) genetic variants and increased SLE risk have been highly replicated [25]. *IFIH1*, encoding an innate immunity viral mRNA detector (early type I IFN-β responsive gene, Helicard), is strongly associated with type 1 diabetes risk [26]. Association of a small risk for type 1 diabetes and *CTLA4* has been consistently replicated [27]–[29]. An insertion polymorphism in *CARD15*, the gene encoding the microbial nucleotide detector Nod2, is a major risk factor for Crohn's disease [30]. An association between SNP's in *IL23R* (IL-23 receptor β-chain) and increased risk of inflammatory bowel disease has been reported for a genome-wide association study and confirmed in three independent populations [31]. A recent meta-analysis demonstrated that an *IL4R* (IL-4 receptor alpha chain) variant modestly increases risk for atopic asthma [32]. A description of these studies and reported associations is provided in **Table 1**. This list is not intended to be comprehensive but rather serves as an example of representative genetic variations which have been well-established to be associated with common inflammatory disorders.

In addition to inflammatory/autoimmune diseases, polymorphisms in inflammation-associated genes may also contribute to risk for diseases in which inflammatory/immune-disorders are not the primary characteristic. There is evidence from the Breast Cancer Association Consortium that *CASP8* (caspase 8) and *TGFB1* (TGF-β1) variants impart risk, albeit low penetrance, for breast cancer [33]. Analysis of two independent populations suggest that a rare polymorphism in *TNF* may also be a low-penetrance risk factor for breast cancer [34]. Several groups report associations between various *PTGS2* (COX-2) genetic variants and colorectal cancer risk [35]–[39]. There is also evidence to suggest that a low frequency *COX2* variant allele decreases prostate cancer risk [40], particularly in subjects who frequently eat fish [41]. Interestingly, in a preliminary study using MegAllele™ I&I panel, including 9,275 SNPs in 1,086 genes involved in immunity and inflammation, more SNPs were found to be significantly associated with prostate cancer risk than expected by chance, which suggests multiple genetic variants in this pathway impart modest risk to prostate cancer [42].

Because genetic associations of disease and genetic variations in inflammatory genes are often relatively modest, it is likely that polymorphisms in multiple inflammatory genes cooperate in an additive or synergistic manner to impact disease risk. Pathway analyses may help to reveal gene-gene interactions or risks imparted independently from other genes in the pathway. The advantages of performing analyses at pathway levels are illustrated by Dinu et al. [43]. Associations between *CFH* (complement factor H) and age-related macular degeneration have been replicated in numerous studies [44]. To test whether genetic variations in the multiple complement pathway genes impact macular degeneration risk, Dinu et al. analyzed the existing genome-wide association study data of Klein et al. [45], restricting the analyses to genes in the complement pathway in subjects carrying the *CFH* risk allele. Significant associations were detected for a *C7* and *MBL2* variants and severity of macular degeneration in the context of the complement pathway analysis and the *CHF* risk allele, and these associations would not have been significant in a genome-wide analysis of the data.

RESULTS

Choosing Genes

Various aspects of immunity contribute to the development of an overall inflammatory immune response. These phases include the development of immune cells, sensing of danger, influx of cells to sites of insult, activation and functional responses of immune and non-immune cells, and resolution of the immune response. To broadly cover most aspects of inflammatory responses, the various *phases of immune responses* were considered in choosing genes for the SNP array panel, outlined in **Table 2**. A schematic representation of most of these phases is presented in **Fig. 1**.

Depicted is a schematic representation of an immune response to a generic pathogenic insult. The phases of immune responses (described in Table 2) are shown in bold. Additional aspects not shown are the involvement of secondary lymphoid tissues for initial T cell and B cell activation by dendritic cells that migrate from the site of inflammation to lymph nodes and other secondary lymphoid structures. The resolution of immune responses, immunological memory, and homeostasis are also not depicted.

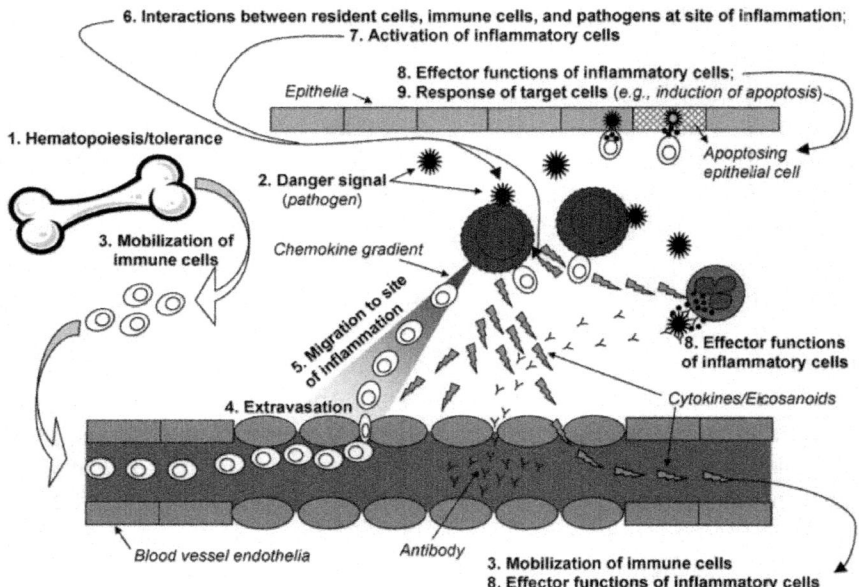

Figure 1: Development of an immune response.

doi:10.1371/journal.pone.0001035.g001

Table 2: Phases of immune response

Phase of immune response	Description
Hematopoiesis/homeosta-sis/tolerance	The generation and differentiation of immune cells and maintenance of their number in circulation and tissues; prevention of self-reactivity.
Danger signal	Innate recognition of and response to pathogenic foreign substances or stress.
Mobilization of immune cells	Systemic soluble mediators informing immune cells in circulation and lymphoid tissues of danger.
Extravasation	The process of circulating immune cells crossing from blood into peripheral tissues and secondary lymphoid tissues.
Migration to site of inflammation	The process of immune cells, after extravasation, reaching the site of inflammatory insult, including chemoattraction, adhesion to substrates, and degradation of extracellular matrix.
Interactions between resident cells, immune cells, and pathogens at site of inflammation	Interactions between resident cells, immune cells, and pathogens at site of inflammation–how infiltrating cells interact with the resident inflammatory cells, non-immune cells (e.g., epithelia), pathogens, and other infiltrating cells, that leads to activation of effector functions.
Activation of inflammatory cells	The signaling pathways and transcription factors stimulated by activating, co-stimulatory, and inhibitory receptors that leads to activation, proliferation, differentiation, and survival of responding immune cells.
Effector functions of inflammatory cells	The factors produced/released by immune cells in attempt to resolve the pathogenic insults, including release of cytotoxic/cytostatic mediators and mediators to enhance or fine-tune the immune response.
Response of target cells	The pathways in non-immune cells (e.g., epithelia) activated in response to the effector functions of immune cells.
Resolution of immune response vs. chronic inflammation	The pathways that lead to the downregulation of immune responses and inflammation after the pathogenic insult is cleared; the factors maintaining late-phase immune responses when the insult is not totally resolved.

doi:10.1371/journal.pone.0001035.t002

Priority was given first to genes of known function in inflammatory responses (in both immune and non-immune cells), and then to genes expressed in immune cells with function implied by homology to other genes but exact function not clear. Ubiquitously expressed genes required for the normal function of most cell types of diverse origin were given lower priority.

However, special emphasis was placed on genes at nodes for signaling to and from multiple pathways, most notably genes in NF-κB, MAPK, and PI3K signaling pathways.

Pathways were built using Ingenuity Pathways Analysis, as described in *Methods* section, using both pre-defined 'canonical pathways' and custom-built pathways based on our own queries for genes/pathways not included in the canonical pathways.

Multiple functional pathways are involved each of the immune response phases, and each functional pathway may contribute to several of the immune response phases. For example (**Fig. 2**), the response of a macrophage responding to a *danger signal* during a gram-negative bacterial infection involves innate pathogen recognition of LPS by the TLR4 complex. TLR4 transduces signals via NF-κB, MAPK, and PI3K signaling pathways, stimulating synthesis of eicosanoids and cytokines to signal other cells of the danger. LPS may also stimulate expression of stress-induced proteins, such as MIC-A and MIC-B (ligands for natural killer cell activating receptors) and T cell co-stimulatory molecules (B7 family proteins).

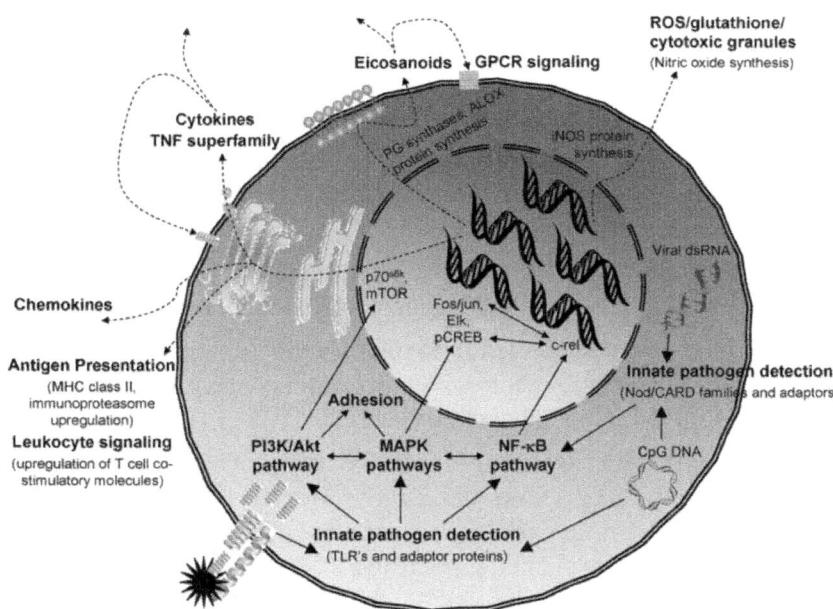

Figure 2: Inflammation subpathways involved in the response to danger signal.

The concerted action of multiple functional subpathways in the initial response of a macrophage to bacteria or virus is depicted. Solid arrows indicate signaling events and dashed arrows stimulated production of proteins and other inflammatory mediators (including autocrine/paracrine responses of the macrophage to the released molecules).

doi:10.1371/journal.pone.0001035.g002

Because immune response phases utilize multiple functional pathways and these pathways are overlapping among phases, the genes chosen for the SNP array panel were assigned to one of the following *functional pathways*:

Adhesion-Extravasation-Migration: adhesion molecules; chemoattractants and chemoattractant receptors; cytoskeletal rearrangement signaling, motility proteins.

Apoptosis signaling: death receptors and ligands and extrinsic apoptosis pathway signaling; mitochondrial-dependent, intrinsic apoptosis pathway signaling; cellular stress signaling.

Calcium signaling: NF-ATs; calcineurins; calcium/calmodulin-dependent kinases.

Complement cascade: components of classical, alternative, and lectin-dependent complement pathways.

Cytokines and cytokine signaling: cytokines; cytokine receptors; cytokine-dependent signaling, including JAK-STAT and interferon-regulatory factor (IRF) pathways; suppressor of cytokine signaling proteins (SOCS).

Eicosanoid synthesis and receptors: enzymes involved in synthesis of prostanoids, leukotrienes, hepoxylins (12-HETE), and lipoxins from arachidonic acid; prostanoid and leukotriene receptors.

Glucocorticoid/PPAR signaling: nuclear receptors for glucocorticoids; steroid-interacting proteins; PPARs and associated proteins.

G-Protein Coupled Receptor Signaling: GPCRs (other than eicosanoid receptors and chemokine receptors); G-protein-dependent signaling pathways, including cAMP-PKA and phospholipase B2.

Innate pathogen detection: Toll-like receptors (TLRs); intracellular nucleotide detectors (e.g., Nod1 and putative members of CARD/Nod family); peptidoglycan recognition proteins; associated signaling molecules linking these detectors to major signaling pathways.

Leukocyte signaling: Signaling molecules, receptors, and adaptors important for regulation of leukocyte activation beyond major signaling pathways (i.e., MAPK, PI3K/Akt, NF-κB, GPCR, and cytokine signaling pathways); including, but not limited to, T cell receptor (TCR) and B cell receptor (BCR) signaling components, B7 family, phosphatases, Foxp3, immunoglobulin receptors, leukocyte inhibitory receptor (CD85) family, scavenger receptors.

MAPK signaling: p38 stress-activated protein kinase, p42/p44 extracellular-regulated kinase (Erk), and Jun kinase (Jnk) signaling pathways.

Natural Killer Cell Signaling: Natural killer (NK) cell-specific activating and inhibitory receptors; adaptors and signaling molecules for transducing NK cell-specific receptor signaling.

NF-kB signaling: Molecules for regulation of NF-κB activation, including adaptors linking. other pathways to and from the NF-κB pathway (e.g., from MAPK, pathogen-detection, and TNF-α signaling pathways).

Antigen presentation: Major Histocompatibility Complex (MHC) molecules and associated proteins; proteins involved in uptake, processing, and loading of peptides on MHC molecules.

PI3K/AKT Signaling: Molecules involved in regulation of PI3K-dependent signaling, including adaptors linking other pathways to and from the PI3K pathway.

ROS/glutathione/cytotoxic granules: Molecules involved in the generation and response to leukocyte-derived cytotoxic agents (reactive oxygen species (ROS), nitric oxide, cytotoxic granules of granulocytes and natural killer cells), including contents of granulocyte and NK cell cytotoxic granules; glutathione peroxidases; peroxiredoxins; catalase; proteinases; superoxide dismutase.

TNF superfamily and signaling: Receptors and ligands of the TNF-α superfamily; adaptors and signaling molecules involved in transducing signals from receptor stimulation to other major signaling pathways (e.g., MAPK and NF-κB pathways).

Examples of the functional subpathways and types of genes chosen for the different phases of immune responses are presented in **Table 3** (proteins encoded by the genes are provided). Most of these immune response phases also utilize common signaling pathways, including elements of MAPK, NF-κB, PI3K/Akt, GCPR, cytokine, and leukocyte signaling pathways. An example of the integration of multiple functional subpathways for one phase of an immune response (danger signal) is depicted in **Fig. 2**. The numbers of genes chosen in each subpathway are listed in **Table 4**, and the complete list of 1027 candidate genes and their primary subpathway (and secondary subpathway for a subset of genes) is provided.

Table 3. Pathways and proteins associated with immune response phases

Phase of immune response	Examples of pathways, proteins, and inflammatory mediators involved in immune response phases
Hematopoiesis/homeostasis/ tolerance	hematopoietic cytokines (M-,G-,GM-CSF;IL-4,-5,-7,-13), stromal factors (c-kit, SCF, Flt3L), regulatory T cell function (Foxp3)
Danger signal	innate pathogen recognition receptors (TLRs, CARDs/NODs, peptidoglycan recognition proteins), scavenger receptors (MSR1), endothelins, adenosine receptors, complement, stress-induced responses (MIC-A,-B), eicosanoid synthesis genes, cytokines, antigen presentation genes
Mobilization of immune cells	systemic inflammatory mediators (IL-1β, IL-6, TNF-a), chemokines, eicosanoids, GPCR signaling (eicosanoids, histamines)
Extravasation	adhesion molecules (integrins, -CAMs), chemokines, vasodilators (eicosanoids/GPCR), cytoskeletal rearrangement singaling molecules (Vav, VASP, MENA), non-muscle myosins
Migration to site of inflammation	adhesion molecules (integrins, -CAMs, maxtrix receptors), chemokines, matrix proteases (MMPs), cytoskeletal rearrangement singaling molecules (Vav,VASP,MENA), focal adhesion proteins (Vav,ROCK), non-muscle myosins
Interactions between resident cells, immune cells, and pathogens at site of inflammation	adhesion molecules, innate detectors of pathogens (TLRs, CARDs/NODs), Fc receptors (FcgRI,II,III; FceRI,II), stress-induced ligands (MIC-A,-B), NK cell-activating receptors, cytokines and receptors, other activating receptors (TCR, BCR complexes; growth factor receptors); co-stimulatory receptors (B7 family, CD2 family), inhibitory receptors (KIRs, LIRs/ILTs), phagocytosis/antigen presentation (XBOX genes, CIITA, TAP, immunoproteasome, HLA molecules)
Activation of inflammatory cells	MAPK pathways (Erk, p38, Jnk), PI3K/Akt signaling, NF-kB signaling, cytokine signaling (JAK/STAT/Tyk, NFIL3, NFIL6, IRFs), GPCR signaling (PKA, PLCb, phosphodiesterases, CREB, Pyk2, Rap1, Src), adaptor signaling proteins (TRAFs, IRAKs, MyD88, DAP10, DAP12, ZAP70, Syk, LAT, SLP76, MyD88, CD3ζ, FcεRγ)
Effector functions of inflammatory cells	cytokines (IFN-γ, IFN-x, TNF-x superfamily, CSFs, interleukins), death receptor ligands (FasL, TRAIL, TNF-a), eicosanoids (prostaglandins, thromboxane, prostacyclin, leukotrienes), cytotoxic mediators (glutathiones/PHOX/ reactive oxygen species, RNS, perforin/granzymes), antibody production, acute phase/fever response (C-reactive protein, factor P)
Response of target cells	cytokine receptors, GPCRs, death receptors, apoptosis signaling, adhesion molecules, growth factor receptors
Resolution of immune response vs. chronic inflammation	apoptosis (death receptor and mitochondrial pathways), TGF-β, IL-10, Foxp3, prostaglandins, phosphatases, inhibitors of cytokine signaling (SOCS, A20/TNFAIP3)

doi:10.1371/journal.pone.0001035.t003

Table 4: Primary subpathways in inflammation panel

Subpathway	Number of genes in subpathway	Number of SNPs in subpathway
Adhesion-Extravasation-Migration	142	1385
Apoptosis Signaling	68	682
Calcium Signaling	14	409
Complement Cascase	40	419
Cytokine signaling	172	1598
Eicosanoid Signaling	39	374
Glucocorticoid/PPAR signaling	21	230
G-Protein Coupled Receptor Signaling	42	1125
Innate pathogen detection	50	457
Leukocyte signaling	121	1743
MAPK signaling	118	1949
Natural Killer Cell Signaling	31	259
NF-kB signaling	33	297
Phagocytosis-Ag presentation	39	286
PI3K/AKT Signaling	37	307
ROS/Glutathione/Cytotoxic granules	22	162
TNF Superfamily Signaling	38	328

doi:10.1371/journal.pone.0001035.t004

Example Application: WFINFLAM tSNP Panel

Because the components comprising inflammation are very numerous and interacting across many pathways, without strong *a priori* evidence it is difficult to choose a handful of candidate genes to fully cover the potential genetic risk factors contributing to the inflammatory component of a particular disease. For this reason, panels of single nucleotide polymorphisms (SNPs) in an array of inflammation-related genes broadly covering most aspects of immunity and inflammation based on our assembled list will be critical in objectively evaluating the impact of genetic variations in inflammation-related genes on an inflammation-dependent outcome.

There is a commercially available product, Affymetrix GeneChip® Human Immune and Inflammation 9K SNP panel, that attempts to serve the purpose. This application -specific panel contains ~9,000 SNPs to cover ~1,000 immunity- and inflammation- related genes (http://www.affymetrix. com/support/technical/datasheets/humanimmune_9k_snp_datasheet.pdf). However, the rationale for choosing the ~1,000 immunity and inflammation genes in this panel is not clear, and the coverage of these genes, regardless of their biological functions, may not be sufficient. In order to investigate the impact of genetic variations in a broad array of inflammation-related genes on disease risk, we created two tagging SNP (tSNP) panels, one each for populations with either Caucasian or African ancestries, for Affimetrix ParAllele genotyping chips. These tSNP panels were designed to capture majority of the genetic variations in these 1027 inflammation candidate genes.

tSNPs for the 1027 inflammation-associated candidate genes were chosen based on a pair-wise r^2 threshold of 0.8 and MAF $\geq 5\%$ using data in the HapMap Phase II database (HapMap Data Release 21a/phaseII; http://hapmap. org/index.html.en). tSNPs were chosen separately for the CEU (representing European ancestry) and YRI populations (representing African ancestry). In order to accommodate as many relevant inflammation genes as possible, less stringent criteria with r^2 threshold >0.5 were employed for intronic regions (excluding 5kb in the beginning and end of these big introns) greater than 50kb in certain large genes with >100 tSNPs. There were seven genes in this category, and the intronic regions for which a less stringent r^2 threshold was applied are listed in . For genes without genotype information from HapMap, additional SNPs in six inflammation genes were chosen to be included based on information from other resources, as described in *Methods* section.

The resulting inflammation tSNP panels, WFINFLAM-CEU for Caucasians and WFINFLAM-YRI for African descent, include 12,011 SNPs and 21,542 SNPs respectively in 1027 inflammation-associated candidate genes. There is an average of 11.7 and 21 SNPs in each candidate gene in WFINFLAM-CEU and WFINFLAM-YRI panels, respectively. **Table 4** briefly summarizes the numbers of genes and tSNPs included in each subpathway. The annotation table for this panel, including the list of 1027 genes, their chromosomal positions, their associated primary and secondary subpathways, and the number of tSNPs chosen for each gene can be found (and also our website: http://www1. wfubmc.edu/Genomics/PublicationsandData/). Additionally, the annotation file for the SNPs included in the WFINFLAM-CEU and WFINFLAM-YRI panels, including the chromosomal locations, their associated genes and the sub-pathways, can also be found (and our website: http://www1.wfubmc.edu/ Genomics/PublicationsandData/). The coverage for the 1027 inflammation candidate genes in CEU is better compared to other widely used genome-wide association panels. For the coverage of the 1027 inflammation-associated candidate genes, 90.4% of the genes have 90% or more SNPs within these genes that can be captured by $r^2 \geq 0.8$ in CEU using WFINFLAM-CEU panel, compared to 78.9% for the Illumina HumanHap 550 genome-wide panel and 45.8% for the Affymetrix 500k genome-wide panel. For populations with African ancestry, the coverage for the 1027 inflammation candidate genes in YRI is greatly improved compared to other widely used genome-wide association panels. For the coverage of the 1027 inflammation-associated candidate genes, 88.2% of the genes have 90% or more SNPs within these genes that can be captured by $r^2 \geq 0.8$ in YRI using WFINFLAM-YRI panel, compared to 27.1% for the Illumina HumanHap 650k genome-wide panel (which was designed to capture more genetic information from YRI population), and 12.5% for the Affymetrix 500k genome-wide panel. The coverage for all these genotyping panels is detailed in .

DISCUSSION

Various components and complex interactions comprise immune and inflammation responses, and numerous genes are involved in this complex network. With a thorough review of various aspects of inflammatory immune responses, and a systematic search for gene-gene interactions using Ingenuity Pathway Analysis, we have provided a comprehensive list of inflammation-associated genes and subpathways for genetic association studies.

Genome-wide association studies have been a very popular approach to test the association between disease phenotypes and genetic variations. However, we believe there are still several advantages for a pathway-focused study. First of all, compared to whole-genome analyses, restricting analyses to SNPs in a specific pathway reduces the number of multiple tests performed in the analysis of a study population, thereby reducing the probability of false positive associations and increasing the effective power of the study.

This kind of study design is particularly effective when inflammation plays an important role in disease etiology and the goal of the studies is to delineate genetic variations in inflammation pathway to disease risk and/or progression. A related second advantage of restricted pathway analysis is in study design. A large proportion of investigators may not have access to the very large number of subjects and multiple confirmation populations needed to overcome false positive associations due to multiple testing in genome-wide association studies. Studies restricted to a pathway analysis permit the use of study populations that are not large enough for use in whole-genome association studies. When target diseases are not prevalent and inflammation is obviously involved in disease etiology, researchers will gain the most out of an inflammation pathway-specific study design. Although some genes not related to inflammation found in whole-genome panels may impart some risk to inflammation-associated diseases, associated genetic variants would be anticipated to be concentrated in a panel of SNPs in inflammation-associated genes. Therefore, the drawback of potentially missing associated non-inflammation genes is offset by the increased probability of detecting true associations in an inflammation-restricted panel. Thirdly, pathway analysis is far less expensive to perform than whole-genome analysis, especially considering the cost for second, and/or third stage confirmation studies needed to follow-up the significant results from an initial screening in order to rule out false positive associations. Lastly, the functional subpathways are also pre-defined with available biological information. This refined information provides investigators with the opportunity to test gene-gene interactions within subpathways in which synergistic interactions are more likely to be concentrated. Additionally, the interplays between subpathways are also clearly defined to enable investigators to test biologically feasible interactions between subpathways.

However, the results from this manuscript also have potential utility for investigators who have more interests in surveying the whole genome. For whole-genome analyses where there is a prior hypothesis for inflammation being associated with the outcome, the inflammation pathway and subpathways defined in this manuscript may provide a framework for testing whether SNPs in the inflammation pathway or subpathways as a whole are overrepresented for significant associations to the outcome. Although pathway networks can be constructed for whole-genome analyses, such networks should be designed *a priori* before beginning the study.

The WFINFLAM and WFINFLAM-YRI SNP array panels for inflammation-associated genes provide a powerful tool for analyzing the contribution of genetic variation in diseases that have inflammatory components. Although whole-genome SNP panels are currently available that include almost all of the genes included in the WFINFLAM panels, the coverage of SNPs in genes included in the WFINFLAM array is superior to the coverage of currently available in whole-genome arrays, especially for populations with African ancestry background. For researchers who would like to use other genotyping platforms, the inflammatory gene list provided here would be a good starting point for designing genotyping assays for other platforms. Additionally, alternate approaches, other than r^2 based method, for choosing SNPs based on the inflammatory gene list provided here could also be considered. For example, researchers may specifically focus on "high-prior" polymorphisms that are known to be functional or have been previously linked to the specific diseases under study, alone or in combination with the tSNPs provided in the WFINFLAM panel. This approach may be more efficient and powerful than the r^2 based method alone, especially if the targeted "high-prior" polymorphisms are causal and their linkage to the nearby tSNPs is incomplete.

In addition, precaution may be warranted for tSNP panels designed based on the HapMap project. The transferability of the LD patterns between populations studied in HapMap project and other study populations may need to be validated. The transferability of HapMap-based selection of tSNPs using the reference CEU population to several other diverse populations of European ancestry has been demonstrated to be almost as effective for overall SNP coverage in selected genomic regions or randomly selected SNPs in the respective populations [46]–[50]. However, the coverage with tSNPs based on HapMap CEU data for a small subset of specific genes or SNPs may be lower in certain populations despite overall similar coverage [51], particularly for isolated indigenous populations [52]. Other than the data provided by the HapMap project, we are not aware of complete sequence variant information available for validation of the transferability of tSNP panels that were designed based on HapMap data to other study populations.

In summary, pathway analysis of inflammation-associated genes is a powerful approach for determining genetic risk factors for both inflammatory diseases and other diseases that may have an under-appreciated modest inflammatory component, such as cancers. The inflammation pathway gene list and functionally-defined subpathways provide useful tools for assessing the impact of genetic variations in inflammation pathways on disease risk, in situations where either pathway-focused studies or genome-wide analyses are employed.

METHODS

Selection of Genes

Networks of genes involved in the regulation of the phases of immune responses (described in*Results* and **Table 2**) were built using Ingenuity Pathways Analysis (Ingenuity Systems:www.analysis.ingenuity.com). Both pre-defined 'canonical pathways' and custom-built pathways based on our own queries for genes/pathways not included in the canonical pathways were used to establish these networks. Canonical pathways included: actin cytoskeleton signaling; antigen presentation pathway; apoptosis signaling; B cell receptor signaling; calcium signaling; cAMP signaling; chemokine signaling; complement and coagulation cascade; death receptor signaling; ERK/MAPK signaling; FcEpsilon receptor signaling; G-coupled protein receptor signaling; GM-CSF signaling; IGF-1 signaling; IL-10 signaling; IL-2 signaling; IL-4 signaling; IL-6 signaling; integrin signaling; interferon signaling; JAK/STAT signaling; leukocyte extravasation signaling; natural killer cell signaling; NF-kB signaling; nitric oxide signaling; notch signaling; p38 MAPK signaling; PI3K/Akt signaling; PPAR signaling; protein ubiquination pathway; PTEN signaling; SPK/JNK signaling; T cell receptor signaling; TGF-β signaling; Toll-like receptor signaling.

These Networks Were Then Arranged Into Inflammation Subpathways By:

combining several networks together (e.g., ERK/MAPK, p38 MAPK, and SAPK/JNK canonical pathways into the 'MAPK signaling' inflammation subpathway; IL-2-, IL-4-, IL-6-, IL-10-, Interferon-, GM-CSF-, IGF-1-, JAK/STAT6-, and TGF-β- signaling canonical pathways into 'cytokine signaling' inflammation subpathway; actin cytoskeleton-, chemokine-, integrin-, and leukocyte extravasation- signaling canonincal pathways into the 'adhesion-extravasation-migration' inflammation subpathway; etc.);

adding additional genes to bridge networks within a subpathway and to include appropriate genes not included in the canonical pathways (e.g., additional cytokines and their receptors were added to the 'cytokine signaling' inflammation subpathway; additional integrins and chemokines/chemoattractant molecules were added to the 'adhesion-extravasation-migration' inflammation subpathway; CD antigens expressed by leukocytes not already included were considered for addition to several subpathways; other missing genes/pathways considered important by the panel of investigators, such as the scavenger receptor network for 'leukocyte signaling' inflammation subpathway and Nod1/CARD family networks for 'innate pathogen detection' inflammation subpathway; etc.);

trimming the networks of genes with low priority for inclusion in the inflammation panel. Genes with lower priority include: genes not expressed in immune cells or not directly involved in cells responding to inflammation, including non-immune cells (e.g., skeletal muscle-specific myosins in the actin cytoskeleton signaling canonical pathway; calsequestrins expressed mainly in various muscle cells, in the calcium signaling canonical pathway; GH1 and GHR, growth hormone expressed in pituitary gland and its receptor, and NGFB and NGFR, nerve growth factor and its receptor, in the NF-κB canonical pathway; etc.; MAPK8IP1, specific for pancreatic cell function, in the PI3K/AKT canonical pathway; etc.); genes with unknown function, though genes with high homology to known inflammatory mediators were considered (e.g., bcl-2 family homologs, IL-1β family homologs). Special emphasis was placed on genes at nodes for signaling to and from multiple pathways, most notably genes in NF-κB, MAPK, and PI3K signaling pathways.

Choosing Tagging SNPs

Tagging SNPs (tSNPs) for candidate genes were chosen using Tagger server (http://www.broad.mit.edu/mpg/tagger/server.html). The target sequences included genomic regions containing the entire candidate genes, 5kb before transcription start site, and 2kb after the transcription end site, based on annotation in NCBI Build 35. When candidate regions for two or more genes overlapped, the combined genomic regions were used for choosing tSNPs. Two separate lists of non-synonymous coding SNPs for the CEPH population (CEU) and for the Yuruba population (YRI) were downloaded from HapMap (rel21a_NCBI_Build35) and forced in as tagging SNPs for ancestry-specific panels. Pair-wise r^2 threshold of 0.8 and MAF$\geq$5% were used. For genes without genotype information from HapMap, SNPs in Affymetrix 500k array, as well as SNPs with frequency data from the Innate Immunity PGA (IIPGA) (http://innateimmunity.net/), were manually chosen to be included in the list based on allele frequencies and inter-SNP distance (the chromosomal positions are based on IIPGA annotation because some of these genes/SNPs were ambiguously mapped in build 35). Some tSNPs were dropped out from the final list because they did not pass the Affymetrix design review due to the following reasons: non-biallelic SNPs, existing SNPs or ambiguous bases too close to the SNP of interest, SNPs exceeding T_m ranges, or SNPs failing BLAST searches. In the attempt to fill in the gaps due to tSNPs failing design criteria, tSNPs for genes with dropped-out tSNPs were chosen again by forcing out the tSNPs failing design review and forcing in the remaining tSNPs when running Tagger. In total, there are 12,011 SNPs in WFINFLAM panel for Caucasians and 21,542 SNPs in WFINFLAM-YRI for African descents in 1027 inflammation-associated candidate genes that passed the Affymetrix design review.

Estimation of genomic coverage by tagging SNPs

We used LdCompare (Hao 2006;http://www.affymetrix.com/support/developer/tools/devnettools.affx) to estimate the coverage of the 1027 inflammation-associated candidate genes with both WFINFLAM panel for Caucasians and WFINFLAM-YRI panel for African descents. We have also computed the coverage of several commercial genotyping arrays, including Affymetrix Mapping 500K, Illumina HumanHap 550 and HumanHap 650, for these 1027 genes. Single-marker coverage (pairwise r^2) and multiple-marker coverage (multiple-marker r^2) were computed and combined to estimate the coverage for the genomic regions containing the entire candidate genes, 5kb before transcription start site, and 2kb after the transcription end site. The summary for the coverage for these genes in all these genotyping panels are detailed in .

AUTHOR CONTRIBUTIONS

Conceived and designed the experiments: JX BC ML. Analyzed the data: BC ML. Wrote the paper: BC ML. Other: Contributed to compilation of data: JX WI LL CM ML BC.

REFERENCES

1. Lawrence T, Gilroy DW (2007) Chronic inflammation: a failure of resolution? Int J Exp Pathol 88: 85–94.

2. Han J, Ulevitch RJ (2005) Limiting inflammatory responses during activation of innate immunity. Nat Immunol 6: 1198–1205.

3. Forrester JS, Bick-Forrester J (2005) Persistence of inflammatory cytokines cause a spectrum of chronic progressive diseases: implications for therapy. Med Hypotheses 65: 227–231.

4. Goronzy JJ, Weyand CM (2005) Rheumatoid arthritis. Immunol Rev 204: 55–73.

5. Ross R (1999) Atherosclerosis–an inflammatory disease. N Engl J Med 340: 115–126.

6. Willerson JT, Ridker PM (2004) Inflammation as a cardiovascular risk factor. Circulation 109: II2–10.

7. Ridker PM, Cushman M, Stampfer MJ, Tracy RP, Hennekens CH (1997) Inflammation, aspirin, and the risk of cardiovascular disease in apparently healthy men. N Engl J Med 336: 973–979.

8. Neuman MG (2007) Immune dysfunction in inflammatory bowel disease. Transl Res 149: 173–186.

9. James SP (2005) Prototypic disorders of gastrointestinal mucosal immune function: Celiac disease and Crohn's disease. J Allergy Clin Immunol 115: 25–30.

10. Coussens LM, Werb Z (2002) Inflammation and cancer. Nature 420: 860–867.

11. Ernst PB, Gold BD (2000) The disease spectrum of Helicobacter pylori: the immunopathogenesis of gastroduodenal ulcer and gastric cancer. Annu Rev Microbiol 54: 615–640.

12. Karin M, Cao Y, Greten FR, Li ZW (2002) NF-kappaB in cancer: from innocent bystander to major culprit. Nat Rev Cancer 2: 301–310.

13. Kuper H, Adami HO, Trichopoulos D (2000) Infections as a major preventable cause of human cancer. J Intern Med 248: 171–183.

14. Shacter E, Weitzman SA (2002) Chronic inflammation and cancer. Oncology (Williston Park) 16: 217–26, 229.

15. Ulrich CM, Bigler J, Potter JD (2006) Non-steroidal anti-inflammatory drugs for cancer prevention: promise, perils and pharmacogenetics. Nat Rev Cancer 6: 130–140.

16. Calder PC (2006) Polyunsaturated fatty acids and inflammation. Prostaglandins Leukot Essent Fatty Acids 75: 197–202.

17. Aganna E, Martinon F, Hawkins PN, Ross JB, Swan DC, et al. (2002) Association of mutations in the NALP3/CIAS1/PYPAF1 gene with a broad phenotype including recurrent fever, cold sensitivity, sensorineural deafness, and AA amyloidosis. Arthritis Rheum 46: 2445–2452.

18. Dode C, Le Du N, Cuisset L, Letourneur F, Berthelot JM, et al. (2002) New mutations of CIAS1 that are responsible for Muckle-Wells syndrome and familial cold urticaria: a novel mutation underlies both syndromes. Am J Hum Genet 70: 1498–1506.

19. Hoffman HM, Mueller JL, Broide DH, Wanderer AA, Kolodner RD (2001) Mutation of a new gene encoding a putative pyrin-like protein causes familial cold autoinflammatory syndrome and Muckle-Wells syndrome. Nat Genet 29: 301–305.

20. Touitou I (2001) The spectrum of Familial Mediterranean Fever (FMF) mutations. Eur J Hum Genet 9: 473–483.

21. Aoki T, Hirota T, Tamari M, Ichikawa K, Takeda K, et al. (2006) An association between asthma and TNF-308G/A polymorphism: meta-analysis. J Hum Genet 51: 677–685.

22. Lee YH, Harley JB, Nath SK (2006) Meta-analysis of TNF-alpha promoter -308 A/G polymorphism and SLE susceptibility. Eur J Hum Genet 14: 364–371.

23. Rahman P, Siannis F, Butt C, Farewell V, Peddle L, et al. (2006) TNFalpha polymorphisms and risk of psoriatic arthritis. Ann Rheum Dis 65: 919–923.

24. Lee YH, Rho YH, Choi SJ, Ji JD, Song GG, et al. (2007) The PTPN22 C1858T functional polymorphism and autoimmune diseases–a meta-analysis. Rheumatology (Oxford) 46: 49–56.

25. Graham RR, Kozyrev SV, Baechler EC, Reddy MV, Plenge RM, et al. (2006) A common haplotype of interferon regulatory factor 5 (IRF5) regulates splicing and expression and is associated with increased risk of systemic lupus erythematosus. Nat Genet 38: 550–555.

26. Smyth DJ, Cooper JD, Bailey R, Field S, Burren O, et al. (2006) A genome-wide association study of nonsynonymous SNPs identifies a type 1 diabetes locus in the interferon-induced helicase (IFIH1) region. Nat Genet 38: 617–619.

27. Howson JM, Dunger DB, Nutland S, Stevens H, Wicker LS, et al. (2007) A type 1 diabetes subgroup with a female bias is characterised by failure in tolerance to thyroid peroxidase at an early age and a strong association with the cytotoxic T-lymphocyte-associated antigen-4 gene. Diabetologia 50: 741–746.

28. Payne F, Cooper JD, Walker NM, Lam AC, Smink LJ, et al. (2007) Interaction analysis of the CBLB and CTLA4 genes in type 1 diabetes. J Leukoc Biol 81: 581–583.

29. Ueda H, Howson JM, Esposito L, Heward J, Snook H, et al. (2003) Association of the T-cell regulatory gene CTLA4 with susceptibility to autoimmune disease. Nature 423: 506–511.

30. Oostenbrug LE, Nolte IM, Oosterom E, van der SG, te Meerman GJ, et al. (2006) CARD15 in inflammatory bowel disease and Crohn's disease phenotypes: an association study and pooled analysis. Dig Liver Dis 38: 834–845.

31. Duerr RH, Taylor KD, Brant SR, Rioux JD, Silverberg MS, et al. (2006) A genome-wide association study identifies IL23R as an inflammatory bowel disease gene. Science 314: 1461–1463.

32. Loza MJ, Chang B (2007) Association between Q551R IL4R genetic variants and atopic asthma risk demonstrated by meta-analysis. J Allergy Clin Immunol In press.

33. Breast Cancer Association Consortium (2006) Commonly studied single-nucleotide polymorphisms and breast cancer: results from the Breast Cancer Association Consortium. J Natl Cancer Inst 98: 1382–1396.

34. Gaudet MM, Egan KM, Lissowska J, Newcomb PA, Brinton LA, et al. (2007) Genetic variation in tumor necrosis factor and lymphotoxin-alpha (TNF-LTA) and breast cancer risk. Hum Genet.

35. Cox DG, Pontes C, Guino E, Navarro M, Osorio A, et al. (2004) Polymorphisms in prostaglandin synthase 2/cyclooxygenase 2 (PTGS2/COX2) and risk of colorectal cancer. Br J Cancer 91: 339–343.

36. Koh WP, Yuan JM, van den BD, Lee HP, Yu MC (2004) Interaction between cyclooxygenase-2 gene polymorphism and dietary n-6 polyunsaturated fatty acids on colon cancer risk: the Singapore Chinese Health Study. Br J Cancer 90: 1760–1764.

37. Lin HJ, Lakkides KM, Keku TO, Reddy ST, Louie AD, et al. (2002) Prostaglandin H synthase 2 variant (Val511Ala) in African Americans may reduce the risk for colorectal neoplasia. Cancer Epidemiol Biomarkers Prev 11: 1305–1315.

38. Siezen CL, van Leeuwen AI, Kram NR, Luken ME, van Kranen HJ, et al. (2005) Colorectal adenoma risk is modified by the interplay between polymorphisms in arachidonic acid pathway genes and fish consumption. Carcinogenesis 26: 449–457.

39. Siezen CL, Bueno-de-Mesquita HB, Peeters PH, Kram NR, van Doeselaar M, et al. (2006) Polymorphisms in the genes involved in the arachidonic acid-pathway, fish consumption and the risk of colorectal cancer. Int J Cancer 119: 297–303.

40. Shahedi K, Lindstrom S, Zheng SL, Wiklund F, Adolfsson J, et al. (2006) Genetic variation in the COX-2 gene and the association with prostate cancer risk. Int J Cancer 119: 668–672.

41. Hedelin M, Chang ET, Wiklund F, Bellocco R, Klint A, et al. (2007) Association of frequent consumption of fatty fish with prostate cancer risk is modified by COX-2 polymorphism. Int J Cancer 120: 398–405.

42. Zheng SL, Liu W, Wiklund F, Dimitrov L, Balter K, et al. (2006) A comprehensive association study for genes in inflammation pathway provides support for their roles in prostate cancer risk in the CAPS study. Prostate 66: 1556–1564.

43. Dinu V, Miller PL, Zhao H (2007) Evidence for association between multiple complement pathway genes and AMD. Genet Epidemiol.

44. Conley YP, Jakobsdottir J, Mah T, Weeks DE, Klein R, et al. (2006) CFH, ELOVL4, PLEKHA1 and LOC387715 genes and susceptibility to age-related maculopathy: AREDS and CHS cohorts and meta-analyses. Hum Mol Genet 15: 3206–3218.

45. Klein RJ, Zeiss C, Chew EY, Tsai JY, Sackler RS, et al. (2005) Complement factor H polymorphism in age-related macular degeneration. Science 308: 385–389.

46. Willer CJ, Scott LJ, Bonnycastle LL, Jackson AU, Chines P, et al. (2006) Tag SNP selection for Finnish individuals based on the CEPH Utah HapMap database. Genet Epidemiol 30: 180–190.

47. De Bakker PI, Graham RR, Altshuler D, Henderson BE, Haiman CA (2006) Transferability of tag SNPs to capture common genetic variation in DNA repair genes across multiple populations. Pac Symp Biocomput 478–486.

48. Montpetit A, Nelis M, Laflamme P, Magi R, Ke X, et al. (2006) An evaluation of the performance of tag SNPs derived from HapMap in a Caucasian population. PLoS Genet 2: e27.

49. Tenesa A, Dunlop MG (2006) Validity of tagging SNPs across populations for association studies. Eur J Hum Genet 14: 357–363.

50. Xu Z, Kaplan NL, Taylor JA (2007) Tag SNP selection for candidate gene association studies using HapMap and gene resequencing data. Eur J Hum Genet.

51. Mueller JC, Lohmussaar E, Magi R, Remm M, Bettecken T, et al. (2005) Linkage disequilibrium patterns and tagSNP transferability among European populations. Am J Hum Genet 76: 387–398.

52. Johansson A, Vavruch-Nilsson V, Cox DR, Frazer KA, Gyllensten U (2007) Evaluation of the SNP tagging approach in an independent population sample-array-based SNP discovery in Sami. Hum Genet 122: 141–150.

Chapter 10

ANALYSIS OF GENE EXPRESSION DATA USING BICLUSTERING ALGORITHMS

Fadhl M. Al-Akwaa[1]

[1]Biomedical Eng. Dept., Univ. of Science & Technology, Sana'a, Yemen

INTRODUCTION

One of the main research areas of bioinformatics is functional genomics; which focuses on the interactions and functions of each gene and its products (mRNA, protein) through the whole genome (the entire genetics sequences encoded in the DNA and responsible for the hereditary information). In order to identify the functions of certain gene, we should able to capture the gene expressions which describe how the genetic information converted to a functional gene product through the transcription and translation processes. Functional genomics uses microarray technology to measure the genes expressions levels under certain conditions and environmental limitations. In the last few years, microarray has become a central tool in biological research. Consequently, the corresponding data analysis becomes one of the important work disciplines in bioinformatics. The analysis of microarray data poses a large number of exploratory statistical aspects including clustering and biclustering algorithms, which help to identify similar patterns in gene expression data and group genes and conditions in to subsets that share biological significance.

What is Clustering?

A large number of clustering definitions can be found in the literature. The simplest definition is shared among all and includes one fundamental concept: the grouping together of similar data items into clusters[1].

Clustering is an important explorative statistical analysis of gene expression data. It aims to identify and group genes that exhibit similar expression patterns over several conditions and also group the conditions based on the expression profiles across set of genes. The successful clustering approach should guarantee two criteria which are homogeneity high similarity between elements in the same cluster, and separation – low similarity between elements from different clusters. When homogeneity and separation are precisely defined, those are two opposing objectives: The better the homogeneity the poorer the separation, and vice versa [2]. Several algorithmic techniques were previously used for clustering gene expression data, including hierarchical clustering [3], self organizing maps [4], and graph theoretic approaches [5].

K-Means

K-means is a classical clustering algorithm [6] invented in 1956 to classify or to group objects (genes) based on attributes or features (experimental conditions) into K number of groups (clusters). K is positive integer number and assumed to be known.

K-means computational approach starts by placing K points into the space represented by the objects that are being clustered. These points represent initial group centroids. We can take any random objects as the initial centroids or the first K objects in sequence can also be used as the initial centroids. Then the K means algorithm will do the four steps below until convergence:

- Determine the centroids coordinate.
- Determine the distance of each object to the centroids using the Euclidean distance.
- Group the objects based on minimum distance.
- Iterate the above steps till no object moves its assigned group.

Each iteration of k-means modifies the current partition by checking all possible modifications of the solution, in which one element is moved to another cluster. This is done by reducing the sum of distances between objects and the centers of their clusters. This procedure is repeated until no further improvement is achieved (No object move the group) and all the objects are grouped into the final required number of clusters.

A disadvantage of K-means algorithm could be perceived in the need to specify the number of clusters K as a parameter value prior to running the algorithm. In cases where there is no expectation about K, user has to make trails with several values of K or use external techniques to guess the no of clusters may be exist.

Hierarchical Clustering (Hcl)

Hierarchical clustering does not partition the genes into subsets. Instead it builds a down-top hierarchy of clusters using agglomerative methods or top - down hierarchy of clusters using divisive methods. The traditional graphical representation of this hierarchy is called dendrogram tree. The divisive method begins at the root and starts to breaks up clusters whose having low similarity. Whereas, the Agglomerative method begins at the leaves of the tree and starts with an initial partition into single element clusters and successively merges clusters until all elements belong to the same cluster [3]. (SeeFigure 1) The agglomerative method is widely used than the divisive one which is not generally available, and rarely has been applied. The idea of the agglomerative method can be summarized as following: Given a set of N items (genes in our case) to be clustered, and an N*N distance (or similarity) matrix [7],

- Assign each item to a cluster, so you have N clusters, each containing just one item.
- Find the closest (most similar) pair of clusters and merge them into a single cluster.
- Compute distances (similarities) between the new cluster and each of the old clusters.
- Repeat steps 2 and 3 until all items are clustered into a single cluster of size N.

In Step 3, distance or similarity measurements between the merged clusters and all the other clusters can be calculated in one of three schemes: single-linkage, complete linkage and average-linkage.

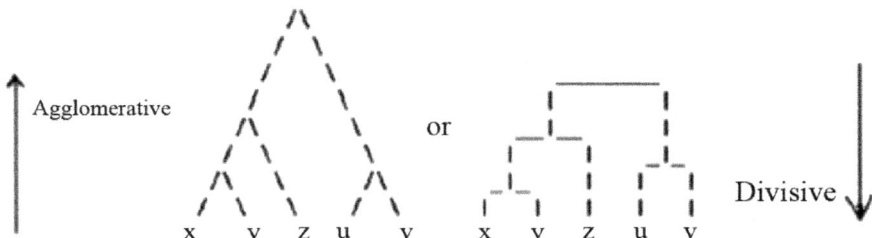

Figure 1: HCL: Agglomerative and Divisive Methods.

Biclustering

Traditional clustering approaches such as k-means and hierarchical clustering put each gene in exactly one cluster based on the assumption that all genes behave similarly in all conditions. However, recent understanding of cellular processes shows that it is possible for subset of genes to be co expressed under certain experimental conditions, and at the same time; to behave almost independently under other conditions. From this context, a new two mode clustering approach called biclustering or co-clustering has been introduced to group the genes and conditions in both dimensions simultaneously.

This allows finding subgroups of genes that show the same response under a subset of conditions, not all conditions. Also, genes may participate in more than one function, resulting in one regulation pattern in one context and a different pattern in another.

Example, if a cellular process is only active under specific conditions and there is a gene participates in multiple pathways that are differentially regulated, one would expect this gene to be included in more than one cluster; and this cannot be achieved by traditional clustering techniques.

Many biclustering methods exist in the literature [8]. Table 1 summarized some of promising biclustering algorithms developed during the last ten years. In brief, we described some of these algorithms according to their prediction strength, their promising results, to what they extend in the community, whether an implementation was available, and the feedback from their authors to explain some ambiguous issues.

Cheng And Church (CC)

CC algorithm[18] is considered to be the first real biclustering implementation after the primary idea has been introduced by Hartigan [19] in 1972.

Table 1: Biclustering Algorithms Comparison

Algorithm	Approach Time	Complicity	Prediction ability
Bivisu [9]	Exhaustive Bicluster Enumeration	$O(m^2 n \log m)a$	Coherent values
MSBE [10]	Greedy Iterative Search	$O((n + m)^2)$	Coherent values
Bimax[11]	Divide-and-Conquer	$O(nm\beta \log \beta)$	Coherent values
ROBA [12]	Matrix algebra	$O(nmLN)$	Coherent Evolution
x-motif [13]	Greedy Iterative Search	$n^{mO(\log(1/\alpha)/\log(1/\beta))}$	Coherent Evolution
SAMBA [14]	Exhaustive Bicluster Enumeration	$O(n2)$	Coherent Evolution
OPSM [15]	Greedy Iterative Search	$O(nm^3 I)$	Coherent Evolution
Plaid[16]	Distribution Parameter Identification	$XX^X b$	Coherent values
ISA [17]	Iterative Signature Algorithm	XXX	Coherent values
CC [18]	Greedy Iterative Search	$O((n + m)nm)$	Coherent values

CC defines a bicluster as a subset of rows and a subset of columns with a high similarity. The proposed similarity score is called mean squared residue (H) and it is used to measure the coherence of the rows and columns in the single bicluster. Given the gene expression data matrix A = (X;Y); a bicluster is defined as a uniform submatrix (I;J) having a low mean squared residue score as following:

The CC Mean Squared Residue:

$$H\left(I,J\right)=\frac{1}{\|I\|\,\|J\|}\sum\nolimits_{i\in I,j\in J}\left(a_{ij}-a_{iJ}-a_{Ij}+a_{IJ}\right)^{2}$$

(1)

- Where: a_{ij} is gene expression level at row i and column j, a_{iJ} is the mean of row i, a_{Ij} is the mean of column j, a_{IJ} is the overall mean. CC algorithm will identify the submatrix as a bicluster if the score is below a level alpha which is a user input parameter to control the quality of the output biclusters. Generally; CC algorithm performs the following major steps:
- Delete rows and columns with a score larger than alpha.
- Adding rows or columns until alpha level is reached.

Iterate these steps until a maximum number of biclusters is reached or no bicluster is found [18].

Iterative Signature Algorithm (ISA)

The ISA algorithm [17, 20] is a novel method for the biclustering analysis of large-scale expression data. It is an efficient algorithm based on the iterative application of the signature algorithm presented in [17]. ISA considers a bicluster to be a transcription module which can be defined as a set of coexpressed genes together with the associated set of regulating conditions (Figure 2). Starting with an initial set of genes, all samples (conditions) are scored with respect to this gene set and those samples are chosen for which the score exceeds a certain threshold (usually defined by the user). In the same way, all genes are scored regarding the selected samples and a new set of genes is selected based on another user-defined threshold. The entire procedure is repeated until the set of genes and the set of samples converge and do not change anymore.

Multiple biclusters can be discovered by running the ISA algorithm on several initial gene sets. This approach requires identification of a reference gene set which needs to be carefully selected for good quality results. In the absence of pre-specified reference gene set, random set of genes is selected at the cost of results quality[17].

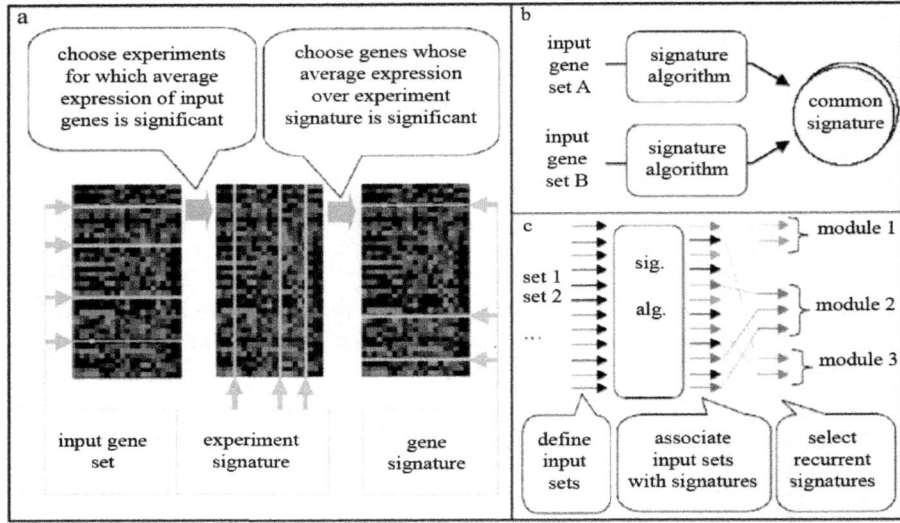

Figure 2: The recurrence signature method. a, The signature algorithm. b, Recurrence as a reliability measure. The signature algorithm is applied to distinct input sets containing different subsets of the postulated transcription module. If the different input sets give rise to the same module, it is considered reliable. c, General application of the recurrent signature method. Copyright © [17].

Biclusters Inclusion Maximal (Bimax)

Bimax[11] is a simple binary model and new fast divide-and-conquer algorithm used to cluster the gene expression data. It is presented in 2006 by Computer Engineering and Networks Laboratory ETH Zurich, Switzerland. Bimax discretized the gene expression data matrix and convert it into a binary matrix by identifying a threshold, so transcription levels (genes expression values) above this threshold become ones and transcription levels below become zeros (or vice versa). Then, it searches for all possible biclusters that contain only ones. This can be done by iterating these steps:

Rearrange the rows and columns to concentrate ones in the upper right of the matrix.

Divide the matrix into two sub matrices.

Whenever in one of the submatrices only ones are found, this sub matrix is returned.

Order Preserving Submatrix(Opsm)

The order-preserving submatrix (OPSM) algorithm [15] is a probabilistic model introduced to discover a subset of genes identically ordered among a subset of conditions. It focuses on the coherence of the relative order of the conditions rather than the coherence of actual expression levels. In other words, the expression values of the genes within a bicluster induce an identical linear ordering across the selected conditions. Accordingly, the authors define a bicluster as a subset of rows whose values induce a linear order across a subset of the columns. The time complexity of this model is $O(nm^3I)$ where n andmare the number of rows and columns of the input gene expression matrix respectively and I is the number of biclusters. A disadvantage of OPSM algorithm is that it takes long time for high dimensional datasets. And this is because its time complexity is cubic with regards to the number of columns (dimensions) of the input matrix [15].

Maximum Similarity Bicluster(Msbe)

MSBE Biclustering algorithm [10] is a novel polynomial time algorithm to find an optimal biclusters with the maximum similarity. The idea behind this algorithm is to find subset of genes that are related to a reference gene. The reference gene is known in advance. MSBE algorithm uses the similarity score for a sub-matrix to find the similar expressions in the microarray datasets. And the threshold of the average similarity score is a user input parameter in order to allow the user to control the quality of the biclustering results.

Clustering or Biclustering

Clustering algorithms [21-23] have been used to analyze gene expression data, on the basis that genes showing similar expression patterns can be assumed to be co-regulated or part of the same regulatory pathway. Unfortunately, this is not always true. Two limitations obstruct the use of clustering algorithms with microarray data. First, all conditions are given equal weights in the computation of gene similarity; in fact, most conditions do not contribute information but instead increase the amount of background noise. Second, each gene is assigned to a single cluster, whereas in fact genes may participate in several functions and should thus be included in several clusters[24].

A new modified clustering approach to uncovering processes that are active over some but not all samples has emerged, which is called biclustering. A bicluster is defined as a subset of genes that exhibit compatible expression patterns over a subset of conditions [11].

During the last ten years, many biclustering algorithms have been proposed (see [8] for a survey), but the important questions are: which algorithm is better? And do some algorithms have advantages over others?

Recently Kevin *et al.*[25]proposed a semantic web algorithm to recommend the best algorithm based on user inputs like: is the dataset contain outliers, is it allowed to get overlapped clusters and the time to retrieve the biclusters.

Generally, comparing different biclustering algorithms is not straightforward as they differ in strategies, approaches, time complicity, number of parameters and prediction ability. In addition, they are strongly influenced by user selected parameter values. For these reasons, the quality of biclustering results is often considered more important than the required computation time. Although there are some analytical comparative studies to evaluate the traditional clustering algorithms[21-23], for biclustering; no such extensive comparison exist even after initial trails have been taken [11]. In the end, Biological merit is the main criterion for evaluation and comparison between the various biclustering methods.

In this chapter we attempt to develope a comparative tool (Bicat-Plus) which is showen in Figure 3 that includes the biological comparative methodology and to be as an extension to the BicAT program[26].

The Goal of BicAT-Plus is to enable researchers and biologists to compare between the different biclustering methods based on set of biological merits and draw conclusion on the biological meaning of the results. In addition, BicAT-Plus help researchers in comparing and evaluating the algorithms results multiple times according to the user selected parameter values as well as the required biological perspective on various datasets.

BicAT-Plus has many features, which could be summarized in the following:-

Algorithms required to be compared could be selected from the biclustering list (left list) to the compared list (right list). External biclustering results for other algorithms could be included in the comparison process. In addition, the organism model, selectable significance level, and GO category should be selected. Finally, Comparison criteria have to be selected based on the user biological metric.

User could perform biclusters functional analysis using the three Gene Ontology (GO) categories (biological process, molecular function and cellular component) (Figure3 with label number 1).

User could evaluate the quality of each biclustering algorithm results after applying the GO functional analysis and display the percentage of the enriched biclusters at different P-values (Figure3 with label number 2).

User could compare between the different biclustering algorithms according to the percentage of the functionally enriched biclusters at the required significance levels, the selected GO category and with certain filtration criteria for the GO terms. (Figure3 with label number 3).

User could evaluate and compare the results of external biclustering algorithms. This gives the BicAT-plus the advantage to be a generic tool that does not depend on the employed methods only. For example, it can be used to evaluate the quality of the new algorithms introduced to the field and compare against the existing ones. (Figure 3 with label number 4).

User could display the results using graphical and statistical charts visualizations in multiple modes (2D and 3D).

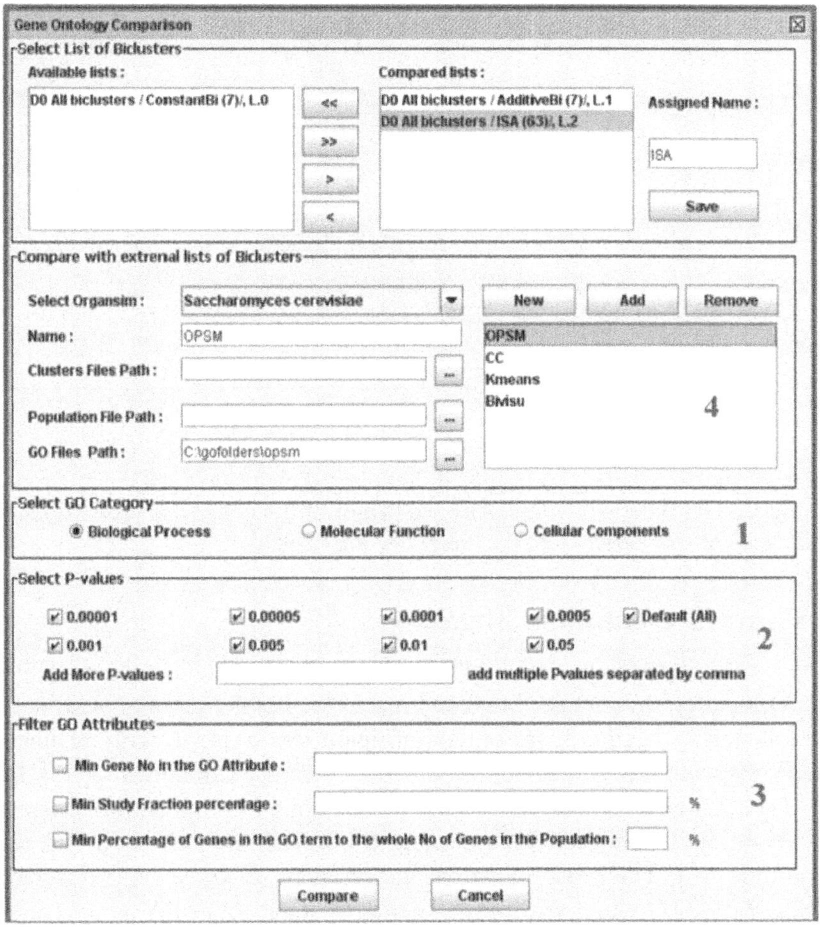

Figure 3: BicAT-Plus Comparison Panel.

MATERIALS AND METHODS

Before using the BicAT-Plus, Active Perl version 5.10 and Java Runtime Environment (JRE) version 6 are required to be installed on your machine. BicAT-Plus has been tested and show good performance on a PC machine with the following configurations: CPU: Pentium 4, 1.5 GHZ, RAM: 2.0 GB, Platform: windows XP professional with SP2.

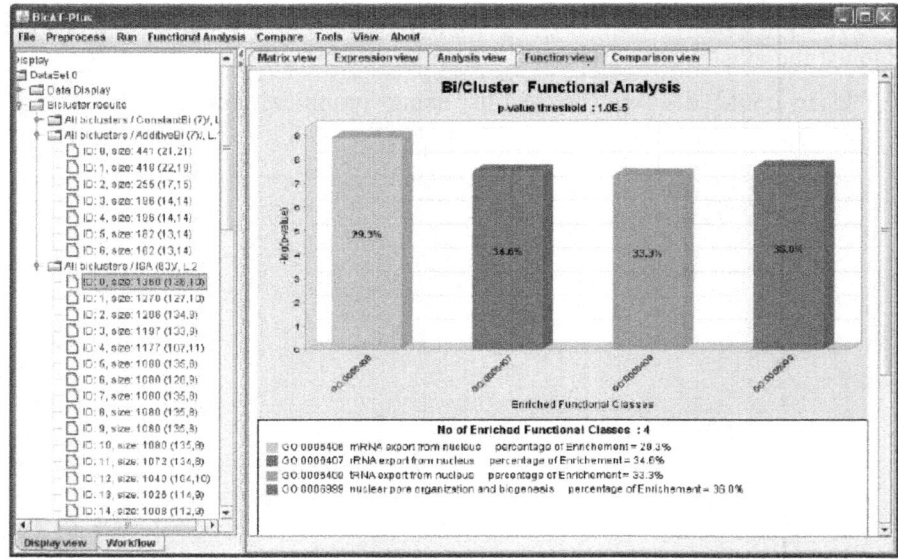

Figure 4: Functional analysis results of the selected bicluster. Each column represents an enriched GO functional class. And the height of the column is proportional to the significance of this enrichment (i.e. height = -log (p-value).

Go Overrepresentation Programs

Many programs like: BINGO[27], FUNCAT[28], GeneMerge[29] and FuncAssociate[30] were used to investigate whether the set of genes discovered by biclustering methods present significant enrichment with respect to a specific GO annotation provided by Gene Ontology Consortium [31]. BicAT-Plus used GeneMerge program as the most popular GO program. GeneMerge provides a statistical test for assessing the enrichment of each GO term in the sample test. The basic question answered by this test is as described by Steven *et al.*[27] "when sampling X genes (test set) out of N genes (reference set, either a graph or an annotation), what is the probability that x or more of these genes belong to a functional category C shared by n of the N genes in the reference set? The hypergeometric test, in which sampling occurs without replacement, answers this question in the form of P-value. It›s counterpart with replacement, the binomial test, which provides only an approximate P-value, but requires less calculation time.»

Comparative Methodologies

BicAT-Plus provides reasonable methods for comparing the results of different biclustering algorithms by:

Identifying the percentage of enriched or overrepresented biclusters with one or more GO term per multiple significance level for each algorithm. A bicluster is said to be significantly overrepresented (enriched) with a functional category if the P-value of this functional category is lower than the preset threshold. The results are displayed using a histogram for all the algorithms compared at the different preset significance levels, and the algorithm that gives the highest proportion of enriched biclusters for all significance levels is considered the optimum because it effectively groups the genes sharing similar functions in the same bicluster.

Identifying The Percentage Of Annotated Genes Per Each Enriched Bicluster

Estimating the predictive power of algorithms to recover interesting patterns. Genes whose transcription is responsive to a variety of stresses have been implicated in a general Yeast response to stress (awkward). Other gene expression responses appear to be specific to particular environmental conditions. BicAT-Plus compares biclustering methods on the basis of their capacity to recover known patterns in experimental data sets. For example, Gasch et al.[32] measure changes in transcript levels over time responding to a panel of environmental changes, so it was expected to find biclusters enriched with one of response to stress (GO:0006950), Gene Ontology categories such as response to heat (GO:0009408), response to cold (GO:0009409) and response to glucose starvation(GO:0042149). The details of this comparison strategy are described in the results and in Table 3.

Comparison Process Steps

The following process diagram shown in Fig 5 summarizes the required steps by the user to compare between the different algorithms using the BicAT-plus:

Download BicAT-Plus from (*www.bioinformatics.org/bicat-plus/*).

Load Gene Expression Data to BicAT-Plus then run the selected five prominent biclustering methods with setting parameters as shown in Table 2.

Run GO comparison tool in the BicAT-Plus and add the available biclustering algorithms to the compared list as shown in Fig 1.

- Select the available GO category e.g. biological process, molecular function and cellular components.
- Select the P-values e.g. 0.00001, 0.0001, 0.01, 0.005, and 0.05.
- Press compare button.
- Press comparison menu, Functional enrichment and select 2D or 3D charts.

Table 2: Default Parameter settings of the compared bi/clustering methods The definitions of these parameters are listed in their original publications [9, 15, 17-18, 20] respectively

Bi/clustering Algorithm	Parameter settings
ISA	$t_g = 2.0$, $t_c = 2.0$, seeds $= 500$
CC	$\delta = 0.5$, $\alpha = 1.2$, $M = 100$
OPSM	$1 = 100$
BiVisu	$E = 0.82$, $N_r = 10$, $N_c = 5$, $P_o = 25$
K-means	K$=100$

RESULTS & DISCUSSION

The above comparison steps is performed on the gene expression data of *S. cerevisiae* provided by Gasch [32]. The dataset contains 2993 genes and 173 conditions of diverse environmental transitions such as temperature shocks, amino acid starvation, and nitrogen source depletion. This dataset is freely available from Stanford University website [33]. For each biclustering algorithm, we used the default parameters as authors recommend in their corresponding publications. See Table 2.

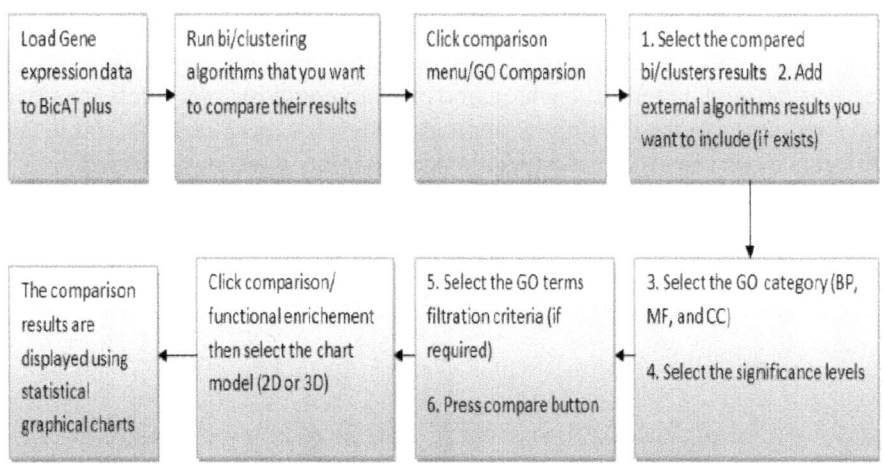

Figure 5: BicAT-Plus Comparison process steps

The Percentage of Enriched Function

After applying the above steps on Gasch data[32], BicAT-plus produce the histogram shown in Fig 6. Investigating this figure, we observed that OPSM algorithm gave a high portion of functionally enriched biclusters at all significance levels (from 85% to 100 %). Next to OPSM, ISA show relatively high portions of enriched biclusters.

In order to evaluate the ability of the algorithms to group the maximum number of genes whose expression patterns are similar and sharing the same GO category, we use the filtration criteria developed in the comparative tool by neglecting those bi/clusters which have study fraction less than 25%. The study fraction of a GO term is the fraction of genes in the study set (bicluster) with this term.

$$Study fraction of a GO term = \frac{No of genes sharing the GO term in a bicluster}{total number of genes in this bicluster} \times 100$$

(2)

Figure 7 shows that OPSM and ISA have highly enriched biclusters/ clusters that have large number of genes per each GO category. On the other hand, Bivisu biclusters are strongly affected by this filtration and they contain a lower number of genes per each category. This filtration will help in identifying the powerful and most reliable algorithms which are able to group maximum numbers of genes sharing same functions in one bicluster.

The Predictability Power to Recover Interested Pattern

The user could compare bi/clusters algorithms based on which of them could recover defined pattern like which one of them could recover bi/clusters which have response to the conditions applied in Gasch experiments. In Table 2, the difference between the biclusters/clusters contents were summarized.

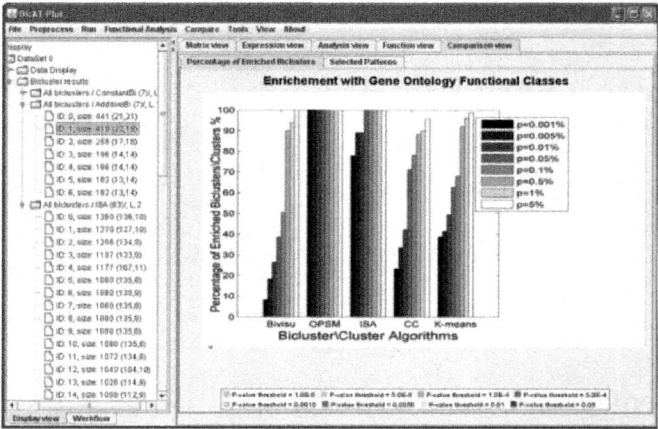

Figure 6: Percentage of biclusters significantly enriched by GO Biological Process category (*S. cerevisiae*) for the five selected biclustering methods and K-means at different significance levels p.

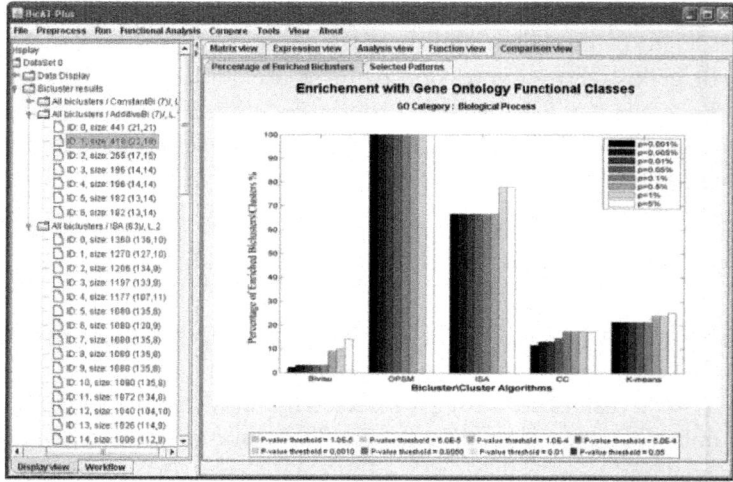

Figure 7: Percentage of significantly enriched biclusters by GO Biological Process category by setting the allowed minimum number of genes per each GO category to 10 and the study fraction to large than 50%.

Although OPSM show high percentage level of enriched biclusters (as shown in Fig 6, 7), its biclusters do not contain any genes within any GO category response to Gasch experiments. The k-means and Bivisu cluster/ bicluster results distinguished a unique GO category, which is GO: 0000304 (response to singlet oxygen), and GO: 0042542 (response to hydrogen peroxide) The powerful usage of these bicluster algorithms is significantly appeared in GO: 0006995 "cellular response to nitrogen starvation" where these algorithms were able to discover 4 out of 5 annotated genes without any prior biological information or on desk experiments.

Table 3: Gene Ontology category per number of annotated genes of the Bicluster/ cluster algorithm results for the experimental condition on Gasch Experiments[32]

GO Term / (number of annotated genes)	K-means	CC	ISA	Bivisu	OPSM
GO:0042493 Response to drug / (118)	4	5	7	6	0
GO:0006970 response to osmotic stress / (83)	3	5	6	3	0
GO:0006979 response to oxidative stress / (79)	2	7	11	0	0
GO:0046686 response to cadmium ion / (102)	2	3	2	2	0
GO:0043330 response to exogenous dsRNA / (7)	2	3	2	2	0
GO:0046685 response to arsenic / (77)	2	0	2	2	0
GO:0006950 response to stress / (532)	9	11	16	2	0
GO:0009408 response to heat / (24)	3	0	2	2	0
GO:0009409 response to cold / (7)	0	0	2	0	0
GO:0009267 cellular response to starvation / (44)	0	2	0	0	0
GO:0006995 cellular response to nitrogen starvation / (5)	4	4	4	0	0
GO:0042149 cellular response to glucose starvation / (5)	0	2	0	0	0

GO:0009651 response to salt stress / (15)	2	7	0	0	0
GO:0042542 response to hydrogen peroxide /(5)	0	0	0	2	0
GO:0006974 response to DNA damage stimulus / (240)	0	22	0	3	0
GO:0000304 response to singlet oxygen / (4)	2	0	0	0	0

CONCLUSION

We have introduced the BicAT-Plus with reasonable comparative methodology based on the Gene Ontology. To the best of our knowledge such an automatic comparison tool of the various biclustering algorithms has not been available in the literature. BicAT-Plus is an open source tool written in java swing and it has a well structured design that can be extended easily to employ more comparative methodologies that help biologists to extract the best results of each algorithm and interpret these results to useful biological meaning.

In other words, the algorithms that show good quality of results (per the dataset) can be used to provide a simple means of gaining leads to the functions of many genes for which information is not available currently (unannotated genes).

Using BicAT-Plus, we can identify the highly enriched biclusters of the whole compared algorithms. This might be quite helpful in solving the dimensionality reduction problem of the Gene Regulatory Network construction from the gene expression data. This problem originates from the relatively few time points (conditions or samples) with respect to the large number of genes in the microarray dataset.

Finally there are several aspects of this research that worth further investigation, according to the Studies carried out so far and also introducing new ideas for consideration

Enrich the BicAT-Plus with more comparative methodologies beside GO. For example, KEGG and promoter analysis by identifying the transcription factors for the clustered genes.

Extend the BicAT-Plus to provide users with multiple export options for the interested enriched biclusters.

Embed the BicAT-Plus as a plug-in in the Cytoscape platform[34] which is open source bioinformatics software for visualizing molecular interaction networks and biological pathways and integrating these networks with annotations, gene expression profiles and other state data. Thus, very promising challenge is to get use of the highly enriched biclusters identified by the BicAT-Plus in solving these integrated networks in the Cytoscape.

REFERENCES

1. G. A. Fung, Overview. Comprehensive, Basic. of, Algorithms. Clustering, Citeseer 20012001137

2. R. Sharan, R. Elkon, R. Shamir, analysis. Cluster, applications. its, gene. to, data. expression, Ernst Schering Res Found Workshop 2002200283108

3. M. B. Eisen, P. T. Spellman, P. O. Brown, D. Botstein, analysis. Cluster, of. display, expression. genome-wide, patterns, Proceedings of the National Academy of Sciences of the United States of America199814863 EOF

4. P. Tamayo, J. Mesirov, Q. Zhu, S. Kitareewan, E. Dmitrovsky, E. S. Lander, T. R. Golub, patterns. Interpreting, gene. of, with. expression, maps. self-organizing, Methods, to. application, dierentiation. hematopoietic, In: Proceedings of the National Academy of Sciences of the United States of America,: 199929072912

5. Sharan RSaR: Click: a clustering algorithm for gene Expression analysis. In: Proceedings of the 8th International Conference on Intelligent Systems for Molecular Biology: 2000307316

6. S. Tavazoie, Campbell. Hughes, R. J. Cho, G. M. Church, determination. Systematic, genetic. of, architecture. network, Nature Genetics1999281 EOF5 EOF

7. S. Johnson, clustering. Hierarchical, Psychometrika. schemes, 1967

8. Madeira SC, Oliveira AL: Biclustering algorithms for biological data analysis: a survey.IEEE/ACM Trans Comput Biol Bioinform 200424 EOF45 EOF

9. Cheng KO, Law NF, Siu WC, Lau TH: BiVisu: software tool for bicluster detection and visualization.Bioinformatics20072342 EOF2344 EOF

10. X. Liu, L. Wang, the. Computing, similarity. maximum, of. bi-clusters, expression. gene, data, Bioinformatics200750 EOF56 EOF

11. A. Prelic, S. Bleuler, P. Zimmermann, A. Wille, P. Buhlmann, W. Gruissem, L. Hennig, L. Thiele, E. A. Zitzler, comparison. Systematic, of. evaluation, methods. biclustering, gene. for, data. expression, Bioinformatics20061122 EOF1129 EOF

12. A. Tchagang, A. Twefik, biclustering. Robust, . R. O. B. A. algorithm, D. N. A. for, data. microarray, analysis, In: IEEE/SP 13thWorkshop on Statistical Signal Processing. 20052005984989

13. T. M. S. K. Murali, conserved. Extracting, expression. gene, from. motifs, expression. gene, data, In: Pac Symp Biocomput. 200320037788

14. A. Tanay, R. S. M. Kupiec, R. Shamir, . Revealing, modularity, in. organization, yeast. the, network. molecular, integrated. by, of. analysis, heterogeneous. highly, data. genomewide, In: Proceedings of the National Academy of Sciences of the United States of America: 200429812986

15. A. Ben-Dor, B. Chor, R. Karp, Z. Yakhini, local. Discovering, in. structure, expression. gene, the. data, submatrix. order-preserving, problem, Journal of Computational Biology2003373 EOF384 EOF

16. H. Wang, W. W. J. Yang, P. S. Yu, . Clustering, Pattern. by, the. p. Similarity, Algorithm. S. I. G. M. O. Cluster, SIGMOD 2002

17. J. Ihmels, G. Friedlander, S. Bergmann, O. Sarig, Y. Ziv, N. Barkai, modular. Revealing, in. organization, yeast. the, network. transcriptional, Nature Genetics2002370 EOF7 EOF

18. Y. Cheng, G. M. Church, of. Biclustering, data. expression, Proceedings of 8th International Conference on Intelligent Systems for Molecular Biology 2000200093103

19. J. Hartigan, Clustering. Direct, a. of, matrix. data, Journal of the American Statistical Association1972123 EOF

20. J. Ihmels, S. Bergmann, N. Barkai, transcription. Defining, using. modules, gene. large-scale, data. expression, Bioinformatics20041993 EOF2003 EOF

21. S. Tavazoie, J. Hughes, M. Campbell, R. Cho, G. Church, determination. Systematic, genetic. of, architecture. network, Nature Genetics1999281 EOF5 EOF

22. R. Guthke, U. Moller, M. Hoffmann, F. Thies, S. Topfer, network. Dynamic, from. reconstruction, expression. gene, applied. data, immune. to, during. response, infection. bacterial, Bioinformatics20051626 EOF1634 EOF

23. P. D'haeseleer, S. Liang, R. Somogyi, network. Genetic, from. inference, clustering. co-expression, reverse. to, engineering, Bioinformatics 2000707 EOF26 EOF

24. D. Reiss, N. Baliga, R. Bonneau, biclustering. Integrated, heterogeneous. of, datasets. genome-wide, the. for, of. inference, regulatory. global, B. M. networks, BMC Bioinformatics2006280 EOF

25. K. Yip, Q. Ya, Peishen, Martin. Schultz, David. W. Cheung, Kei. Cheung-Hoi, Biosphere. A. Sem, Web. Semantic, to. Approach, Microarray.

Recommending, Services. Clustering, In: The Pacific Symposium on Biocomputing. 20062006188199

26. S. Barkow, S. Bleuler, A. Prelic, P. Zimmermann, E. Zitzler, A. T. a. Bic, analysis. biclustering, toolbox, Bioinformatics20061282 EOF1283 EOF

27. S. Maere, K. Heymans, M. Kuiper, N. G. O. a. Bi, plugin. Cytoscape, assess. to, of. overrepresentation, Ontology. Gene, in. categories, Networks. Biological, Bioinformatics 2005

28. A. Ruepp, A. Zollner, D. Maier, K. Albermann, J. Hani, M. Mokrejs, I. Tetko, U. Guldener, G. Mannhaupt, M. Munsterkotter, et al.Fun. The, a. Cat, annotation. functional, for. scheme, classification. systematic, proteins. of, whole. from, genomes, Nucl Acids Res 2004

29. Castillo-Davis CI, Hartl DL: GeneMerge- post-genomic analysis, data mining, and hypothesis testing. Bioinformatics2003

30. G. F. Berriz, O. D. King, B. Bryant, C. Sander, F. P. Roth, gene. Characterizing, with. sets, Associate. Func, Bioinformatics20032502 EOF2504 EOF

31. M. Ashburner, C. Ball, J. Blake, D. Botstein, H. Butler, J. Cherry, A. Davis, K. Dolinski, S. Dwight, J. Eppig, et al.ontology. Gene, for. tool, unification. the, biology. of, The Gene Ontology Consortium. Nat Genet 2000

32. A. P. Gasch, P. T. Spellman, C. M. Kao, O. Carmel-Harel, M. B. Eisen, G. Storz, D. Botstein, P. O. Brown, Expression. Genomic, in. Programs, Response. the, Yeast. of, to. Cells, Changes. Environmental, Mol Biol Cell 20004241 EOF57 EOF

33. http://genome-www.stanford.edu/yeast/_stress

34. P. Shannon, A. Markiel, O. Ozier, N. Baliga, J. Wang, D. Ramage, N. Amin, B. Schwikowski, T. Ideker, a. Cytoscape, environment. software, integrated. for, of. models, interaction. biomolecular, networks, Genome Res 2003

Chapter 11

GENOME SEQUENCE AND GENETIC DIVERSITY OF THE COMMON CARP, CYPRINUS CARPIO

Peng Xu[1,10,] Xiaofeng Zhang[2,10], Xumin Wang[3,10], Jiongtang Li[1,10], Guiming Liu[3,10,] Youyi Kuang[2,10], Jian Xu[1,10], Xianhu Zheng[2,10], Lufeng Ren[3], Guoliang Wang[3], Yan Zhang[1], Linhe Huo[3], Zixia Zhao[1], Dingchen Cao[2], Cuiyun Lu[2], Chao Li[2], Yi Zhou[4], Zhanjiang Liu[1,5], Zhonghua Fan[3], Guangle Shan[3], Xingang Li[3], Shuangxiu Wu[3], Lipu Song[3,] Guangyuan Hou[1], Yanliang Jiang[1], Zsigmond Jeney[6], Dan Yu[3], Li Wang[3], Changjun Shao[3], Lai Song[3], Jing Sun[3], Peifeng Ji[1], Jian Wang[1], Qiang Li[1], Liming Xu[1], Fanyue Sun[5], Jianxin Feng[7], Chenghui Wang[8], Shaolin Wang[9], Baosen Wang[1], Yan Li[1], Yaping Zhu[1], Wei Xue[1], Lan Zhao[1], Jintu Wang[1], Ying Gu[2], Weihua Lv[2], Kejing Wu[3], Jingfa Xiao[3], Jiayan Wu[3], Zhang Zhang[3], Jun Yu[3] & Xiaowen Sun[1,2]

[1]Centre for Applied Aquatic Genomics, Chinese Academy of Fishery Sciences, Beijing, China

[2]Heilongjiang River Fisheries Research Institute, Chinese Academy of Fishery Sciences, Harbin, China

[3]Chinese Academy of Sciences Key Laboratory of Genome Sciences and Information, Beijing Institute of Genomics, Chinese Academy of Sciences, Beijing, China

[4]Stem Cell Program, Division of Hematology and Oncology, Boston Children's Hospital and Dana-Farber Cancer Institute, Harvard Medical School, Boston, Massachusetts, USA

[5]Fish Molecular Genetics and Biotechnology Laboratory, Department of Fisheries and Allied Aquacultures, Auburn University, Auburn, Alabama, USA

[6]Research Institute for Fisheries, Aquaculture and Irrigation, Szarvas, Hungary

[7]Henan Academy of Fishery Science, Zhengzhou, China

[8]College of Fisheries and Life Science, Shanghai Ocean University, Shanghai, China

[9]Department of Psychiatry and Neurobiology Science, University of Virginia, Charlottesville, Virginia, USA

[10]These authors contributed equally to this work. Correspondence should be addressed to X.S

ABSTRACT

The common carp, *Cyprinus carpio*, is one of the most important cyprinid species and globally accounts for 10% of freshwater aquaculture production. Here we present a draft genome of domesticated *C. carpio* (strain Songpu), whose current assembly contains 52,610 protein-coding genes and approximately 92.3% coverage of its paleotetraploidized genome ($2n = 100$). The latest round of whole-genome duplication has been estimated to have occurred approximately 8.2 million years ago. Genome resequencing of 33 representative individuals from worldwide populations demonstrates a single origin for *C. carpio* in 2 subspecies (*C. carpio Haematopterus* and *C. carpio carpio*). Integrative genomic and transcriptomic analyses were used to identify loci potentially associated with traits including scaling patterns and skin color. In combination with the high-resolution genetic map, the draft genome paves the way for better molecular studies and improved genome-assisted breeding of *C. carpio* and other closely related species.

INTRODUCTION

Carp (cyprinids) contribute over 20 million metric tons to fish production worldwide and account for approximately 40% of total global aquaculture production and 70% of total freshwater aquaculture production. They have emerged as the most economically important teleost family. In comparison to other major aquaculture species, such as salmon and shrimp, carp are recognized as an ecofriendly fish because most are omnivorous filter-feeders and thus consume much less fish meal and fish oil. As one of the dominant cyprinid species, *C. carpio* (the common carp) is cultured in over 100 countries worldwide and accounts for up to 10% (over 3 million metric tons) of global annual freshwater aquaculture production[1, 2]. In addition to its value as a food source, *C. carpio* is also an important ornamental fish species. One of its variants, koi, is the most popular outdoor ornamental fish because of its distinctive color and scale patterns.

Most teleosts have undergone a teleost-specific genome duplication (TSGD) and contain 24 to 25 chromosomes in their haploid genome. The haploid genome of *C. carpio* has 50 chromosomes[3], and molecular evidence suggests that an additional whole-genome duplication (WGD) event tetraploidized the genome[4, 5, 6, 7]. Although cytogenetic evidence of the allotetraploidization of *C. carpio* has suggested that 50 bivalents rather than 25 quadrivalents are formed during meiosis[6], genome-scale validation is of great importance. Owing to its economic value in aquaculture, *C. carpio* has been intensively studied in terms of its physiology, development, immunology, disease resistance, selective breeding and transgenic manipulation. In addition, it is also considered an alternative vertebrate fish model to zebrafish (*Danio rerio*). A variety of *C. carpio* genome resources have been developed over the past decade, including a large number of genetic markers[8, 9], genetic maps[10, 11, 12, 13], a BAC-based physical map[14, 15], a large number of ESTs[16, 17, 18] and cDNA microarrays[19]. Recently, a comparative exomic study of *C. carpio* and *D. rerio* has been reported, providing additional genome resourcing data for the research community[20].

Using a whole-genome shotgun strategy and combining data from several next-generation sequencing platforms, we have produced a high-quality genome assembly for *C. carpio* (strain Songpu) and completed the genomic resequencing of 33 *C. carpio* accessions that represent major domesticated strains and populations. In addition to comparative and evolutionary studies of *C. carpio* and its closely related species using the genome sequences, we also demonstrate the genetic basis of phenotypic traits on scale patterning and body color determination, on the basis of data from two distinct domesticated strains (Songpu and Hebao). This study on the *C. carpio* genome provides a valuable resource for the molecular-guided breeding and genetic improvement of the common carp.

RESULTS

Sequencing and Assembly

We prepared genomic DNA from a homozygous double-haploid clonal line from the domesticated strain Songpu, which has a documented breeding history. We performed whole-genome shotgun sequencing on three next-generation sequencing platforms, Roche 454, Illumina and SOLiD, using both single-end and paired-end or mate-pair libraries of various insert size ranging from 250 bp to 8 kb (Supplementary Table 1). Our contig assembly was based on single-end pyrosequencing data (CABOG, Celera Assembler with Best Overlap Graph), and the scaffold assembly was based on paired-end and mate-pair sequences from different sequencing platforms and 29,046 paired BAC-end sequences[14, 15]. After gap filling, the contig and scaffold N50 lengths reached 68.4 kb and 1.0 Mb, respectively. The total length of all scaffolds was 1.69 Gb (Table 1). We estimated the genome size to be 1.83 Gb on the basis of *K*-mer analysis, which is consistent with estimates based on cytogenetic methods[3, 21] (Supplementary Fig. 1). Thus, the scaffolds covered at least 92.3% of the genome (90.2% if we excluded sequence gaps of 40 Mb in length) and 90% of the assembly containing 2,503 large scaffolds, for a total length of 1.53 Gb.

Table 1: Summary of genome assembly

Genome assembly		N50 (size/number)	N90 (size/number)	Total length
	Contigs	68.4 kb/7,171	16.5 kb/25,070	1.65 Gb
	Scaffolds	1.0 Mb/491	96.6 kb/2,503	1.69 Gb
	Chromosomes	50 chromosomes (from 1,456 scaffolds)		875 Mb
Noncoding RNAs		Copies		Length
	rRNAs	1,012		100.2 kb
	miRNAs	914		20.4 kb
	tRNAs	3,622		268.1 kb
TEs		Total length		Percent of genome
	Total	529.4 Mb		31.23
	DNA transposons	297.2 Mb		17.53
	Retroelements	169.3 Mb		9.99
Protein-coding genes	Total number	Annotated		Unannotated
	52,610	47,795		4,815

To validate the genome assembly, we mapped all paired-end and mate-pair reads from different sequencing platforms to the assembly and found that an average of 80.3% of the reads (78.1% of Illumina reads, 74.6% of Life Technologies SOLiD reads, 98% of Roche 454 reads and 98.8% of BAC-end reads) could be mapped (Supplementary Fig. 2 and Supplementary Table 2). To assess the accuracy of the assembly, we aligned our assembly to an assembled BAC and five large scaffolds from previously published genome sequences[20]; the result demonstrates high consistency between the two data sets (Supplementary Fig. 3). To assess gene coverage, we mapped assembled transcriptome sequences, including publically available ESTs and new mRNA reads from multiple tissues, to the assembly. The effort yielded ~88.8% coverage of these transcripts by nucleotide sequence similarity (Supplementary Table 3); of all of the mapped genes, 90% were common among the sequenced teleosts (Supplementary Table 4). Owing to the multiple rounds of GWD, *C. carpio* genes are rich in paralogs, which are thought to interfere with assembly. Therefore, we mapped 19 duplicated genes that were shared among teleosts to the genome and found that 16 of the gene pairs mapped to distinct locations whereas the other 3 collapsed into single genes, given the high similarity among the paralogs (Supplementary Table 5).

Genetic Map and Markers

Our attempt to anchor the genome assembly onto a newly updated high-resolution genetic map, constructed using a genetic mapping panel of 107 full siblings produced from the cross of a Songpu pair (Supplementary Note), succeeded in placing a total of 3,470 high-quality SNPs and 773 microsatellite markers (Supplementary Table 6), which clustered into 50 linkage groups (Supplementary Fig. 4) and covered a genetic distance of 3,946.7 cM. The physical coverage of these linkage groups contained 1,456 of the longest scaffolds (~875 Mb in total length and 16.7 Mb of gaps) (Fig. 1 and Table 1), and the ratio of the median genetic distance to the physical distance was 0.2 Mb/cM (Supplementary Figs. 5 and 6).

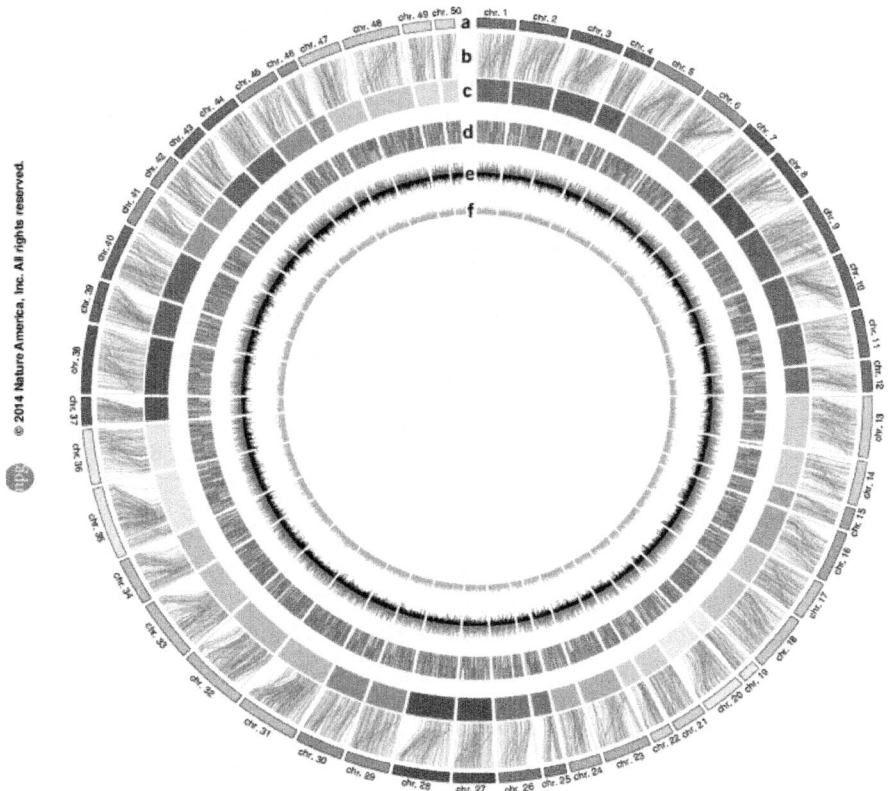

© 2014 Nature America, Inc. All rights reserved.

Figure 1: The genome landscape of the 50 assembled chromosomes of *C. carpio*.

(a)   The genetic linkage map. (**b**) Anchors between the genetic markers and the assembled scaffolds. (**c**) Assembled chromosomes. (**d**) Gene distribution on each chromosome; red lines indicate genes on the plus strand, and blue lines indicate genes on the minus strand. (**e**) GC content within a 10-kb sliding window; blue indicates GC content that is higher than average, and black indicates GC content that is lower than average. (**f**) Repeat content within a 10-kb sliding window.

Genome Characterization

The *C. carpio* genome has a GC content of 37.0%, slightly higher than that of *D. rerio* but much lower than that of other sequenced teleost genomes (Supplementary Fig. 7 and Supplementary Note). To identify transposable elements (TEs)[22], we constructed a *C. carpio*–specific repeat database. We found that 529 Mb of the assembled contigs (31.3% of the genome assembly) could be attributed to TEs (Table 1 and Supplementary Tables 7 and 8). This proportion of the content is higher than that for most of the sequenced teleost genomes (7.1% in *Takifugu rubripes*[23], 5.7% in *Tetraodon nigroviridis*[24], 30.68% in *Oryzias latipes*[25] and 13.48% in *Gasterosteus aculeatus*[26]) but lower than that for the *D. rerio* genome (59.78%) (Supplementary Table 9). Of the TEs, the fraction of class I TEs (retroelements) was 9.99% of the total genome assembly (4.90% long interspersed nuclear elements (LINEs), 4.35% long terminal repeats (LTRs) and 0.47% short interspersed nuclear elements (SINEs)), whereas that of the class II TEs (DNA transposons) was 17.53%. The most abundant DNA transposon family identified in the *C. carpio* genome was the hAT superfamily, which had approximately 463,000 copies and accounted for 33% of all identified DNA transposons (consistent with our previous findings from the analysis of BAC-end sequences[15]). The distribution of the divergence rates for the TEs peaked at 6% in *C. carpio* and at 8% in *D. rerio*, suggesting a more recent expansion of these elements in the *C. carpio* genome (Supplementary Fig. 8).

We used a comprehensive strategy to annotate *C. carpio* genes by combining *ab initio* gene prediction (FGENESH and AUGUSTUS), protein-based homology (Supplementary Note) and transcript-based evidence (transcriptomes from multiple tissues and developmental stages) (Supplementary Table 10). All predicted gene structures were integrated with EVidenceModeler (EVM)[27] to yield a consensus gene set containing a total of 52,610 protein-coding genes, of which 91.4% were proven to be expressed (Table 1, Supplementary Fig. 9 and Supplementary Tables 11 and 12). This gene number is almost twice that found in *D. rerio*, confirming the fact that the tetraploid genome retained a large portion of its gene duplicates after the latest WGD. The average gene and coding sequence lengths were 12,145 bp and 1,487 bp, respectively, and *C. carpio* genes had an average of 7.48 exons per gene (Supplementary Fig. 10 and Supplementary Table 13). In addition, the non-protein-coding genes included 1,012 rRNA, 3,622 tRNA and 914 microRNA (miRNA) genes (Table 1 and Supplementary Table 14).

Genome Evolution

C. carpio has 100 chromosomes, approximately twice as many as are found in other cyprinid fish species. Many studies have corroborated the occurrence of either TSGD or the third round of WGD in most ray-finned fishes[28, 29, 30, 31, 32] and have predicted that TSGD has facilitated the evolutionary radiation and phenotypic diversification of the teleost fishes[29, 31]. The *C. carpio* genome is believed to have undergone an additional round of genome duplication (4R) and to have thus tetraploidized[4, 5, 6, 33]. We have identified approximately 50 chromosome bivalents rather than quadrivalents in the meiotic nuclei of *C. carpio*, suggesting that it is not a true tetraploid species according to karyotyping. The tetraploidy observed in *C. carpio* seems to result from allotetraploidization (species hybridization) rather than autotetraploidization (genome doubling)[6]. We aligned 52,610 high-confidence gene models to the 50 *C. carpio* chromosomes and the *D. rerio* genome (*n* = 25 chromosomes) and identified 8,002 orthologous gene pairs with a clear two-to-one orthologous relationship between the two species, respectively (Fig. 2a). The major obscure synteny found on the long arm of *D. rerio* chromosome 4 is actually in accordance with a recent report that highlighted unique features of this region[34]: the region shows little orthology with other sequenced teleost genomes and harbors zebrafish-specific gene duplication and a high-density small nuclear RNA (snRNA) cluster that accounts for 53.2% of all snRNAs in the genome. This region most likely emerged in *D. rerio* after *Danio-Cyprinus* divergence. In addition, we also observed a number of minor chromosome rearrangements on the carp chromosomes, including on the long arm of chromosome 8 (showing weakened orthology with *D. rerio* chromosome 4) and the region containing orthologs with *D. rerio* chromosome 17. We also identified 2,114 best-match reciprocal paralogous gene pairs and built ohnologous blocks on 25 paired chromosomes. A circular representation of ohnolog pairs clearly demonstrates their one-to-one syntenic relationship (Fig. 2b), consistent with previous observations for genome tetraploidization.

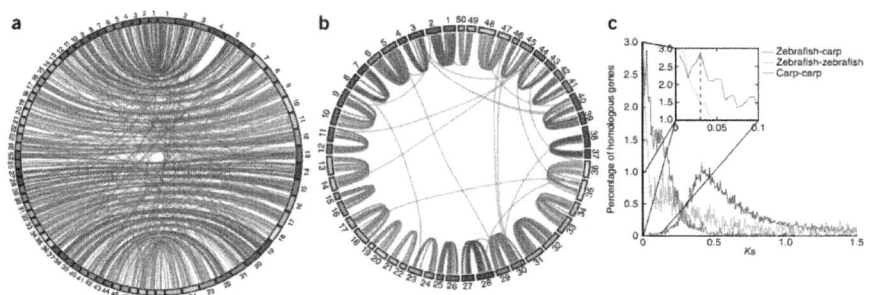

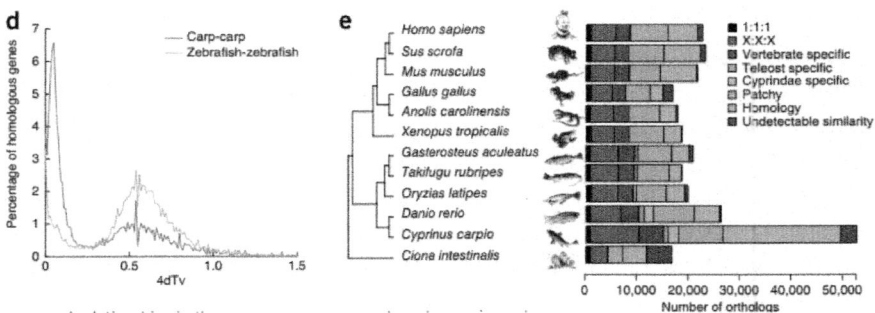

Figure 2: Comparative analyses of the *C. carpio* **genome.**

(a)   Maps of the 50 common carp chromosomes and of the 25 zebrafish chromosomes based on the positions of 8,002 orthologous pairs demonstrate highly conserved synteny for the 2 species and the tetraploidization of the common carp genome. (**b**) Schematic of major interchromosomal relationships in the common carp genome based on reciprocal best-match relationships. Chromosomes are represented as colored blocks. The positions of duplicated genes on the chromosomes are linked by gray lines. (**c**) The distribution of the synonymous substitution rates (*K*s) of homologous gene groups for intraspecies and interspecies comparisons. The peak (*K*s = 0.03) of the paralogous *K*s distribution (inset) indicates the recent WGD. (**d**) The third and fourth genome duplications in the common carp genome were identified by 4dTv analyses. (**e**) Comparison of the gene repertoire of the *Ciona* species with those of five teleost and six tetrapod genomes, ranging from the highly conserved chordate genes (the black fraction on the left) to species-specific genes (the rust red fraction on the right). "1:1:1" indicates universal single-copy genes, and "X:X:X" indicates any other orthologous group (missing in one species), where X means one or more orthologs per species. "Patchy" indicates the existence of other orthologs that are present in at least one teleost and one tetrapod genome. The species tree on the left was built on the basis of 941 single-copy orthologs, for which the *Ciona* species served as the outgroup.

To further provide insight into the tetraploid nature of the genome at the gene level after the 4R WGD event, we investigated the *hox* gene clusters in *C. carpio*. This species has almost twice the number of *hox* clusters as *D. rerio*[35] and the same number of *hox* gene clusters as the Atlantic salmon (*Salmo salar*)[36], which is an autotetraploid species[37] (Supplementary Figs. 11 and 12, andSupplementary Note).

To determine the date of the *C. carpio* WGD event (4R), we used a total of 5,783 gene families and calculated their synonymous substitution rates (*K*s values) (Fig. 2c and Supplementary Note). On the basis of a *K*s rate of 3.51 × 10⁻⁹ substitutions per synonymous site per year[5] and the obtained*K*s value of 0.03, we estimated that the latest WGD (4R) happened 8.2 million years ago, a date more recent than the predictions suggested in previous reports[5, 7]. The carp-zebrafish paralogous genes displayed a distinct peak (*K*s = 0.45) that corresponded to a divergence time of 128 million years ago. In combination with the duplication time and divergence time predictions, these data suggest that the latest WGD event (4R) occurred long after *C. carpio* and *D. rerio* split. Similarly, an analysis of fourfold synonymous third-codon transversion (4dTv) provided additional evidence for an extra round of WGD (Fig. 2d). *C. carpio* and *D. rerio* had a peak in common (4dTv = 0.58), which corresponds to the TSGD event (3R). An extra 4dTv peak within the *C. carpio* paralogous genes (4dTv = 0.1) corresponds to the latest carp-specific WGD (4R).

The annotated gene models of the *C. carpio* genome are substantially better than those of other completely sequenced fish genomes. To understand the evolutionary relationship of the 52,610 gene models with those of other vertebrates, we performed systematic cross-species comparative analysis and classified the genes according to their similarities. We first used five teleosts (*C. carpio*, *D. rerio*, *T. rubripes*, *O. latipes* and *G. aculeatus*), six tetrapods (*Homo sapiens*, *Mus musculus*, *Sus scrofa*, *Gallus gallus*, *Anolis carolinensis* and *Xenopus tropicalis*) and *Ciona intestinalis* (outgroup) for the comparison (Fig. 2e). We identified 941 single-copy orthologs that were conserved among all investigated species, which only accounted for 1.8% of the predicted gene models of *C. carpio*. Second, we constructed the species phylogeny using a maximum-likelihood approach with multiple alignments of single-copy orthologs. The remaining gene models (98.2%) were more complex and included many-to-many orthologs (28.0%), non-uniformly occurring, patchy orthologs (11.0%) and undetectable models (6.1%). Third, the predicted *C. carpio* gene models corresponded to orthologous genes in *D. rerio* (8,002 orthologous genes), including 2,037 (3.9%) cyprinid-specific gene models. Fourth, we also identified 3,212 species-specific gene models of *C. carpio* that did not have any homologs in the 10 other vertebrates and the *Ciona* species examined. This number is higher than the number of species-specific genes in the *D. rerio* genome, suggesting that a significant number of novel genes were generated in *C. carpio* after the divergence of *C. carpio* and *D. rerio*, likely owing to the latest WGD and independent gene evolution (Supplementary Table 15 and Supplementary Note).

Genetic Diversity

C. carpio, as a genetically diverse and successful species, has adapted to various environments across a broad ecological spectrum in Eurasia and has been domesticated for more than 2,000 years. This species has been bred into numerous strains and local populations, producing distinct phenotypic changes in its growth rate, temperature and hypoxia tolerance, body color, scale pattern and body shape, which are partially attributable to genome diversity due to its two WGD (3R and 4R) events[31]. To investigate its genetic variation, we selected 33 representative *C. carpio* accessions for genome resequencing, which included 13 accessions of 4 wild populations from the Danube River, the Yellow River, the Heilongjiang (Amur) River and the Chattahoochee River and 20 accessions of 6 domesticated strains from Asia and Europe (including Songpu, Xingguo red, Oujiang color, Hebao, Szarvas 22 and koi) (Supplementary Table 16). With a total of 4,176 million paired-end reads (101-bp read length, 417.6 Gb in total length and 229-fold coverage of the genome; Supplementary Table 17), we identified 18,949,596 candidate SNPs and 1,694,102 small insertion-deletions (indels) (Supplementary Table 18).

To investigate the divergence of the representative *C. carpio* accessions from diverse geographical habitats and domestic histories, we constructed phylogenetic trees on the basis of the sequence variations (Fig. 3a and Supplementary Fig. 13). It was obvious that the European and Asian accessions formed two distinct clades. One of the strains, Songpu, was also grouped into the European clade as it was bred from mirror carp originally introduced from Europe in the 1950s. Our principal-component analysis (PCA) yielded a similar result (Fig. 3b), showing Asian accessions as a tight cluster that was separate from the European accessions. We further analyzed the population structure using the Bayesian clustering program STRUCTURE[38]. Because the values of ln likelihood were distinctively high for the models $K = 3$ and 4 (Supplementary Fig. 14), we show the clusters of $K = 3$ and 4 in Figure 3c. Almost all the accessions either had common ancestry or showed a single origin of the Eurasian population, and the results agree with the hypothesis that modern *C. carpio* evolved from the Caspian Sea ancestor and spread into Europe and the eastern mainland of Asia[39]. There were no uniform patterns covering all the populations, with the exception of two extremely isolated wild populations, Heilongjiang and Oujiang; in other words, extensive genetic admixture has been occurring in both the wild and domesticated *C. carpio* populations. For instance, Songpu carried admixture from the Asian population ($K = 3$), a finding supported by the recent history of introgression after its introduction to China. We also observed that the US accessions separated into both the European and Asian clades, and the trend indicates multiple introductions to North America from both Europe and Asia. This observation is also supported by our PCA and population structure analyses.

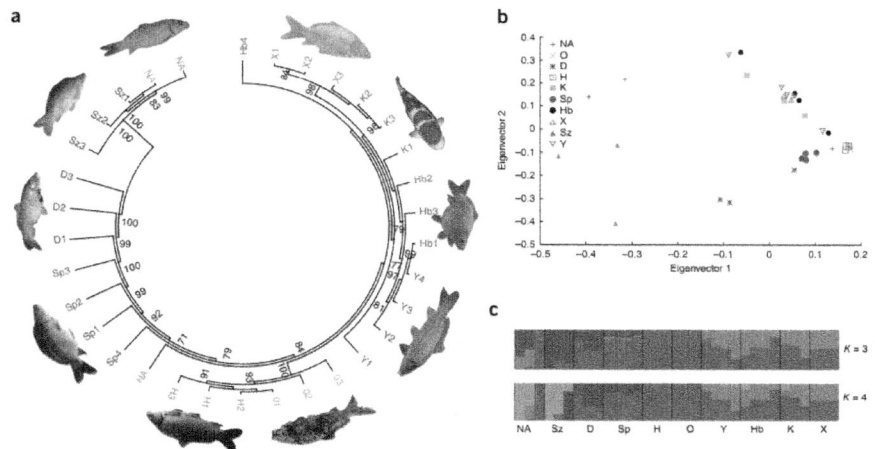

Figure 3: Genetic analysis of ten *C. carpio* **strains using genome resequencing data.**

(**a**) A maximum-likelihood phylogenetic tree of ten strains of common carp generated on the basis of SNPs. Strain abbreviations: Sp, Songpu; D, Danube; Sz, Szarvas; NA, North American; Y, Yellow River; H, Heilongjiang; O, Oujiang color; Hb, Hebao; X, Xingguo; K, koi. (**b**) PCA of *C. carpio* strains. (**c**) The population structure of common carp strains. Each color represents one ancestral population; each strain is represented by a vertical bar, and the length of each colored segment in each vertical bar represents the proportion contributed by ancestral populations. $K = 3$ and 4 was used for analysis with the highest ln value.

We performed a further genetic diversity scan comparing the Hebao and Songpu genomes to identify highly different genomic regions. Hebao is one of the typical strains derived from East Asian subspecies (*C. carpio haematopterus*), whereas Songpu is the strain derived from mirror carp of European subspecies (*C. carpio carpio*) (Fig. 4a). We predict that these two varieties retain substantial genetic differences, given their distinct body shapes, scale morphogenesis and patterns, and skin color phenotypes. We identified a total of 205 genome regions with the highest (top 1% of $\varpi_{Hebao}/\varpi_{Songpu}$) genetic diversity (12.67 Mb in length) containing 326 candidate genes (Supplementary Tables 19 and 20). Gene Ontology (GO) and Kyoto Encyclopedia of Genes and Genomes (KEGG) analysis indicated that a significant portion of these candidate genes were associated with epithelial morphogenesis, hair follicle morphogenesis, pigmentation and immune response, including adherens junction signaling, signaling by Rho family GTPases, tight junction signaling, prolactin signaling, fibroblast growth factor (FGF) signaling, interleukin (IL)-6 signaling and other functional pathways (Fig. 4b and Supplementary Table 21). The results are consistent with the phenotype differences in scale pattern and skin color observed for Hebao and Songpu. We investigated the candidate genes and detected 82 genes and 106 genes that harbored nonsynonymous SNPs in the Songpu and Hebao genomes, respectively (Supplementary Tables 22 and 23). We also identified two Songpu genes (*fgfr1a1* and *lrrc72*) and two Hebao genes (*zpld1*and *nlk*) that harbored deletions in coding regions (Supplementary Figs. 15 and 16). All these instances of sequence diversity altering protein-coding sequences provide candidate loci for assessing phenotypic differences between Hebao and Songpu. Notably, the *fgfr1a1* gene (encoding FGF receptor 1 a1) on chromosome 34 of the Songpu genome contained a 306-bp specific deletion in intron 10 (228-bp deletion) and exon 11 (78-bp deletion) (Fig. 4c). The deletion had previously been reported as the causative mutation for scale loss and reduction in the mirror carp[40]. Extensive investigation on large samples from four strains confirmed that the deletion was only found in Songpu (Supplementary Fig. 16).

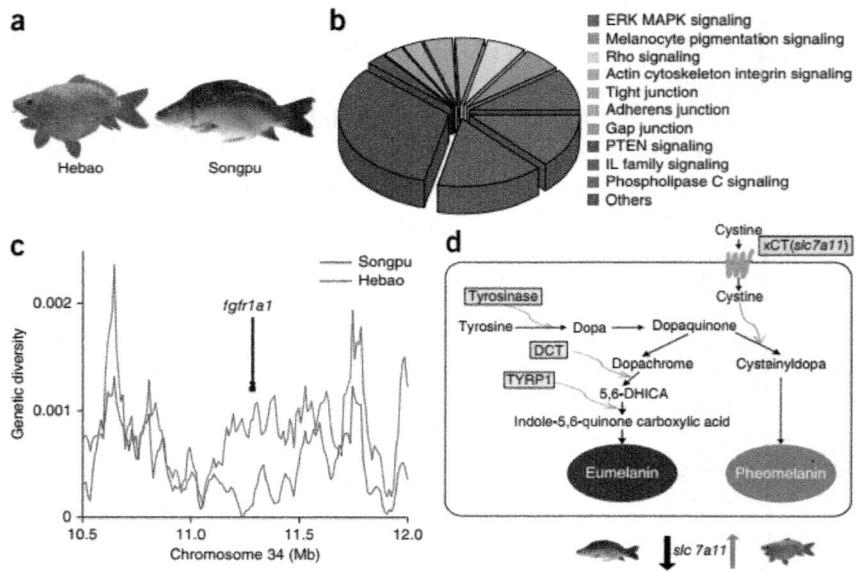

Figure 4: Comparison of the diversity distributions of two *C. carpio* **strains.**

(a)   Typical Songpu and Hebao carp. (b) The distribution of genes involved in KEGG pathways in the regions with the top 1% of $\varpi_{\text{Hebao}}/\varpi_{\text{Songpu}}$.

(c) The diversity (ϖ) distribution on chromosome 34 for Songpu and Hebao showed a 78-bp deletion in exon 11 of the *fgfr1a1* gene in Songpu; this deletion results in a reduced-scale phenotype. (d) The pheomelanin synthesis pathway. Transcriptome analysis showed that the *slc7a11* gene is significantly more upregulated in the skin of Hebao than that of Songpu. *slc7a11*encodes the transmembrane cystine/ glutamate exchanger (xCT), which transports cystine into melanocytes to synthesize pheomelanin (yellow to red pigment).

Comparative Analysis of the Skin Transcriptome

To further elucidate the differences in the scale pattern and skin color of Hebao and Songpu, we performed a comparative analysis of the skin transcriptome in both strains using a deep RNA sequencing (RNA-seq) approach. We identified 894 differentially expressed transcripts, including 567 upregulated genes in Hebao and 327 upregulated genes in Songpu. The experiment was validated with quantitative RT-PCR (qRT-PCR) on selected genes (Supplementary Fig. 17). Further analysis showed distinct expression patterns in Hebao and Songpu for many genes associated with the Wnt/β-catenin signaling pathway (Supplementary Table 24), which is an essential pathway in initiating hair follicle formation[41]. Both mammalian hair and teleost scales are skin appendages, and their formation involves similar developmental pathways. We inferred that the gene expression differences in these two different carp populations were correlated with the reduced-scale phenotype in Songpu and the full-scale phenotype in Hebao.

We also observed a difference in the expression of the *slc7a11* gene (encoding solute carrier family 7 member 11), the plasma membrane cystine/ glutamate exchanger (xCT) that transports cystine into melanocytes. In the melanogenesis pathway, tyrosine is oxidized to form dopaquinone, which is then intracellularly catalyzed to become eumelanin (brown to black pigment) through polymerization and oxidation reactions. However, cystine and dopaquinone can switch off the eumelanin synthesis pathway and promote the synthesis of pheomelanin (yellow to red pigment)[42,43]. *slc7a11* was expressed at a higher level in Hebao than in Songpu, suggesting that more cystine is transported into the pigment cells in Hebao, resulting in the preponderant synthesis of pheomelanin in the skin of Hebao while eumelanin synthesis is suppressed. Higher pheomelanin accumulation in the pigment cells gives Hebao its red skin appearance (Fig. 4d). However, the genetic basis for differences in *slc7a11* expression remains unclear, and further investigation will be necessary to understand the overall role of *slc7a11* in color variation.

DISCUSSION

As one of the most representative carp species, *C. carpio* had a value and global production in 2011 of $5.31 billion and 3.73 million tons, respectively (FAO statistics; see URLs), and the importance of *C. carpio* has been increasing over the past decade. The species is also widely cultured as an ornamental fish because of its various color and scale patterns. We sequenced and assembled the *C. carpio* genome from the genome of a gynogenetic individual using multiple next-generation sequencing platforms and a hybrid assembly strategy. The draft genome provides an important genomic resource to study the genetic basis of economically important traits in carp and to facilitate genome-based genetic breeding technologies in common carp aquaculture. The draft genome also provides insight into the latest WGD event of allotetraploidization that occurred approximately 8.2 million years ago, doubling the chromosome number and gene content of *C. carpio*. The whole-genome resequencing of selected accessions also offers a glimpse into the phylogenetic relationship and population structure of major global accessions of the *C. carpio*population. Comparison of the genomic diversity of two distinct strains, Songpu and Hebao, coupled with additional transcriptomic studies, has allowed us to identify genetic loci and to determine the molecular basis of scale patterns and skin colors, providing a foundation for further studies using comprehensive approaches to completely define the mechanisms underlying these phenotypes. Thus, the draft genome assembly presented here provides a valuable resource for genetic, genomic and biological studies of *C. carpio* and for improving the aquaculturally important traits of farmed *C. carpio* and other key cyprinid species in aquaculture.

METHODS

Ethics Statement.

This study was approved by the Animal Care and Use Committee of the Centre for Applied Aquatic Genomics at the Chinese Academy of Fishery Sciences.

Genome Sequence and Assembly.

A gynogenetic Songpu *C. carpio* was selected as the genomic DNA source for whole-genome sequencing. We constructed 21 shotgun libraries and an 8K mate-pair library according to Roche 454 standard operating procedures. The 22 libraries were sequenced on a Roche 454 genome sequencer using GS FLX Titanium chemistry. We also constructed six paired-end libraries by following the Illumina procedure. Paired-end sequencing of each library was performed on an Illumina HiSeq 2000 instrument to produce the raw data. We then filtered out low-quality and short reads to obtain a set of usable reads. Another library with an 8-kb jumping distance was generated and was sequenced on the SOLiD platform. The published 65,720 clean BAC-end sequences were collected for genome assembly. We assembled the Roche 454 read data set and the Sanger BAC-end sequences into contigs using the Celera assembler[44]. Reads from the Illumina libraries, the SOLiD libraries and the 8,000 Roche 454 mate-pair libraries were aligned to the genomic sequences, and paired-end relationships between the reads were used to construct scaffolds. BAC-end sequences were mapped to the scaffolds and were used for further scaffolding. Finally, we used the paired-end information from the short paired-end reads to fill the gaps between the scaffolds with Gapcloser[45].

Linkage Mapping and Map Integration

Microsatellite and SNP markers were used for genotyping analysis and linkage map construction. A tailed primer protocol was used to amplify microsatellite alleles[46, 47]. PCR products were analyzed on a 3130xl Genetic Analyzer. Restriction site–associated DNA (RAD) technology[48] was used to develop polymorphic SNP markers. JoinMap4.0 software was used to perform the linkage analysis. Linkage between markers was examined by estimating the logarithm of odds (LOD) scores for the recombination rate, and map distances were calculated using the Kosambi mapping function. We then used BLAT (with alignment length coverage of >70%) to align the molecular markers to scaffolds. We linked the scaffolds onto chromosomes with a string of 100 Ns representing the gap between 2 adjacent scaffolds on the basis of a high-resolution genetic map.

Repeat Analysis.

Both homology-based and *de novo* prediction analyses were used to identify the repeat content in the carp genome. For the homology-based analysis, we used Repbase (version 20120418) to perform a TE search with RepeatMasker (3.3.0) and the WuBlast search engine. For the *de novo*prediction analysis, we used RepeatModeler to construct a TE library. Elements within the library were then classified using a homologous search with Repbase and a Support Vector Machine (SVM) method (TEClass).

Gene Prediction and Functional Annotation.

We used three approaches for gene prediction: *ab initio* gene prediction, sequence homology–based prediction and expression evidence–based prediction. Briefly, two *ab initio* prediction software programs, AUGUSTUS[49] and FGENESH[50], were used to predict genes in the repeat-masked genome sequences. Gene model parameters for the programs were trained from long genes and known teleost genes processed by PASA[51]. Sequence homology–based gene prediction included both raw and precise alignments. First, protein sequences from the NCBI non-redundant (nr) database and 68 species sequences in Ensembl (version 68)[52] were collected to build a database. Assembled genome sequences were aligned to their corresponding protein sequences in the database using BLASTX. Identified homologous proteins were selected and then aligned to the genome with TBLASTN. Adjacent and overlapping matches were merged using Perl scripts, building the longest protein for each genomic sequence region. Each target region in the genome was then extended by 10 kb from both ends of the aligned region to cover potential UTRs. Protein sequences were then aligned to those genome fragments by Genewise[53]. Transcriptome reads were generated using the Roche Genome Sequencer FLX (previously released data; available from the Sequence Read Archive (SRA) under accessions SRA009366 and SRA050545) and the Illumina HiSeq 2000 (Supplementary Note). Reads were mapped to genomic sequences by TopHat[54], and Cufflinks[55] was used to produce transcript assemblies. For a gene locus with several alternatively spliced transcripts generated by Cufflinks, the transcript with the longest exon length was chosen. All evidence was merged to form a comprehensive consensus gene set using EVM[27]. PASA was used to update the EVM consensus predictions by adding UTR annotations. To obtain gene function annotations, BLAST searches were conducted against the NCBI nr, SwissProt and TrEMBL protein databases, and homologs were called with E values of $<1 \times 10^{-5}$. The functional classification of GO categories was performed using the InterProScan[56] program. Pathway analysis was performed using the KEGG[57] annotation service, the KEGG Automatic Annotation Server (KAAS)[58].

Comparative Genomic Analysis

Protein-coding genes and coding DNA sequences from 11 species (*D. rerio, G. aculeatus, O. latipes, T. rubripes, T. nigroviridis, X. tropicalis, A. carolinensis, H. sapiens, M. musculus, S. scrofa, G. gallus* and *C. intestinalis*) were downloaded from Ensembl (version 68)[52]. For genes with alternatively spliced variants, only the longest transcript was selected. Any genes encoding proteins of fewer than 30 amino acids were discarded. The OrthoMCL pipeline[59] was used to define gene families in the common ancestor of the species. All-against-all similarities were performed using BLASTP, with an E-value cutoff of 1×10^{-5}. The well-aligned regions of each gene family, aligned using MUSCLE[60], were extracted with Gblocks[61]. Phylogenetic analysis of the superalignments was performed using a maximum-likelihood method implemented in PhyML[62] with the Jones-Taylor-Thornton (JTT) model. *C. intestinalis* was selected as the outgroup. We used MCScanX[63]to identify syntenic blocks for *C. carpio* and *D. rerio*, with the gap size set to 15 genes and at least 5 syntenic genes. Circos[64] was used for visualization.

To detect the conserved synteny blocks generated by the fourth round of genome duplication, we identified the reciprocal best-match paralogs from the above all-against-all BLASTP comparisons. Two chromosome regions with the gap size set to 15 genes and at least 5 syntenic genes were considered to have been duplicated.

Genome Evolutionary Analysis

We used two methods to detect genome duplication signatures. All-against-all BLASTP comparisons (E value $< 1 \times 10^{-5}$) were used to identify pairs of homologous genes. For each homologous gene pair in *C. carpio*, the synonymous site divergence value (Ks) was calculated using the CodeML program (run mode -2) from the PAML package[65]. The distributions of Ks values for *D. rerio* paralogous pairs and pairs between *D. rerio* and the common carp were analyzed using the same pipeline. We calculated the 4dTv values of paralogous pairs within species and of orthologous pairs between species to give the distribution of the 4dTv value to estimate the speciation and WGD event that occurred during evolutionary history.

Whole-Genome Resequencing and Phylogenetic Analysis

The 10 strains of *C. carpio*, consisting of 33 individuals, were randomly collected across Europe, North America and China. Danube River carp and Szarvas 22 were collected from the carp live gene bank of the Research Institute for Fisheries, Aquaculture and Irrigation of Hungary (HAKI). North American carp were collected from the Chattahoochee River in the United States. All other strains or wild populations were collected from China, including Songpu carp from the Heilongjiang Fishery Research Institute; Yellow River carp from the Henan Academy of Fishery Sciences; Heilongjiang River carp from Fuyuan county, Heilongjiang province; Hebao carp from Wuyuan county, Jiangxi province; Xingguo red carp from Xingguo county, Jiangxi province; Oujiang color carp from Longquan county, Zhejiang province; and koi from the breeding population of the Beijing Fishery Research Institute. Fin chips and blood samples were collected, and DNA was extracted using the DNeasy Blood and Tissue kit (Qiagen). Genome resequencing was conducted using the Illumina HiSeq 2000 platform. Paired-end reads from each accession were aligned to the reference genome using the Burrows-Wheeler Aligner

(BWA)[66]. After mapping, SNPs were identified on the basis of the mpileup files generated with SAMtools[67]. The filtering threshold was set to require a read depth of ≥ 10 and a quality score of ≥ 20. Genotypes supported by at least two reads and with a minor allele frequency of ≥ 0.1 were assigned to each genomic position. We performed all-against-all BLASTP for genes in 5 teleosts (*C. carpio, D. rerio, T. rubripes, O. latipes* and *G. aculeatus*) to determine the similarity for each gene pair and to identify single-copy genes, obtaining 8,375 homozygous SNPs from 7,709 single-copy genes. A maximum-likelihood tree was constructed with PhyML[68] and displayed with MEGA[69]. PCA was performed with EIGENSOFT[70], and homozygous SNPs were used to investigate the population structure using STRUCTURE[39]with 2,000 iterations and 2–8 clusters (*K*). The result of the structure matrix was plotted using DISTRUCT[71] software.

Genome Diversity Analysis and Comparison

We calculated the ϖ distribution for each linkage group using a sliding window method with Tajima›s D test in Variscan[72] software. The window width was set to 50 kb, and the stepwise distance was 10 kb. ϖ values were compared, and the ratios were sorted. Using the ratio values, we identified the regions with the 1% highest and lowest diversity, and annotated genes were analyzed. Putative SNPs and deletions in the coding regions were identified by mapping the RNA-seq reads to annotated reference genes using BWA and SAMtools. PCR was performed on the deletion regions to verify the identified gene deletions. PCR products were analyzed via electrophoresis on a 2% agarose gel.

RNA Sequencing Analysis

RNA was extracted from the skin tissues of 18 Hebao and 18 Songpu individuals and pooled for each strain. RNA-seq reads were generated using the Illumina HiSeq 2000 platform (Supplementary Note). Reads with a low quality score and a read length of less than 10 bp were removed. All cleaned reads were mapped to the assembled reference with Bowtie[73]. Then, RSEM (RNA-Seq by Expectation Maximization)[74323 (2011)] was used to estimate and quantify gene and isoform abundance. Gene expression was measured in fragments per kilobase of exon per million fragments mapped (FPKM)[55]. Finally, edgeR[75] was used to normalize the expression levels in both strains to identify the differentially expressed transcripts by pairwise comparisons. For qRT-PCR validation, total RNA was isolated and purified from all of the samples using the RNeasy kit (Qiagen) and was quantified using a NanoDrop and a Bioanalyzer 2100 (Agilent Technologies). qRT-PCR was performed on the ABI PRISM 7500 Real-Time PCR System with three replicates using the QuantiTect SYBR Green PCR kit

(Qiagen). The *actb* gene was used as the internal reference. Primer information is provided in Supplementary Table 25. Two-sided *t* tests were used to compare expression levels. GO annotation of the genes was performed on the basis of orthologous relationships with the gene set of *D. rerio*. Pathway analysis was performed using Ingenuity Pathway Analysis (IPA) tools (Ingenuity Systems).

REFERENCES

1. FAO Fisheries and Aquaculture Department. The State of World Fisheries and Aquaculture 2006 (Food and Agriculture Organization of the United Nations, Rome, 2007).

2. Bostock, J. et al. Aquaculture: global status and trends. Phil. Trans. R. Soc. B 365,2897–2912 (2010).

3. Hinegardner, R. & Rosen, D.E. Cellular DNA content and the evolution of teleostean fishes.Am. Nat. 106, 621–644 (1972).

4. Wang, J.T., Li, J.T., Zhang, X.F. & Sun, X.W. Transcriptome analysis reveals the time of the fourth round of genome duplication in common carp (Cyprinus carpio). BMC Genomics 13,96 (2012).

5. David, L., Blum, S., Feldman, M.W., Lavi, U. & Hillel, J. Recent duplication of the common carp (Cyprinus carpio L.) genome as revealed by analyses of microsatellite loci. Mol. Biol. Evol. 20, 1425–1434 (2003).

6. Ohno, S., Muramoto, J., Christian, L. & Atkin, N.B. Diploid-tetraploid relationship among old-world members of the fish family Cyprinidae. Chromosoma 23, 1–9 (1967).

7. Larhammar, D. & Risinger, C. Molecular genetic aspects of tetraploidy in the common carpCyprinus carpio. Mol. Phylogenet. Evol. 3, 59–68 (1994).

8. Ji, P. et al. High throughput mining and characterization of microsatellites from common carp genome. Int. J. Mol. Sci. 13, 9798–9807 (2012).

9. Xu, J. et al. Genome-wide SNP discovery from transcriptome of four common carp strains.PLoS ONE 7, e48140 (2012).

10. Zhang, X. et al. A consensus linkage map provides insights on genome character and evolution in common carp (Cyprinus carpio L.). Mar. Biotechnol. (NY) 15, 275–312 (2013).

11. Zheng, X. et al. A genetic linkage map and comparative genome analysis of common carp (Cyprinus carpio L.) using microsatellites and SNPs. Mol. Genet. Genomics 286, 261–277(2011).

12. Zhang, Y. et al. Genetic linkage mapping and analysis of muscle fiber–related QTLs in common carp (Cyprinus carpio L.). Mar. Biotechnol.

(NY) 13, 376–392 (2011).

13. Sun, X.W. & Liang, L.Q. A genetic linkage map of common carp (Cyprinus carpio L.) and mapping of a locus associated with cold tolerance. Aquaculture 238, 165–172 (2004).

14. Li, Y. et al. Construction and characterization of the BAC library for common carp Cyprinus carpio L. and establishment of microsynteny with zebrafish Danio rerio. Mar. Biotechnol. (NY) 13, 706–712 (2011).

15. Xu, P. et al. Genomic insight into the common carp (Cyprinus carpio) genome by sequencing analysis of BAC-end sequences. BMC Genomics 12, 188 (2011).

16. Christoffels, A., Bartfai, R., Srinivasan, H., Komen, H. & Orban, L. Comparative genomics in cyprinids: common carp ESTs help the annotation of the zebrafish genome. BMC Bioinformatics 7 (suppl. 5), S2 (2006).

17. Zhang, Y. et al. Identification of common carp innate immune genes with whole-genome sequencing and RNA-Seq data. J. Integr. Bioinform. 8, 169 (2011).

18. Ji, P. et al. Characterization of common carp transcriptome: sequencing, de novo assembly, annotation and comparative genomics. PLoS ONE 7, e35152 (2012).

19. Williams, D.R. et al. Genomic resources and microarrays for the common carp Cyprinus carpio L. J. Fish Biol. 72, 2095–2117 (2008).

20. Henkel, C.V. et al. Comparison of the exomes of common carp (Cyprinus carpio) and zebrafish (Danio rerio). Zebrafish 9, 59–67 (2012).

21. Ojima, Y. & Yamamoto, K. Cellular DNA contents of fishes determined by flow cytometry. La Kromosomo II 57, 1871–1888 (1990).

22. Kidwell, M.G. & Lisch, D.R. Transposable elements and host genome evolution. Trends Ecol. Evol. 15, 95–99 (2000).

23. Aparicio, S. et al. Whole-genome shotgun assembly and analysis of the genome of Fugu rubripes. Science 297, 1301–1310 (2002).

24. Van de Peer, Y. Tetraodon genome confirms Takifugu findings: most fish are ancient polyploids. Genome Biol. 5, 250 (2004).

25. Kasahara, M. et al. The medaka draft genome and insights into vertebrate genome evolution. Nature 447, 714–719 (2007).

26. Jones, F.C. et al. The genomic basis of adaptive evolution in threespine sticklebacks. Nature484, 55–61 (2012).

27. Haas, B.J. et al. Automated eukaryotic gene structure annotation using EVidenceModeler and the Program to Assemble Spliced Alignments. Genome Biol. 9, R7 (2008).

28. Meyer, A. & Van de Peer, Y. From 2R to 3R: evidence for a fish-specific genome duplication (FSGD). Bioessays 27, 937–945 (2005).

29. Hoegg, S., Brinkmann, H., Taylor, J.S. & Meyer, A. Phylogenetic timing of the fish-specific genome duplication correlates with the diversification of teleost fish. J. Mol. Evol. 59,190–203 (2004).

30. Santini, F., Harmon, L., Carnevale, G. & Alfaro, M. Did genome duplication drive the origin of teleosts? A comparative study of diversification in ray-finned fishes. BMC Evol. Biol. 9, 194(2009).

31. Crow, K.D. & Wagner, G.P. Proceedings of the SMBE Tri-National Young Investigators Workshop 2005. What is the role of genome duplication in the evolution of complexity and diversity? Mol. Biol. Evol. 23, 887–892 (2006).

32. Jaillon, O. et al. Genome duplication in the teleost fish Tetraodon nigroviridis reveals the early vertebrate proto-karyotype. Nature 431, 946–957 (2004).

33. Zhang, Y. et al. Genome evolution trend of common carp (Cyprinus carpio L.) as revealed by the analysis of microsatellite loci in a gynogentic family. J. Genet. Genomics 35, 97–103(2008).

34. Howe, K. et al. The zebrafish reference genome sequence and its relationship to the human genome. Nature 496, 498–503 (2013).

35. Amores, A. et al. Zebrafish hox clusters and vertebrate genome evolution. Science 282,1711–1714 (1998).

36. Mungpakdee, S. et al. Differential evolution of the 13 Atlantic salmon Hox clusters. Mol. Biol. Evol. 25, 1333–1343 (2008).

37. Davidson, W.S. et al. Sequencing the genome of the Atlantic salmon (Salmo salar). Genome Biol. 11, 403 (2010).

38. Pritchard, J.K., Stephens, M. & Donnelly, P. Inference of population structure using multilocus genotype data. Genetics 155, 945–959 (2000).

39. Balon,E.K. Origin and domestication of the wild carp, Cyprinus carpio: from Roman gourmets to the swimming flowers. Aquaculture 129, 3–48 (1995).

40. Rohner, N. et al. Duplication of fgfr1 permits Fgf signaling to serve as a target for selection during domestication. Curr. Biol. 19, 1642–1647 (2009).

41. Millar, S.E. Molecular mechanisms regulating hair follicle development. J. Invest. Dermatol.118, 216–225 (2002).

42. Hoekstra, H.E. Genetics, development and evolution of adaptive pigmentation in vertebrates.Heredity 97, 222–234 (2006).

43. Ito, S. & Wakamatsu, K. Human hair melanins: what we have learned and have not learned from mouse coat color pigmentation. Pigment Cell Melanoma Res. 24, 63–74 (2011).

44. Miller, J.R. et al. Aggressive assembly of pyrosequencing reads with mates. Bioinformatics24, 2818–2824 (2008).

45. Li, R. et al. De novo assembly of human genomes with massively parallel short read sequencing. Genome Res. 20, 265–272 (2010).

46. Neilan, B.A., Wilton, A.N. & Jacobs, D. A universal procedure for primer labelling of amplicons. Nucleic Acids Res. 25, 2938–2939 (1997).

47. Schuelke, M. An economic method for the fluorescent labeling of PCR fragments. Nat. Biotechnol. 18, 233–234 (2000).

48. Sun, X. et al. SLAF-seq: an efficient method of large-scale de novo SNP discovery and genotyping using high-throughput sequencing. PLoS ONE 8, e58700 (2013).

49. Stanke, M., Schoffmann, O., Morgenstern, B. & Waack, S. Gene prediction in eukaryotes with a generalized hidden Markov model that uses hints from external sources. BMC Bioinformatics 7, 62 (2006).

50. Salamov, A.A. & Solovyev, V.V. Ab initio gene finding in Drosophila genomic DNA. Genome Res. 10, 516–522 (2000).

51. Campbell, M.A., Haas, B.J., Hamilton, J.P., Mount, S.M. & Buell, C.R. Comprehensive analysis of alternative splicing in rice and comparative analyses with Arabidopsis. BMC Genomics 7, 327 (2006).

52. Flicek, P. et al. Ensembl 2013. Nucleic Acids Res. 41, D48–D55 (2013).

53. Birney, E., Clamp, M. & Durbin, R. GeneWise and Genomewise. Genome Res. 14, 988–995(2004).

54. Trapnell, C., Pachter, L. & Salzberg, S.L. TopHat: discovering splice junctions with RNA-Seq. Bioinformatics 25, 1105–1111 (2009).

55. Trapnell, C. et al. Transcript assembly and quantification by RNA-Seq reveals unannotated transcripts and isoform switching during cell differentiation. Nat. Biotechnol. 28, 511–515(2010).

56. Zdobnov, E.M. & Apweiler, R. InterProScan—an integration platform for the signature-recognition methods in InterPro. Bioinformatics 17, 847–848 (2001).

57. Ogata, H. et al. KEGG: Kyoto Encyclopedia of Genes and Genomes. Nucleic Acids Res. 27,29–34 (1999).

58.   Moriya, Y., Itoh, M., Okuda, S., Yoshizawa, A.C. & Kanehisa, M. KAAS: an automatic genome annotation and pathway reconstruction server. Nucleic Acids Res. 35, W182–W185(2007).

59.   Li, L., Stoeckert, C.J. Jr. & Roos, D.S. OrthoMCL: identification of ortholog groups for eukaryotic genomes. Genome Res. 13, 2178–2189 (2003).

60.   Edgar, R.C. MUSCLE: multiple sequence alignment with high accuracy and high throughput.Nucleic Acids Res. 32, 1792–1797 (2004).

61.   Talavera, G. & Castresana, J. Improvement of phylogenies after removing divergent and ambiguously aligned blocks from protein sequence alignments. Syst. Biol. 56, 564–577(2007).

62.   Guindon, S., Lethiec, F., Duroux, P. & Gascuel, O. PHYML Online—a web server for fast maximum likelihood–based phylogenetic inference. Nucleic Acids Res. 33, W557–W559(2005).

63.   Wang, Y. et al. MCScanX: a toolkit for detection and evolutionary analysis of gene synteny and collinearity. Nucleic Acids Res. 40, e49 (2012).

64.   Krzywinski, M. et al. Circos: an information aesthetic for comparative genomics. Genome Res. 19, 1639–1645 (2009).

65.   Yang, Z. PAML 4: phylogenetic analysis by maximum likelihood. Mol. Biol. Evol. 24,1586–1591 (2007).

66.   Li, H. & Durbin, R. Fast and accurate short read alignment with Burrows-Wheeler transform.Bioinformatics 25, 1754–1760 (2009).

67.   Li, H. et al. The sequence alignment/map format and SAMtools. Bioinformatics 25,2078–2079 (2009).

68.   Guindon, S. & Gascuel, O. A simple, fast, and accurate algorithm to estimate large phylogenies by maximum likelihood. Syst. Biol. 52, 696–704 (2003).

69.   Tamura, K. et al. MEGA5: molecular evolutionary genetics analysis using maximum likelihood, evolutionary distance, and maximum parsimony methods. Mol. Biol. Evol. 28,2731–2739 (2011).

70.   Price, A.L. et al. Principal components analysis corrects for stratification in genome-wide association studies. Nat. Genet. 38, 904–909 (2006).

71.   Rosenberg, N.A. DISTRUCT: a program for the graphical display of population structure.Mol. Ecol. Notes 4, 137–138 (2004).

72.   Vilella, A.J., Blanco-Garcia, A., Hutter, S. & Rozas, J. VariScan: analysis of evolutionary patterns from large-scale DNA sequence polymorphism data. Bioinformatics 21, 2791–2793(2005).

73. Langmead, B. & Salzberg, S.L. Fast gapped-read alignment with Bowtie 2. Nat. Methods 9,357–359 (2012).

74. Li, B. & Dewey, C.N. RSEM: accurate transcript quantification from RNA-Seq data with or without a reference genome. BMC Bioinformatics 12, 323 (2011).

75. Robinson, M.D., McCarthy, D.J. & Smyth, G.K. edgeR: a Bioconductor package for differential expression analysis of digital gene expression data. Bioinformatics 26, 139–140

Chapter 12

COMPARATIVE GENOMIC ANALYSIS OF SOYBEAN FLOWERING GENES

Chol-Hee Jung, Chui E. Wong Mohan B. Singh , Prem L. Bhalla

Plant Molecular Biology and Biotechnology Laboratory, ARC Centre of Excellence for Integrative Legume Research, Melbourne School of Land and Environment, The University of Melbourne, Parkville, Victoria, Australia

ABSTRACT

Flowering is an important agronomic trait that determines crop yield. Soybean is a major oilseed legume crop used for human and animal feed. Legumes have unique vegetative and floral complexities. Our understanding of the molecular basis of flower initiation and development in legumes is limited. Here, we address this by using a computational approach to examine flowering regulatory genes in the soybean genome in comparison to the most studied model plant, Arabidopsis. For this comparison, a genome-wide analysis of orthologue groups was performed, followed by an *in silico* gene expression analysis of the identified soybean flowering genes. Phylogenetic analyses of the gene families highlighted the evolutionary relationships among these candidates. Our study identified key flowering genes in soybean and indicates that the vernalisation and the ambient-temperature pathways seem to be the most variant in soybean. A comparison of the orthologue groups containing flowering genes indicated that, on average, each Arabidopsis flowering gene has 2-3 orthologous copies in soybean. Our analysis highlighted that the *CDF3*, *VRN1*, *SVP*, *AP3* and *PIF3* genes are paralogue-rich genes in soybean. Furthermore, the genome mapping of the soybean flowering genes showed that these genes are scattered randomly across the genome. A paralogue comparison indicated that the soybean genes comprising the largest orthologue group are clustered in a 1.4 Mb region on chromosome 16 of soybean. Furthermore, a comparison with the undomesticated soybean (*Glycine soja*) revealed that there are hundreds of SNPs that are associated with putative soybean flowering genes and that there are structural variants that may affect the genes of the light-signalling and ambient-temperature pathways in soybean. Our study provides a framework for the soybean flowering pathway and insights into the relationship and evolution of flowering genes between a short-day soybean and the long-day plant, Arabidopsis.

INTRODUCTION

Plants switch to the reproductive phase of development when environmental and endogenous factors are the most favourable for reproductive success and seed production. This proper timing is the result of elaborate regulatory networks that coordinate the external stimuli with endogenous cues, inducing the expression of genes that initiate the floral transition at the shoot apical meristem (SAM).

Much of our current understanding of the floral initiation process is derived from studies usingArabidopsis thaliana as the model system. More than 180 Arabidopsis genes have been identified that play a role in regulating flowering time, and these genes have been organised into six major pathways (reviewed by Fornara et al. [1]). Although the photoperiod and vernalisation pathways monitor seasonal changes in day length or temperature and, hence, initiate flowering in response to exposure to long days or prolonged cold temperatures, the ambient temperature pathway coordinates the response to daily growth temperatures. The autonomous pathway together with those involving age or gibberellin constitutes the rest of the floral pathways, which function more independently of external stimuli. These pathways are integrated by downstream target genes including LEAFY (LFY), FLOWERING LOCUS T (FT) and SUPPRESSOR OF CONSTANS1 (SOC1), with their resulting outcomes conveyed to floral meristem identity genes such as APETALA1 (AP1) at the SAM that triggers the flowering process [2], [3].

Flowering is one of the most important agronomic traits influencing crop yield. There is thus a great necessity for research that examines the molecular control of this fundamental process in important crop species. This knowledge is critical for the breeding of climate change resilient crop varieties. Soybean, a major food crop, is also a member of the large and diverse legume family, which has the unique capability of forming nitrogen-fixing symbioses with soil microorganisms and has thus been used as part of sustainable agricultural practices for thousands of years. Soybean is distributed broadly across latitudes and is cultivated as different maturity groups, with each having a narrow range of latitudinal adaptation. Unlike Arabidopsis, soybean can undergo a reversion of flowering when plants are shifted from flowering inductive to non-inductive conditions [4]. In addition, soybean also follows a floral developmental plan that is distinct from that of Arabidopsis [5]. Therefore, an understanding of the molecular mechanisms underlying these soybean traits is of fundamental and practical interest. Recent studies have begun to shed light on the molecular adaptation of different soybean cultivars to a wide range of photoperiodic conditions [6], [7]. These studies have highlighted similarities as well as differences in the roles of flowering time genes between soybean and Arabidopsis. The blue light receptor CRYPTOCHROME2 (CRY2) regulates photoperiodic flowering in Arabidopsis; however, in soybean, GmCRY1a but not GmCRY2a is the major regulator of photoperiodic flowering. On the other hand, Kong et al. (2010) revealed that, although soybean contains several FT homologues, the dynamic expression of only two of them is responsible for the adaptation of soybean to diverse photoperiodic environments [6]. Nevertheless, it is still unclear if similar downstream target genes are activated in soybean as in Arabidopsis or how the floral pathway is modified to generate the outputs that reconcile the differences in floral development between the two species.

In view of the recent availability of the soybean genome sequence, we have undertaken a genome-wide analysis for the identification of all soybean orthologues for the corresponding Arabidopsis genes, particularly those involved in flowering. As a paleopolyploid, soybean contains duplicate copies of most genes, and these duplicates may have undergone sub- or neo-functionalisation. We identified 491 putative soybean flowering regulatory genes that are scattered randomly throughout the genome, and then we performed phylogenetic analyses of these gene families to acquire an understanding of the evolutionary relationships among these candidate genes. The identified putative soybean flowering genes were further subjected to an in silico gene expression analysis using two independent transcriptome datasets [8], [9]. Although the distributions of the soybean genes in the paralogue-rich groups are not correlated with the recently duplicated regions in the genome, soybean chromosome 16, especially in the ~1.4 Mb region around 34–35 Mb, is highly enriched for genes within paralogue-rich groups. Our study provides an essential genomic resource for functional analyses of the soybean flowering pathway, facilitating future research and efforts into breeding robust high-yielding crop varieties.

RESULTS

Identification of Soybean Homologues of Arabidopsis Genes

The most recent genome annotation of soybean lists 46,367 genes with high confidence from the current draft genome sequence of soybean [10]. The current soybean annotation (G.max 1.09) identifies the closest Arabidopsis homologue of nearly all of the predicted soybean genes. However, soybean genes are associated with only 55% of the total Arabidopsis genes in the TAIR9 annotation. Thus, we combined the information from the TAIR9 annotation together with the orthologue-based method, which clusters soybean genes and Arabidopsis genes independently into pre-defined orthologue groups (OGs) in the OrthoMCL database (release 5.0) [11]. Then, we matched soybean and Arabidopsis genes under the same OGs as putative orthologues (see Methods). This combined analysis for homologue identification connected 20,730 Arabidopsis genes in 11,344 OGs to 45,175 soybean genes (Dataset S1).

Soybean Flowering-related Genes

In this study, we focus on the 183 Arabidopsis genes that are known to take part in flowering regulatory pathways from previous studies [9], [12], [13], [14], [15], [16], [17], [18], [19], [20],[21], [22]. The orthologue identification analysis found 491 soybean genes that are putative flowering genes (Figure 1 and Table S1). The majority of the Arabidopsis flowering genes have putative soybean orthologues (163 out of 183). However, the soybean orthologues for 20 Arabidopsis flowering genes are not identified by the orthologue-based method used in this study. These Arabidopsis genes include TARGET OF EARLY ACTIVATION TAGGED 2 (TOE2), TOE3 and the vernalisation-insensitive genes VERNALIZATION5/VIN3-LIKE 2(VEL2), and VEL3 (Table S1).

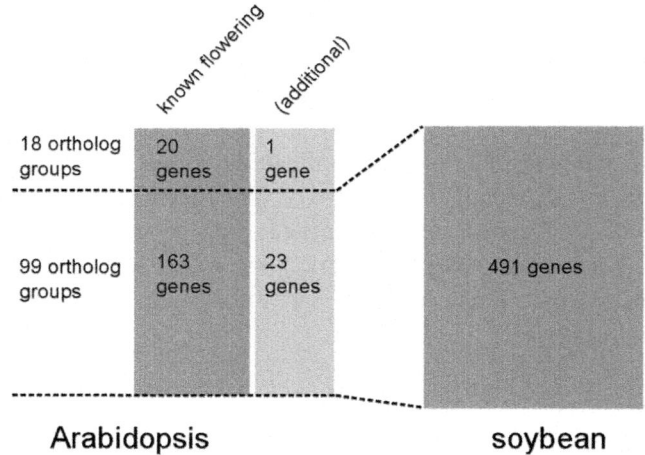

Figure 1: Number of orthologue groups and genes related to flowering pathways.183 Arabidopsis flowering genes are classified into 117 orthologue groups along with 24 additional genes. Function of these additional genes has not been investigated so far. Out of the 117 OGs, 99 OGs contain 491 soybean genes that are putative soybean flowering genes. The numbers of genes are indicated within each box.

Nevertheless, this lack of orthologue identification does not necessarily mean that these 20 Arabidopsis flowering genes are absent in soybean, as they still have similar soybean genes based on a direct BLAST analysis. In addition, the orthologue-based method identified 24 additional Arabidopsis genes that are grouped into the same OGs as known flowering genes (Table S1) but have not been investigated for their role in floral initiation. doi:10.1371/journal. pone.0038250.g001

Subsequent analyses of phylogenetic trees generated from the multiple sequence alignments of the soybean and Arabidopsis genes within each OG estimated that 322 genes are located in the same clades as the Arabidopsis genes known to be involved in flowering pathways (seeMaterials and Methods), indicating that they are likely the true orthologues of their corresponding Arabidopsis flowering genes. The simplified pathway diagram, which contains most of the Arabidopsis flowering-time genes and their putative soybean orthologues, is shown in Figure 2.

Arabidopsis flowering pathway depicted by Higgins et al. (2010) [12] was adapted for this study. Arabidopsis gene symbols are shown in upper box of each node along with the corresponding OG ID and the total number of soybean genes in the same OG. Only the soybean genes that are closer to the Arabidopsis flowering genes in each OG are listed in each node. Arabidopsis genes in grey shades are those that are not assigned with putative soybean orthologues. However, based on BLAST analysis these genes still have homologous soybean genes, the best matching soybean gene in BLAST analysis is shown below these genes. Dashed-lines were used for the arrows or T-bars that involve grey shades. Soybean genes marked with * are those that share the same clade with Arabidopsis genes but have not been investigated so far as flowering genes. OG IDs starting with 'OG5_AT' are arbitrarily generated in this study and do not exist in OrthoMCL 5.0 database. Other conventions are same as those used in the Figure 1A by Higgins et al. (2010) [12]. Arrows show promoting effects, T-bars show repressing effects. Environmental cues are shown as lower case letters in square boxes; 'v' is extended cold (vernalization); 'ld' is long days; 'sd' is short days; 'am' is ambient (non-vernalizing) temperature. Genes are shown in italics and proteins in non-italics in ovals. 'Pfr' indicates P_{fr} phytochrome signaling. Arabidopsis genes assigned to specific pathways are color-coded (photoperiod pathway in green, vernalization in blue and autonomous pathway in purple). Flowering pathway integrators are shown in red. Triple headed arrows indicate activation by red or blue light.

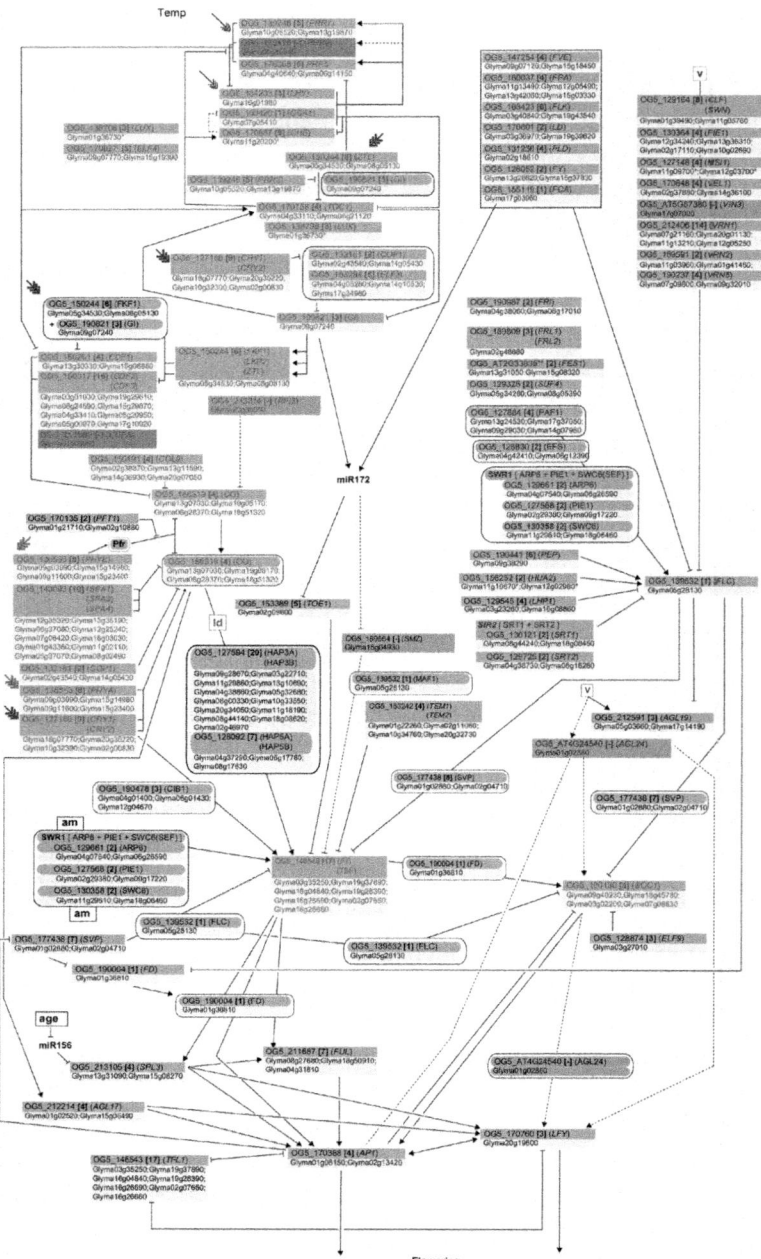

Figure 2: An outline of flowering pathway showing soybean orthologues for Arabidopsis flowering genes.

Expression of Soybean Homologues of Arabidopsis Flowering Genes

The transcriptional activities of the putative soybean floral regulatory genes were examined to gather further evidence for their involvement in flowering. To this end, we utilised the soybean gene expression data from two recent transcriptome analyses [8], [9]. We found that the expression of most of the putative soybean floral genes (449 out of 491 genes; 91.4%) is supported by these two datasets (Figure S1 and Dataset S2). Furthermore, the vast majority of the expressed putative soybean flowering genes, 403 out of 449, exhibited transcriptional activities in flowers (Dataset S2), among which 19 genes are preferentially or specifically expressed in flowers [9]. These 19 genes are spread across 10 OGs, which contain the Arabidopsis MADS box genes, including AP1, PISTILLATA (PI) and SEPALLATA1 (SEP1) (Table S2).

Key Pathways and Gene Families for Flowering

Among the genes for flowering pathways, the key players are those that are involved in the light-signalling pathway, the vernalisation pathway, the autonomous pathway and the ambient temperature pathway, along with genes for meristem identity and flowering pathway integrators[12]. In Arabidopsis, 120 genes, which are grouped into 69 OGs, are known to be key players in flowering [12]. Among these 120 Arabidopsis genes, 112 of them are found in 62 OGs having 314 putative soybean orthologues, (Table 1). Table 1 shows the number of OGs associated with the key pathways or groups of genes for flowering. As one gene can take part in two or more different pathways, one OG can participate in multiple pathways or groups. The 237 soybean genes that are orthologous to key flowering genes of Arabidopsis are scattered throughout the genome rather than clustered within certain regions (Figure 3).

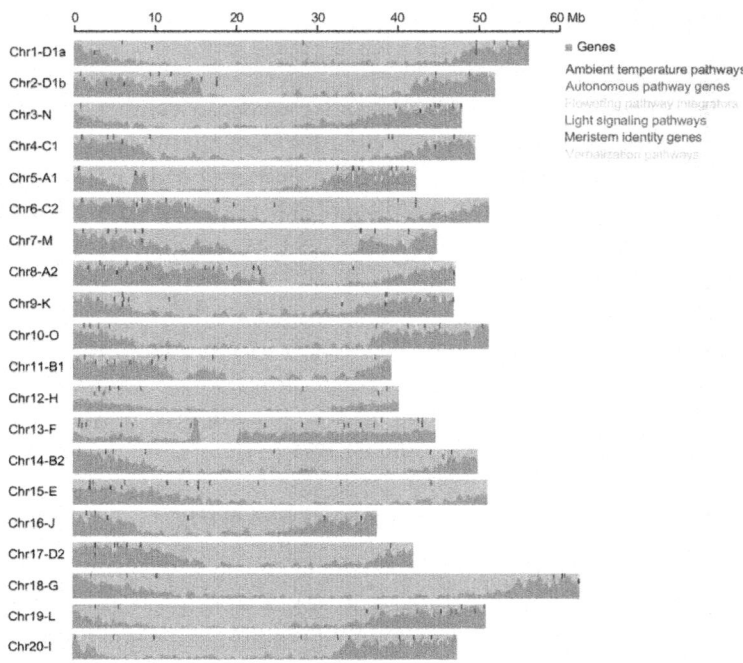

Figure 3: Genomic loci of soybean genes homologous to Arabidopsis floral regulatory genes.

Soybean genes in Table 1 that are homologous to Arabidopsis floral regulatory genes are indicated with solid bars of different colours while the purple shade represents gene density. These soybean genes are randomly spread over the whole genome. Gene density depiction is adapted and modified from Figure 1 by Schmutz et al. (2010) [10].

doi:10.1371/journal.pone.0038250.g003

Table 1: Key flowering pathways and the numbers of associated genes in Arabidopsis, soybean, Medicago, A. lyrata **and** Brachypodium

Pathway	OGs	Arabidopsis"	Soybean*	Medicago	A lyrata	Brachypodium
Light signaling	25	48/(53)	121/(115)	47	59	44
Vernalization	23	32/(36)	81/(71)	46	33	25
Autonomous	16	17/(23)	49/(46)	33	24	23
Ambient temperature	8	16/(19)	38/(30)	28	18	9
Meristem identity	5	7/(7)	19/(18)	6	7	3
Flowering pathway integrators	11	36/(39)	82/(69)	32	40	33

As one gene can take part in two or more different pathways, one OG can participate in multiple pathways or groups.
"Numbers within parentheses indicate the total number of Arabidopsis genes belonging to the OGs associated with the given pathways.
*Numbers within parentheses indicate the number of soybean genes showing transcriptional activity detected by using either of the two recent transcriptome datasets [8,9].
doi:10.1371/journal.pone.0038250.t001

Light signalling pathways.

Light is one of the main environmental regulators of flowering in plants. Plants sense the time of day and season of year by monitoring the light environment through light signalling pathways[23]. Soybean is a facultative short-day crop, but soybean cultivars also belong to different maturity groups depending upon their photoperiod sensitivity. This strong latitudinal cline is also observed in its undomesticated wild relative, Glycine soja (G. soja). In Arabidopsis, photoperiod pathway genes together with photoreceptor genes and circadian clock components take part in light signalling pathways. The number of known Arabidopsis flowering genes involved in these pathways is 48, which are clustered into 25 OGs. However, these OGs contain 53 Arabidopsis genes in total, suggesting that the additional 5 genes may also be involved in floral initiation (Table 1). In total, 121 soybean genes are identified as putative orthologues of 48 Arabidopsis flowering genes in 25 OGs (Table 1). The multiple sequence alignments followed by phylogenetic tree analyses for the Arabidopsis and soybean gene sequences in each of the 25 OGs revealed that 66 of the soybean genes are more closely located to their corresponding Arabidopsis genes than other soybean genes in the same OGs (Dataset S3). Furthermore, an in silico gene expression analysis of the identified soybean flowering genes determined that 115 of the 121 soybean orthologues are expressed, including 109 genes expressed in flowers [8],[9] (Figure S1 and Dataset S2).

The key Arabidopsis genes involved in the light signalling pathway include the CONSTANS(CO), PHYTOCHROME (PHY) and CRYPTOCHROME (CRY), CIRCADIAN CLOCK ASSOCIATED 1 (CCA1), LATE ELONGATED HYPOCOTYL (LHY) and PSEUDO-RESPONSE REGULATOR 1 [PRR1, also called TIMING OF CAB EXPRESSION 1 (TOC1)] genes. CO, along with CONSTANS-LIKE 1 (COL1) and CONSTANS-LIKE 2 (COL2), are contained in OG5_156319, which also contains four soybean genes as soybean orthologues (Glyma08g28370, Glyma13g07030, Glyma18g51320 and Glyma19g05170) (See Table S1). All four soybean-orthologue candidates of Arabidopsis CO are expressed in tested tissues/developmental stages in the two recent transcriptome datasets [8], [9], but only two candidates are expressed in flowers (Dataset S2). The CRY genes CRY1 and CRY2 are grouped into OG5_127186, which contains nine soybean genes (Table S1). The UV REPAIR DEFECTIVE 3 (UVR3) gene is also grouped into OG5_127186. In the phylogenetic tree of genes contained in OG5_127186, CRY1, CRY2 and UVR3 are all located in the same clade, along with 5 soybean genes (Figure 4A). Among these soybean genes, Glyma08g22400 is the closest orthologue of Arabidopsis UVR3, while Glyma18g07770,

Glyma20g35220, Glyma10g32390 and Glyma02g00830 are closer to CRY2 (Figure 4A). Phylogenetic trees in Figure 4 include putative orthologues in Arabidopsis lyrata (A. lyrata), Medicago truncatula(Medicago) as well as a monocot Brachypodium distachyon (Brachypodium). All three Brachypodium genes and one Medicago genes clustered in the same OG are also found in theCRY1 clade, leaving four soybean genes and one Medicago gene in separate clades, indicating that these may have diverged functions (Figure 4A). Five PHY genes of Arabidopsis (PHYA,PHYB, PHYC, PHYD and PHYE) have eight soybean orthologue candidates, which are contained within OG5_136555 (Table S1). All of these soybean genes, except for Glyma15g23400, are expressed in flowers in one or both of the two transcriptome gene expression analyses integrated in this study [8], [9]. The MYB-transcription factor genes CCA1and LHY are among the key circadian clock components in Arabidopsis and are regulated byTOC1 (also known as PRR1) [24]. CCA1 has a single soybean gene orthologue candidate (Glyma07g05410), while LHY and TOC1 have three and four soybean orthologue candidates, respectively (Table S1). All of the putative soybean orthologues of CCA1, LHY and TOC1 are expressed in the samples tested, including flowers, when analysed for their in silico gene expression [8], [9]. The GIGANTEA (GI) gene in OG5_190821 is a part of the evening loop in Arabidopsis and performs different functions through its interactions with other genes, including the FLAVIN-BINDING, KELCH REPEAT, F BOX 1 (FKF1), LOV KELCH PROTEIN 2 (LKP2) and ZEITLUPE (ZTL) genes contained within OG5_150244, which contains six soybean genes in total (Table S1) [12]. Higgins et al. (2011) reported that GI is a highly conserved single copy gene in Arabidopsis, rice, Brachypodium and barley [12], but it has three orthologous soybean genes (Glyma09g07240, Glyma10g36600, Glyma20g30980) (Table S1).

Arabidopsis genes that are known as flowering genes are shown in red along with their closest soybean homologues in the corresponding phylogenetic trees, which are most likely to be orthologues of Arabidopsis flowering genes. Numbers in each branch indicate the confidence value calculated from 1000 times of bootstrapping. Putative orthologues of A. lyrata, Medicago and Brachypodium are also included in the phylogenetic trees. Gene names starting with 'AT', 'Glyma', 'Alyr_' and 'Bradi' are for genes of A. thaliana, soybean, A. lyrata and Brachypodium, respectively. Medicago gene names start with either 'Medtr' or 'AC'. To clarify relationships between different nodes diverse tree formats for OGs are used. The Arabidopsis genes marked with '*' have not been identified and investigated as flowering genes. Phylogenetic tree for OG5_127186 (A), OG5_139532 (B), OG5_163423 (C), OG5_131236 (D), OG5_147254 (E), OG5_146543 and OG5_126706 (G) is shown.

doi:10.1371/journal.pone.0038250.g004

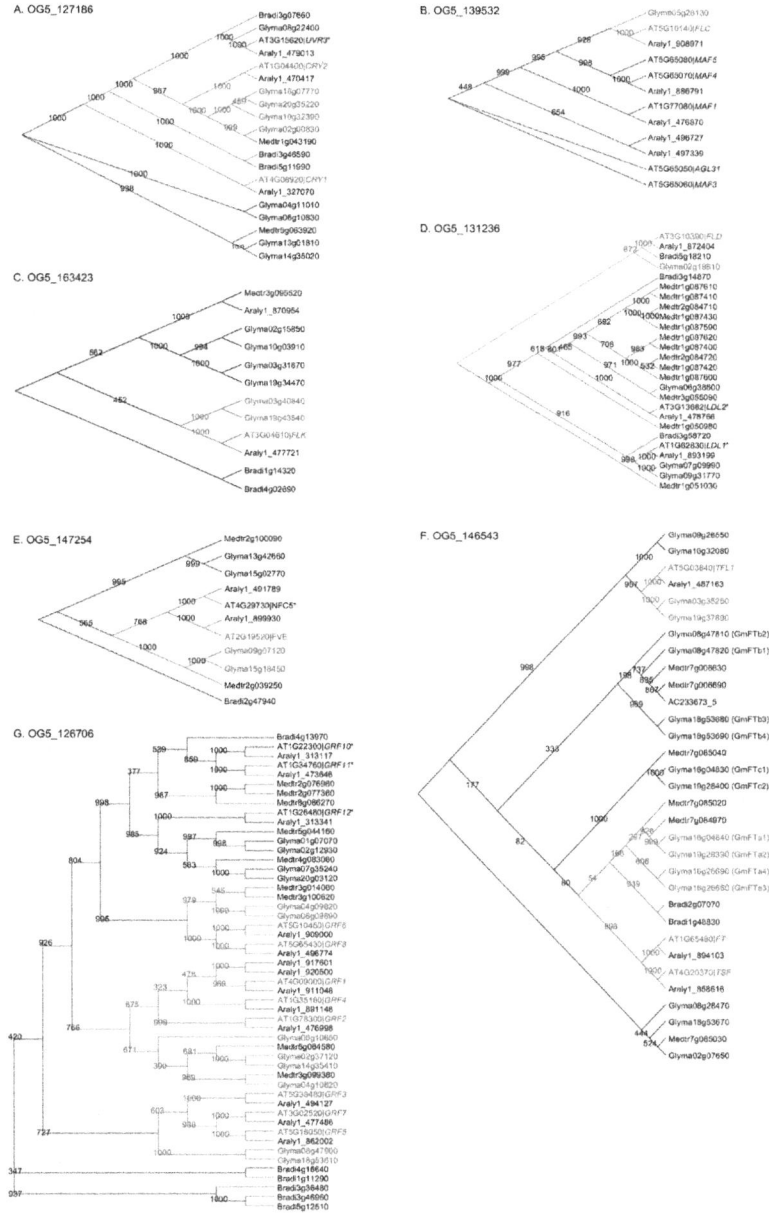

Figure 4: Phylogenetic relationship between soybean and Arabidopsis genes in each orthologue group.

Vernalisation pathway.

Vernalisation involves plants that require prolonged periods of low temperature to initiate flowering. The vernalisation pathway in Arabidopsis involves 32 genes clustered into 23 OGs (Table 1). Among these, 30 Arabidopsis genes in 21 OGs have 81 soybean orthologue candidates (Table 1), of these 81 genes, 71 show evidence of transcription (Dataset S2). However, the orthologous counterparts of the Arabidopsis VERNALISATION INSENSITIVE 3(VIN3) gene in OG5_AT5G57380 and AGAMOUS-LIKE24 (AGL24) gene in OG5_AT4G24540 were not identified in soybean (Table S1) by this method. Nonetheless, a BLAST analysis suggests the potential existence of their soybean orthologues (see below and the Discussion). Among the OGs containing Arabidopsis genes associated with the vernalisation pathway, the ratio of the number of soybean genes to that of Arabidopsis genes is highest in OG5_212406, in which the ratio is 14 soybean genes to 1 Arabidopsis gene, REDUCED VERNALISATION RESPONSE 1 (VRN1) (Table S1). In contrast, the six Arabidopsis genes in OG5_139532, which includes a MADS-box transcription factor gene, FLOWERING LOCUS C (FLC), that negatively regulates flowering [25], share only one soybean gene as a putative orthologue (Glyma05g28130), resulting in the lowest soybean-to-Arabidopsis gene count ratio among the vernalisation-related OGs (Table S1). In the phylogenetic tree of OG5_139532, Glyma05g28130 is most closely related to FLC (Figure 4B). Interestingly, no Medicago and Brachypodium genes are found in this OG. As mentioned above, VIN3 (in OG5_AT5G57380), which is a repressor of FLC in cold temperatures [26], and a flowering promoter gene, AGL24[27], [28], [29] [reviewed by Alexandre and Hennig (2008) [30]] in OG5_AT4G24540, are not assigned with putative soybean orthologues (Table S1), but share closely related soybean genes with other flowering genes (see below and Discussion).

Autonomous pathway.

Autonomous pathways in plants are activated in response to endogenous changes that are independent from the environmental cues leading to flowering [31]. There are 17 genes, grouped into 16 OGs, involved in the Arabidopsis autonomous pathway (Table 1). Each OG has a single Arabidopsis gene that is known to be functional during floral initiation, except for OG5_129164, which contains two Arabidopsis flowering genes: CURLY LEAF (CLF) andSWINGER (SWN) (Table S1). Three other OGs (OG5_147254, OG5_127148 and OG5_131236) also include one or two additional Arabidopsis genes, raising the total number of Arabidopsis genes in the autonomous pathway-related OGs to 23 (Table 1). The total number of orthologous soybean genes to the 17 Arabidopsis genes (or 23 if the additional genes are included) is 49, of which 46 genes are transcriptionally active (Table 1). OG5_163423 has six soybean genes that are orthologous to AT3G04610 [FLOWERING LOCUS KH DOMAIN(FLK)], a repressor of FLC expression [32], which is the highest soybean-to-Arabidopsis gene count ratio among the OGs for autonomous pathways. The subsequent phylogenetic tree analyses revealed that only two soybean genes (Glyma03g40840 and Glyma19g43540) are located in the same clade with Arabidopsis FLK, indicating that they are likely true orthologues of FLK (Figure 4C). Similarly, in OG5_131236 and OG5_147254, only one (Glyma02g18610) and two (Glyma15g18450 and Glyma09g07120) soybean genes, respectively, are found and thus are also likely to be true orthologues of their Arabidopsis counterparts involved in autonomous pathways (Figure 4D,E). OG5_131236 has three Arabidopsis genes, includingFLOWERING LOCUS D (FLD), which down-regulates FLC and has Glyma02g18610 as its closest orthologue according to the phylogenetic tree (Figure 4D), and FVE [also known asMULTICOPY SUPPRESSOR OF IRA1 4 (MSI4)] in OG5_147254, which also down-regulatesFLC and has Glyma15g18450 and Glyma09g07120 as its closest orthologues (Figure 4E). In comparison, four soybean orthologue candidates of Arabidopsis FPA, which has a redundant role with FLD, FVE, and LD [33], are equally distant from their Arabidopsis counterpart (data not shown). Because the minimum number of sequences for the generation of a phylogenetic tree is four, we are unable to generate phylogenetic trees for four OGs (OG5_128052, OG5_155119, OG5_169591 and OG5_170601) (Table S1). Therefore, all of the soybean genes in these OGs are regarded as the closest homologues of the Arabidopsis genes contained in the corresponding groups. Each of VEL2, VEL3 and VIN3 are grouped into a singleton OG and are not assigned orthologous counterparts in soybean (Table S1) but do have homologous genes in soybean according to the direct BLAST analysis (see below andDiscussion).

Ambient temperature pathway.

Plants respond to ambient temperature changes to modulate their flowering times [34]. The ambient temperature pathway in Arabidopsis involves 16 genes that are clustered into 8 OGs that have 38 soybean genes in total (Table 1). Three OGs (OG5_131236, OG5_147254 and OG5_155119) are also involved in autonomous pathways, and the Arabidopsis genes contained in OG5_139532, OG5_129661 and OG5_177438 are also involved in its vernalisation pathway (Table S1). In most of the OGs related to the ambient temperature pathway, the numbers of soybean genes are greater than those of

Arabidopsis genes; however, the opposite findings are observed in the cases for OG5_139532 and OG5_190004. OG5_139532 contains six Arabidopsis genes (including FLC) that are orthologous to only one soybean gene, Glyma05g28130 (see above and Figure 4B). Similarly, Glyma01g36810 is the only soybean orthologue of the Arabidopsis genes AT4G35900 (FD) and AT2G17770 (FDP) in OG5_190004, which encode for the basic leucine zipper (bZIP) domain protein and positively regulate flowering [35]. Arabidopsis AT4G16280 (FCA) in OG5_155119 has one putative soybean orthologue (Glyma17g03960) (Table S1).

Meristem identity genes.

Meristem identity genes are activated by upstream pathways and initiate floral development by triggering the transition of the apical meristem from the vegetative phase to the reproductive phase [12]. Seven Arabidopsis genes, including FD, LFY, SQUAMOSA PROMOTER BINDING PROTEIN-LIKE 3 (SPL3), AP1 and AGL8 [also known as FRUITFULL (FUL)], are involved in this role and are clustered in five OGs (Table 1 and Table S1). The total number of soybean genes clustered within these OGs is 19, 18 of which were expressed in either the transcriptome data [8] or [9] (Table 1). SPL3 in OG5_213105, which positively regulates FT, AP1 and LFY in Arabidopsis [36], has four putative soybean orthologues (Glyma07g31880, Glyma13g24590, Glyma13g31090 and Glyma15g08270). Arabidopsis LFY (AT5G61850) in OG5_170760 has three orthologous counterparts in soybean (Glyma04g37900, Glyma06g17170 and Glyma20g19600) (Table S1). AP1 and CAULIFLOWER (CAL), which are important in initiating flowering, are grouped into OG5_170388, which contains four soybean genes (Table S1).AGL8 (or FUL) in OG5_211687, which is also important for the initiation of flowering [37], [38], has 7 putative soybean orthologues (Table S1).

Flowering pathway integrators.

Genes of flowering pathway integrators integrate signals from several related pathways and determine the exact timing of flowering [2], [3]. In this study, 36 Arabidopsis flowering pathway integrator genes, including FT, LFY, FLC, CO and SOC1, were grouped into 11 OGs (Table 1and Table S1). Among the 82 soybean genes grouped into these same 11 OGs, 69 genes demonstrated expression in at least one of the recent transcriptome datasets of SoyBase [8] or Libault et al. [9] (Table 1). As FT, FLC, LFY and CO are also involved in other flowering pathways, their relationships with soybean counterparts are described above. OG5_169591, which contains the REDUCED VERNALIZATION RESPONSE 2 (VRN2) gene, includes two soybean genes (Glyma01g41460 and Glyma11g03960). OG5_146543 consists of three Arabidopsis genes [FT, TWIN SISTER of FT (TFT) and TERMINAL FLOWER 1 (TFL1)] and 17 soybean genes. These 17 soybean genes include all 10 genes identified as soybean counterparts of Arabidopsis FT in two recent studies [6], [39], validating our approach for the identification of orthologues. The phylogenetic tree for OG5_146543 is also in agreement with that produced by Hecht and colleagues, which sub-classed 10 soybean FT homologues intoFTa, FTb and FTc (Figure 4F) [39]. Eleven members of the NUCLEAR FACTOR Y transcription factors separate into three OGs (OG5_127594, OG5_128092 and OG5_152404) and have 34 soybean homologues (Table S1). The SUPPRESSOR OF OVEREXPRESSION OF CO 1 (SOC1) in OG5_190130 and EARLY FLOWERING 9 (ELF9) in OG5_128874 have four and three putative soybean orthologues, respectively (Glyma03g02200, Glyma07g08830, Glyma09g40230 and Glyma18g45780 for SOC1; Glyma03g27010, Glyma10g31450 and Glyma20g36110 for ELF9) (Table S1). OG5_126706 contains 11 Arabidopsis genes that encode 14-3-3 proteins that are involved in various processes, including signal transduction[40]; 8 of these 11 genes have demonstrated proven roles as flowering pathway integrators in previous studies (Table 1). In soybean, 12 genes are found in the same OG as the putative orthologues of these 11 Arabidopsis genes. In the phylogenetic tree analysis, the GENERAL REGULATORY FACTOR 10 (GRF10), GRF11 and GRF12 genes branch out on their own along with four soybean genes (Figure 4G). Interestingly, these have not yet been investigated for their role as flowering genes even though they group together with eight other 14-3-3 protein genes (that are known as flowering genes) in OG5_126706.

Comparison with Other Species

As Medicago truncatula is another important model species of the legume family that has also been extensively sequenced and studied, we applied the same methods used for the identification of orthologues of Arabidopsis genes to the annotated protein sequences of Medicago (Mt 3.5 annotation) (Dataset S1). Along with Medicago, we also applied the orthologue identification method to Arabidopsis lyrata (A. lyrata) and Brachypodium distachyon(Brachypodium). The numbers of putative orthologues of Arabidopsis genes involved in key flowering pathways in each species are shown in Table 1. Although the number of Medicago genes in the Mt3.5 annotation (version 3) is similar to that of soybean genes in G.max 1.09, smaller numbers of Medicago genes were grouped into the same OGs that contain key flowering genes of Arabidopsis. In particular, the numbers of putative Medicago orthologues of Arabidopsis genes involved in the light signalling pathways, meristem identity and coding for flowering pathway integrators were less than half of the number of soybean genes in these same pathways (Table 1). The number of Brachypodium and A. lyrata genes in each of OGs is also comparable with that of A. thaliana, even though the genome size varies from ~130 Mb (A. thaliana) to ~270Mb (Brachypodium) [41]. (Table 1).

Comparison with Glycine soja

Recently, the draft genome sequence of Glycine soja (G. soja), wild soybean, was released[42]. As Kim et al. identified structural variations, such as large deletions, inversions and insertions, between G. max and G. soja genomes by mapping short reads of G. soja against the G. max genome sequence [42], we determined how many of the G. max genes involved in flowering pathways harbour such structural variations. Of the 1,538 genes in G. max associated with large deletions when promoter regions (the 1 kb region upstream) are taken into account, 10 genes are involved in flowering, 7 of which are involved in key flowering pathways. Similarly, 11 out of the 689 genes associated with large insertions are involved in flowering, but only 4 of them are associated with key flowering pathways (Table S3).Kim et al. also compared the single nucleotide polymorphisms (SNPs) in G. soja to those in the G. max genome [42]. A total of 4,187 SNPs were found within the genic regions of 405 G. soja genes that are counterparts of putative flowering genes of G. max. The number of flowering genes in G. soja that contain SNPs increases to 458 when the 1 kb upstream is included as the promoter region. However, a substantial number of these flowering genes, 182 genes (39.8% of 458 genes), have SNPs only in non-protein coding regions and/or promoters. For example, among the 17 G. soja genes corresponding to G. max genes that are homologous to the Arabidopsis FT, TSF and TFL1(OG5_146543) genes, only five have SNPs in CDSs: Glyma08g47820, Glyma16g04840, Glyma03g35250, Glyma09g26550 and Glyma16g32080. Of these, Glyma08g47820 and Glyma16g04840 were named as GmFT6 and GmFT3a, respectively, by Kong et al. [6]. However, neither GmFT6 nor GmFT3a expression is detected in either of the transcriptome datasets [8], [9].

A subsequent analysis of the proportions of genes containing structural variations or SNPs indicated that these mutations are not particularly enriched or depleted in flowering genes (data not shown).

Soybean Genes with More or Less Paralogues than in Arabidopsis

Soybean flowering genes with more or less paralogues than in Arabidopsis. We subsequently focused on the soybean OGs that are potentially involved in flowering pathways and have significantly more or less paralogues than the corresponding Arabidopsis genes (paralogue-rich and paralogue-less groups, respectively). As soybean underwent additional rounds of whole genome duplication events compared to Arabidopsis, the ratio of soybean gene counts against Arabidopsis gene counts per group is 2.5 on average, which suggests that each OG has 2-3 times more soybean genes than Arabidopsis genes (Figure 5). However, in 8 OGs, the number of soybean genes is far greater than that of Arabidopsis genes (Table 2). The Arabidopsis genes in these groups include PIF3, CDF2/3, EEL, AREB3, AP2,AP3, SVP, AGL8 (or FUL) and VRN1 (Table 2). The majority of the soybean genes in the paralogue-rich groups are transcriptionally active according to the two recent transcriptome datasets for soybean [8], [9] (Dataset S2). However, the hierarchical clustering of the expression profiles of the 82 soybean genes in these 8 paralogue-rich OGs is not in agreement with the gene grouping based on the sequence similarity (i.e., the grouping of OGs), suggesting that the paralogues have diverged in terms of function (Figure 6).

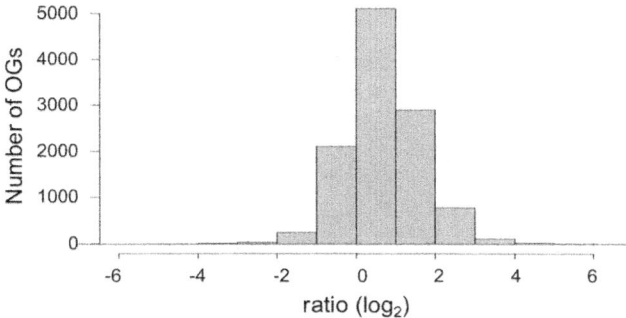

Figure 5: Ratio of soybean gene count to Arabidopsis gene count per OG.

On average, the number of soybean genes per OG is 2-3 times of Arabidopsis genes, but 300 OGs have far more soybean genes (more than 2-standard deviation from average), hence designated as paralogue-rich OGs. doi:10.1371/journal.pone.0038250.g005.

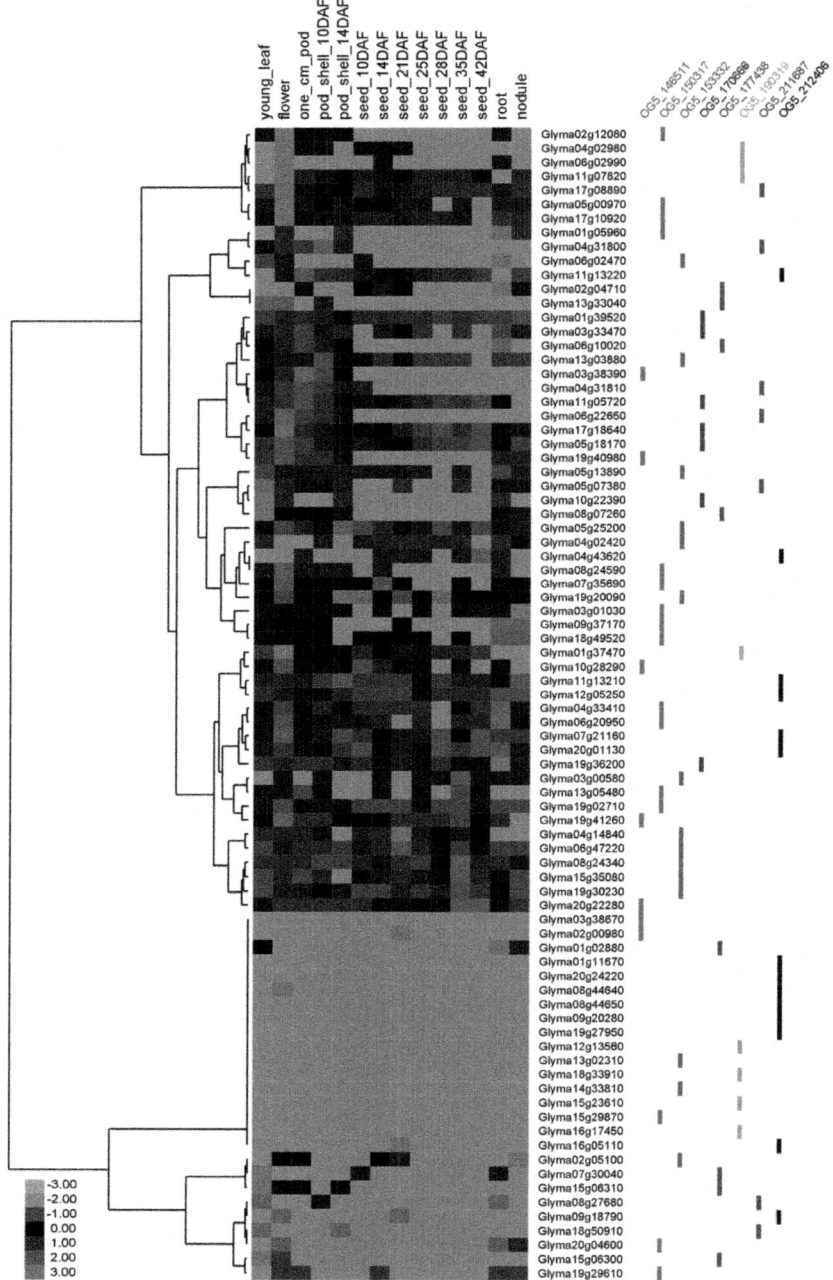

Figure 6: Expression profiles of soybean genes in eight paralogue-rich ortho-logue groups.

The hierarchical clustering of the expression profiles for 82 soybean genes shows that genes in the same group do not always have similar expression patterns, indicating functional divergence among paralogues. The expression data were extracted from the soybean transcriptome data in SoyBase [8]. DAF: Days After Flowering.doi:10.1371/journal.pone.0038250.g006

Table 2: List of paralogue-rich OGs containing Arabidopsis flowering genes

OrthoMCL ID/soybean gene	Arabidopsis gene/description	PFAM	Panther
OG5_139532/ Glyma05g28130	AT1G77080: MAF1 (MADS AFFECTING FLOWERING 1); transcription factor	K-box region	MADS BOX PROTEIN
	AT5G10140: FLC (FLOWERING LOCUS C); specific transcriptional repressor/transcription factor	K-box region	MADS BOX PROTEIN
	AT5G65050: AGL31 (AGAMOUS LIKE MADS-BOX PROTEIN 31); transcription factor	K-box region	MADS BOX PROTEIN
	AT5G65060: MAF3 (MADS AFFECTING FLOWERING 3); transcription factor	K-box region	MADS BOX PROTEIN
	AT5G65070: MAF4 (MADS AFFECTING FLOWERING 4); transcription factor	K-box region	MADS BOX PROTEIN
	AT5G65080: MAF5 (MADS AFFECTING FLOWERING 5); transcription factor	K-box region	MADS BOX PROTEIN
OG5_189849/ Glyma20g39140	AT1G50680: AP2 domain-containing transcription factor, putative	B3 DNA binding domain	-
	AT1G51120: AP2 domain-containing transcription factor, putative	B3 DNA binding domain	-
OG5_190004/ Glyma01g26810	AT2G17770: ATBZIP27; transcription factor	-	-

In contrast to the results for the paralogue-rich OGs, the soybean gene counts are smaller than the Arabidopsis gene counts for the three paralogue-less OGs (OG5_139532, OG5_189849 and OG5_190004) (Table 3). Furthermore, there are 20 Arabidopsis flowering genes in 18 OGs that are not assigned putative soybean orthologues by our method (Table S1). Among these genes are Arabidopsis VEL2, VEL3 and AGL24. VEL2 and VEL3 belong to a small gene family of plant homeodomain (PHD) finger-containing proteins that coordinate flowering through epigenetic regulation [43], [44], while AGL24 is one of the MADS-box genes found to promote flowering by integrating flowering signals from several floral pathways [27], [28], [29]. However, BLAST searches of these Arabidopsis flowering genes against all of the annotated soybean genes reveal that they do have homologous soybean genes. In spite of this, the soybean genes that best match these Arabidopsis genes are putative soybean orthologues of other Arabidopsis genes (Table 4). Table 4 includes Arabidopsis genes in OGs that do not contain any other soybean gene members besides the best-matching soybean gene in the BLAST results. All of the soybean genes bearing a sequence similarity with Arabidopsis genes in OGs lacking other soybean gene members (with BLAST e-values less than 1e-10) are provided in the Dataset S4.

Table 3: List of OGs for flowering genes containing less number of paralogues in soybean.

OrthoMCL ID/soybean gene	Arabidopsis gene/description	PFAM	Panther
OG5_139532/ Glyma05g28130	AT1G77080: MAF1 (MADS AFFECTING FLOWERING 1); transcription factor	K-box region	MADS BOX PROTEIN
	AT5G10140: FLC (FLOWERING LOCUS C); specific transcriptional repressor/transcription factor	K-box region	MADS BOX PROTEIN
	AT5G65050: AGL31 (AGAMOUS LIKE MADS-BOX PROTEIN 31); transcription factor	K-box region	MADS BOX PROTEIN
	AT5G65060: MAF3 (MADS AFFECTING FLOWERING 3); transcription factor	K-box region	MADS BOX PROTEIN
	AT5G65070: MAF4 (MADS AFFECTING FLOWERING 4); transcription factor	K-box region	MADS BOX PROTEIN
	AT5G65080: MAF5 (MADS AFFECTING FLOWERING 5); transcription factor	K-box region	MADS BOX PROTEIN
OG5_189849/ Glyma20g39140	AT1G50680: AP2 domain-containing transcription factor, putative	B3 DNA binding domain	-
	AT1G51120: AP2 domain-containing transcription factor, putative	B3 DNA binding domain	-
OG5_190004/ Glyma01g36810	AT2G17770: ATBZIP27; transcription factor	-	-
	AT4G35900: FD; DNA binding/protein binding/transcription activator/transcription factor	-	CYCLIC-AMP-DEPENDENT TRANSCRIPTION FACTOR ATF-6

doi:10.1371/journal.pone.0038250.t003

Table 4: Arabidopsis flowering genes in OGs with no soybean genes members and their best BLAST-hit soybean genes

Arabidopsis genes	Symbol/annotation	Best BLAST-hit soybean gene	OG ID for soybean gene	Arabidopsis genes in OG
AT5G67180	AP2 domain-containing transcription factor, putative	Glyma17g18640	OG5_170666	AP2
AT2G35670	FERTILIZATION INDEPENDENT SEED 2 (FIS2)	Glyma01g41460	OG5_169591	VRN2
AT5G27220	protein transport protein-related	Glyma05g21790	OG5_170932*	AT5G48385
AT5G62040	BFT (brother of FT and TFL1 protein)	Glyma16g32080	OG5_146543	FT;TSF;TFL1
AT2G46790	ARABIDOPSIS PSEUDO-RESPONSE REGULATOR 9 (APRR9)	Glyma04g40640	OG5_178368	APRR5
AT4G34000	ABSCISIC ACID RESPONSIVE ELEMENTS-BINDING FACTOR 3 (ABF3)	Glyma02g14880	OG5_144915	ABF1;ABF2;ABF4
AT2G39250	SCHNARCHZAPFEN (SNZ)	Glyma15g04930	OG5_153389	TOE1
AT3G54990	SCHLAFMUTZE (SMZ)	Glyma15g04930	OG5_153389	TOE1
AT1G26790	Dof-type zinc finger domain-containing protein	Glyma18g49520	OG5_150317	CDF3;CDF2
AT1G69570	Dof-type zinc finger domain-containing protein	Glyma01g05960	OG5_150317	CDF3;CDF2
AT2G24790	CONSTANS-LIKE 3 (COL3)	Glyma06g06300	OG5_144994	ATCOL5;ATCOL4
AT2G47700	zinc finger (C3HC4-type RING finger) family protein	Glyma20g38050	OG5_178422?	AT3G05545
AT2G18870	VERNALIZATION5/VIN3-LIKE (VEL3)	Glyma07g09800	OG5_190237	VRN5
AT2G18880	VERNALIZATION5/VIN3-LIKE (VEL2)	Glyma17g07000	OG5_170648	VEL1
AT3G30260	AGAMOUS-LIKE 79 (AGL79)	Glyma16g13070	OG5_170388	AP1;CAL
AT4G16810	VEFS-Box of polycomb protein	Glyma01g41460	OG5_169591	VRN2
AT4G24540	AGAMOUS-LIKE 24 (AGL24)	Glyma01g02880	OG5_177438	SVP
AT5G42910	basic leucine zipper transcription factor (BZIP15)	Glyma04g04170	OG5_144915	ABF1;ABF2;ABF4
AT5G57380	VERNALIZATION INSENSITIVE 3 (VIN3)	Glyma17g07000	OG5_170648	VEL1
AT5G60120	TARGET OF EARLY ACTIVATION TAGGED (EAT) 2 (TOE2)	Glyma12g07800	OG5_153389	TOE1

*Not among the OGs containing Arabidopsis flowering genes.
doi:10.1371/journal.pone.0038250.t004

Genomic distribution of genes for paralogue-rich groups and one genomic region harbouring numerous homologues of specific Arabidopsis genes.

After we observed that some of the Arabidopsis flowering genes have more or less copies in soybean than average, we particularly expanded the investigation of paralogue-rich groups to the genomic scale. The 300 paralogue-rich OGs that have significantly more numbers of soybean genes (more than 2 standard deviations above the average) contain 4,236 soybean genes, which are spread evenly across the genome. On average, these genes comprise 6–16% of the total number of genes on each chromosome, and their fraction does not have any obvious correlation with the percentage of recently duplicated segments collected from the PHYTOZOME website (www.phytozome.net) (Pearson's Correlation Coefficient: -0.08) (Figure 7A). However, soybean chromosome 16 (Chr16) is exceptionally enriched for genes belonging to paralogue-rich OGs, containing 292 genes, which is 16.2% of the total of genes on Chr16. In particular, 76 of these 292 genes are condensed within a 1.4 Mb region on Chr16 (Figure 7B). These 76 genes are associated with only two OGs: OG5_134835 and OG5_170470. The numbers of total paralogues of these OGs in soybean are 115 and 20, respectively, while Arabidopsis has only one (AT2G34930) and two genes (AT2G44290 and AT2G44300) in OG5_134835 and OG5_170470, respectively. AT2G34930 in OG5_134835 is known as a disease-resistance family protein and is involved in a defence response to fungus and signal transduction (TAIR, www.arabidopsis.org/servlets/TairObje ct?id=32293&type=locus).

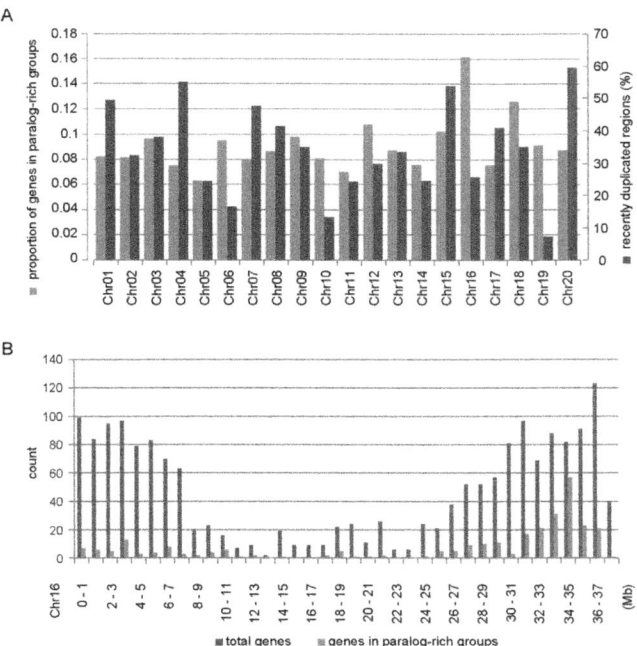

Figure 7: Distribution of genes in paralogue-rich groups in whole genome (A) and chromosome 16 (B).(A) Paralogue-rich orthologue groups and the percentage of recently duplicated segments per chromosome seems to have no apparent correlation.

(B) In chromosome 16, a large number of the genes in paralogue-rich groups, mostly from OG5_134835 or OG5_170470, are condensed in the region around 34-35Mb (see the main text).

doi:10.1371/journal.pone.0038250.g007

Two Arabidopsis genes in OG5_170470 (AT2G44290 and AT2G44300) are annotated as protease inhibitor/seed storage/lipid transfer protein (LTP) family proteins (AT2G44290: TAIR,www.arabidopsis.org/servlets/TairObject?id=33037&type=locus; AT2G44300: TAIR,www.arabidopsis.org/servlets/TairObject?id=33039&type=locus).

DISCUSSION

In this study, we identified potential soybean orthologues of most of the Arabidopsis genes with an emphasis on those that are involved in flowering pathways by associating soybean and Arabidopsis genes based on the current soybean annotation (G.max 1.09) in conjunction with the pre-defined Arabidopsis OGs in the OrthoMCL database [11]. Although the current soybean annotation assigned a closest Arabidopsis gene to nearly all of the predicted soybean genes, only 55% of the total Arabidopsis genes in the TAIR9 annotation are associated with soybean genes. However, the combined methodology used in this study, which incorporates the orthologue-group based results and the current soybean annotation, increased the number of Arabidopsis genes that are assigned with putative soybean orthologues to 20,730, which is more than 75% of the total Arabidopsis genes in the TAIR9 annotation. Nevertheless, it is difficult to determine the true soybean orthologues of Arabidopsis genes, especially when an OG contains multiple Arabidopsis genes. Thus, we resolved this by constructing phylogenetic trees for the candidates concerned. Conversely, the 25% of Arabidopsis genes that failed to have putative soybean orthologues by our analysis are not necessarily absent in soybean. In fact, all of the 20 Arabidopsis flowering genes that are not assigned with potential soybean counterparts have homologous soybean genes with significant e-values (<1e-10) when we subsequently performed BLASTP analyses with them (Table 4 and Dataset S4). However, the orthologue identification method used in this study determined that all of the top-matching soybean genes for these 20 Arabidopsis flowering genes (by BLASTP) were also candidate orthologues of other Arabidopsis genes. This finding suggests that the sequences concerned may have diverged beyond the sensitivity of our orthologue detection method; therefore, future functional analyses of these genes are necessary to confirm their orthology. In addition, it should be also noted that the false positive and false negative rates of OrthoMCL algorithm are 0.17 and 0.06, respectively, even though the OrthoMCL is among the best performing orthology identification tools [45]. Thus, failure of the orthologue identification for a subset of Arabidopsis genes may also due to the error in OrthoMCL-DB.

Using an approach that associates a group of soybean genes with a group of Arabidopsis genes that has the same OG ID also enabled us to investigate which soybean genes have a statistically higher or lower number of paralogues (or copies) in comparison to their Arabidopsis counterparts. For example, the analysis of paralogue-rich and paralogue-less soybean genes can be expanded to whole genes. Based on the near log-normal distribution of soybean-Arabidopsis gene count ratio per OG (mean: 1.03 and standard-deviation: 0.88 in $\log_2$), the number of soybean genes are much higher for 300 OGs (7 time or more than Arabidopsis genes) and smaller for 304 OGs (half or less than Arabidopsis genes) (Figure 5). A preliminary Gene Ontology analysis on the Arabidopsis genes in paralogue-rich and paralogue-less OGs (i.e., those that have far more and less copies in soybean than average, respectively) showed interesting features. Three hundred and ninety nine (399) Arabidopsis genes in 300 paralogue-rich OGs are enriched for 'response to auxin stimulus (GO:0009733)', 'defense response (GO:0006952)', 'response to wounding (GO:0009611)' and 'lipid transport (GO:0006869)' (Table S4). While the 1,946 Arabidopsis genes that have much less copies in soybean (paralogue-less OGs) are also enriched for 'defense response (GO:0006952)', they are mainly enriched for other GO terms, such as 'intracellular signaling cascade (GO:0007242)' (Table S4). The outcome raises questions as to whether the differential accumulation of gene copies between soybean and Arabidopsis is a possible evolutionary innovation that distinguishes the two species from one another.

It is known that duplicated genes (i.e., genes in the same OG) do not necessarily retain the same functions. In our analysis, we found paralogues with diverged expression patterns as well as some more highly related paralogues located within the same clades in the phylogenetic tree and exhibiting similar expression patterns and, hence, more likely to retain similar functions (Figure 6 and Figure 8). Intriguingly, there exist some OGs that consist of soybean genes that are all non-tandem duplicates but with similar expression patterns (for example OG5_150317 inFigure 8). This begs the question as to how the soybean genome keeps copies of the same genes with presumably the same function in different places in the genome, especially because we observed no correlation between the fraction of genes in paralogue-rich OGs in each chromosome and the percentage of recently duplicated segments in the respective chromosomes (Figure 7A).

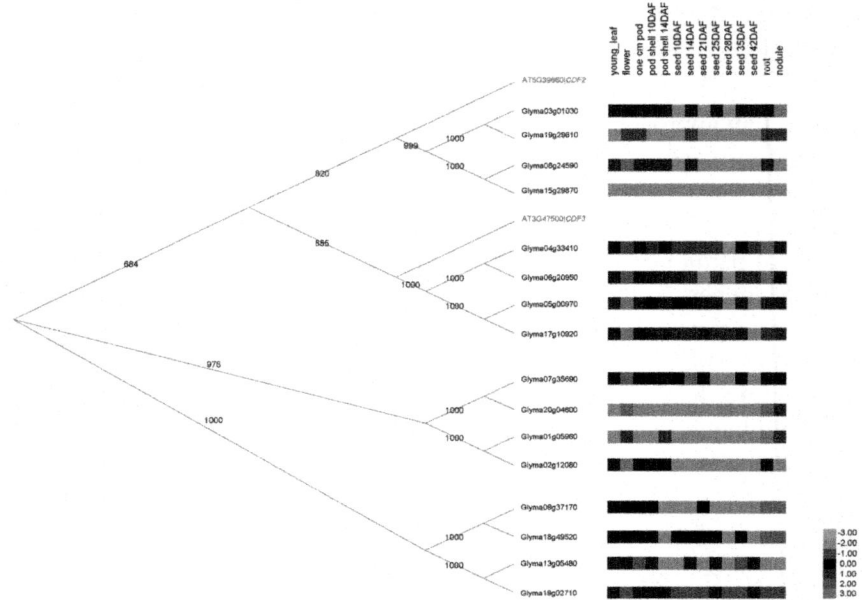

Figure 8: Expression patterns of soybean genes in OG5_150317.While the expression patterns of the 16 soybean genes in OG5_150317 were spread in different clusters as shown in Figure 6, but those that are in the same clade of the phylogenetic tree tend to have similar expression profiles. The expression data were extracted from the soybean transcriptome data in SoyBase [8]. DAF: Days After Flowering.

doi:10.1371/journal.pone.0038250.g008.

Soybean is a short-day species that does not require vernalisation to induce flowering [46]. It is, therefore, intriguing to observe that some of the Arabidopsis genes involved in the vernalisation pathway (e.g., FLC, VRN1 and VRN2) are represented in soybean. A previous study failed to identify any FLC genes, a key regulator of the vernalisation pathway, in Medicago truncatula, soybean, or Lotus japonicus [18], which can be attributed to the more incomplete genome sequence at the time; however, our analysis grouped Arabidopsis FLC along with 5 other Arabidopsis genes and one soybean gene (Glyma05 g28130) into the same OG (OG5_139532), in which the soybean gene and FLC have the closest relationship (Figure 4B). VRN1 forms part of a chromatin-modifying polycomb that is involved in the methylation of histone 3 lysine 9 (H3K9) and histone 3 lysine 27 (H3K27) and, hence, the repression of FLCexpression [47], [48]. VRN1 belongs to a paralogue-rich OG that has far more soybean orthologue candidates than the majority of Arabidopsis genes, although other Arabidopsis genes that mediate vernalisation responsiveness, such as VIN3, are not assigned any putative soybean orthologues. SVP in OG5_177438, which encodes a MADS box transcription factor and is a negative regulator of flowering in Arabidopsis [49], has 8 putative soybean orthologues, while its close homologue AGL24 is not assigned a candidate soybean orthologue. It is likely that the soybean genes in OG5_177438 may have acquired and replaced the function of AGL24 because the most homologous soybean gene of Arabidopsis AGL24 is one of the putative soybean orthologues of Arabidopsis SVP (Table 4).

Two microRNAs (miRNAs), miR156 and miR172, are involved in the regulation of SQUAMOSA PROMOTER BINDING PROTEIN LIKE family genes and APETALA2-LIKE (AP2-like) transcription factors in Arabidopsis [50] and are conserved in soybean [51]. Zhang et al. also concluded that the SPL family genes and AP2-like transcription factors are among the predicted target genes of miR156 and miR172, respectively, in soybean [51]. However, whether or not miR172 plays a similar role in soybean flowering still needs to be investigated.

The functional analysis of genes through reverse genetics approaches is more complicated for soybean than Arabidopsis. Computational analyses such as our study can therefore pinpoint the putative soybean orthologues of Arabidopsis genes with known functions. Indeed, putative soybean orthologues of FT and TFL1 were first identified via computational analysis [18], [52]. This study determines all of the soybean genes that are putative orthologues of Arabidopsis genes by first grouping Arabidopsis and soybean genes into pre-defined orthologue groups and then associating genes in the same group from each species. This method determined not only the inter-species relationship of genes between soybean and Arabidopsis but also the intra-species relationships of genes in terms of their sequence similarities. Subsequently, the most probable soybean orthologues for Arabidopsis genes, especially those involved in key flowering pathways, were inferred through phylogenetic tree analyses. These inferences were also strengthened by referring to publicly available transcriptome datasets for the expression profile of the soybean genes. As more soybean transcriptome data becomes available in the future, it is expected that the putative soybean orthologues of Arabidopsis flowering genes can be more precisely compared by combining the data on their sequence similarities and expression patterns. Additionally, our methods found that 24 Arabidopsis genes, which had not been previously investigated for their roles in flowering, belong to OGs with known Arabidopsis flowering genes. Although sequence similarity does not always indicate functional similarity, these Arabidopsis genes may well be involved in the initiation of flowering. In summary, our study has identified numerous floral regulatory candidate genes in soybean.

Further studies of the genes identified here will provide a new perspective on the molecular processes underlying the floral transition process in soybean.

Materials and Methods

Sequence Data

The protein sequences of annotated soybean genes (G. max 1.09) were downloaded from the PHYTOZOME website (www.phytozome.net). The protein sequences of annotated Arabidopsis genes (TAIR9 release) were downloaded from The Arabidopsis Information Resource (TAIR) website (www.arabidopsis.org). The protein sequences of Medicago genes (Mt3.5 version 3) were downloaded from J. Craig Venter Institute website (www. jcvi.org). For those that have splicing variants, the longest isoforms were selected. Peptide sequences of Brachypodium distachyon were downloaded from PHYTOZOME website, then the longest peptide sequence for each locus was extracted for the analysis [41]. Peptide sequence of filtered Arabidopsis lyrata gene model which best represent each locus is downloaded from PHYTOZOME website[53].

Soybean Homologue Identification and Grouping Orthologous Genes

The current soybean annotation information (G.max 1.09) associates 44,818 out of 46,367 soybean genes to 15,113 Arabidopsis genes. In order to identify soybean orthologues of more Arabidopsis genes, we assigned orthologue group (OG) IDs pre-defined in OrthoMCL database (release 5.0) [11] to Arabidopsis genes. OrthoMCL-DB is a list of OG IDs and the genes under same OG IDs from multiple species. OrthoMCL algorithm examines the all-versus-all BLAST search result and use the Markov Clustering algorithm to find interspecies homologues (orthologues) and intraspecies homologues (paralogues) [11]. Soybean genes that have homologous Arabidopsis genes in the current annotation information were given the same OG ID with the corresponding Arabidopsis genes, and those that do not were assigned OG IDs by OrthoMCL. An OG IDs was assigned to each gene, if possible, via the web-based tool in the OrthoMCL website (http://www.orthomcl.org), which considers the BLAST-hit quality of the input protein sequence to the protein sequences in the OrthoMCL-DB to find the OG containing the closest protein to the input. For the 317 Arabidopsis genes that have homologous soybean genes according to the current soybean annotation but do not have OG IDs assigned by OrthoMCL, arbitrary OG IDs were given, which are 'OG5_' followed by the Arabidopsis gene locus name (e.g., OG5_AT2G33835 for an Arabidopsis gene AT2G33835). Among these, 9 groups have Arabidopsis flowering genes, which are AT2G33835, AT2G18880, AT2G18870, AT3G30260, AT4G16810, AT4G24540, AT5G57380, AT5G42910 and AT5G60120. Then, for each OG, the soybean gene members were regarded as putative soybean orthologues of Arabidopsis gene members.

Transcriptional Activity and Gene Expression Data Analysis

The expression profiles of the 491 putative soybean orthologues of Arabidopsis flowering genes were extracted from SoyBase [8] and the integrated transcriptome atlas of soybean generated by Libault et al. (2010) [9]. For the SoyBase data, a transcriptionally active gene is a gene that has at least two or more sequence reads at one or more of the tested tissues/developmental stages. For the dataset of Libault et al. (2010) [9] any gene that has normalized read counts greater than 0 at least in one tissue/developmental stage is regarded as a transcriptionally active gene. The hierarchical clustering of the expression patterns was performed by CLUSTER 3.0 using the Z-scores of the normalized read counts (bonsai.ims.u-tokyo.ac.jp/~mdehoon/software/cluster), and the clustered results were visualized by Java Treeview (http://rana.lbl.gov/downloads/TreeView/TreeView_vers_1_60.exe).

Multiple Alignment and Phylogenetic Tree Generation

Multiple sequence alignment was carried out by MUSCLE 3.8.31 [54], and phylogenetic trees were generated by CLUSTALW 2.0.12 [55] with the bootstrap option. Dendroscope 2.7.4 [56]was used for the graphical representation of the phylogenetic trees.

Identification of Closer Soybean Orthologues of Arabidopsis Flowering Genes within OGs

The phylogenetic tree information for between all Arabidopsis and soybean genes within the same OG in Newick format was parsed to decide which soybean genes are closer homologues to Arabidopsis genes involved in flowering pathways using in-house Python scripts. Any soybean gene located in the same clade with Arabidopsis flowering genes is regarded as a soybean homologue involved in flowering pathways, unless the soybean gene is equally close or closer to another Arabidopsis gene that are not explicitly involved in flowering pathways. For the small OGs (less than 4 sequences in total), all soybean genes are regarded as close orthologues of corresponding Arabidopsis flowering genes if the OG has only Arabidopsis flowering genes.

AUTHOR CONTRIBUTIONS

Conceived and designed the experiments: CJ PLB MBS. Performed the experiments: CJ. Analyzed the data: CJ. Contributed reagents/materials/ analysis tools: CJ. Wrote the paper: CJ CEW PLB MBS.

REFERENCES

1.   Fornara F, de Montaigu A, Coupland G (2010) SnapShot: Control of flowering in Arabidopsis. Cell 141: 550, 550 e551–552:

2.   Parcy F (2005) Flowering: a time for integration. Int J Dev Biol 49: 585–593.

3.   Simpson GG, Dean C (2002) Arabidopsis, the Rosetta stone of flowering time? Science 296: 285–289.

4.   Washburn CF, Thomas JF (2000) Reversion of flowering in Glycine Max (Fabaceae). Am J Bot 87: 1425–1438.

5.   Tucker SC (2003) Floral development in legumes. Plant Physiol 131: 911–926.

6.   .Kong F, Liu B, Xia Z, Sato S, Kim BM, et al. (2010) Two coordinately regulated homologs of LOWERING LOCUS T are involved in the control of photoperiodic flowering in soybean. Plant Physiol 154: 1220–1231.

7.   Zhang Q, Li H, Li R, Hu R, Fan C, et al. (2008) Association of the circadian rhythmic expression of GmCRY1a with a latitudinal cline in photoperiodic flowering of soybean. Proc Natl Acad Sci U S A 105: 21028–21033.

8.   .Severin AJ, Woody JL, Bolon YT, Joseph B, Diers BW, et al. (2010) RNA-Seq Atlas of Glycine max: A guide to the soybean transcriptome. BMC Plant Biol 10: 160.

9.   .Libault M, Farmer A, Joshi T, Takahashi K, Langley RJ, et al. (2010) An integrated transcriptome atlas of the crop model Glycine max, and its use in comparative analyses in plants. Plant J 63: 86–99.

10. .Schmutz J, Cannon SB, Schlueter J, Ma J, Mitros T, et al. (2010) Genome sequence of the palaeopolyploid soybean. Nature 463: 178–183.

11. Li L, Stoeckert CJ Jr, Roos DS (2003) OrthoMCL: identification of ortholog groups for eukaryotic genomes. Genome Res 13: 2178–2189.

12. .Higgins JA, Bailey PC, Laurie DA (2010) Comparative genomics of flowering time pathways using Brachypodium distachyon as a model for the temperate grasses. PLoS One 5: e10065.

13. .Barton MK, Poethig RS (1993) Formation of the shoot apical meristem

in Arabidopsis thaliana - an analysis of development in the wild-type and in the shoot meristemless mutant. Development 119: 823–831.

14.   Endrizzi K, Moussian B, Haecker A, Levin JZ, Laux T (1996) The SHOOT MERISTEMLESS gene is required for maintenance of undifferentiated cells in Arabidopsis shoot and floral meristems and acts at a different regulatory level than the meristem genes WUSCHEL and ZWILLE. Plant J 10: 967–979.

15.   Laux T, Mayer KF, Berger J, Jurgens G (1996) The WUSCHEL gene is required for shoot and floral meristem integrity in Arabidopsis. Development 122: 87–96.

16.   .Mayer KF, Schoof H, Haecker A, Lenhard M, Jurgens G, et al. (1998) Role of WUSCHEL in regulating stem cell fate in the Arabidopsis shoot meristem. Cell 95: 805–815.

17.   .Robles P, Pelaz S (2005) Flower and fruit development in Arabidopsis thaliana. Int J Dev Biol 49: 633–643.

18.   .Hecht V, Foucher F, Ferrandiz C, Macknight R, Navarro C, et al. (2005) Conservation of Arabidopsis flowering genes in model legumes. Plant Physiol 137: 1420–1434.

19.   .Koo SC, Bracko O, Park MS, Schwab R, Chun HJ, et al. (2010) Control of lateral organ development and flowering time by the Arabidopsis thaliana MADS-box Gene AGAMOUS-LIKE6. Plant J 62: 807–816.

20.   .Nakaminami K, Hill K, Perry SE, Sentoku N, Long JA, et al. (2009) Arabidopsis cold shock domain proteins: relationships to floral and silique development. J Exp Bot 60: 1047–1062.

21.   .Chen M, Ni M (2006) RFI2, a RING-domain zinc finger protein, negatively regulates CONSTANS expression and photoperiodic flowering. Plant J 46: 823–833.

22.   .Han P, Garcia-Ponce B, Fonseca-Salazar G, Alvarez-Buylla ER, Yu H (2008) AGAMOUS-LIKE 17, a novel flowering promoter, acts in a FT-independent photoperiod pathway. Plant J 55: 253–265.

23.   Liscum E, Hodgson DW, Campbell TJ (2003) Blue light signaling through the cryptochromes and phototropins. So that's what the blues is all about. Plant Physiol 133: 1429–1436.

24.   Alabadi D, Oyama T, Yanovsky MJ, Harmon FG, Mas P, et al. (2001) Reciprocal regulation between TOC1 and LHY/CCA1 within the Arabidopsis circadian clock. Science 293: 880–883.

25.   Michaels SD, Amasino RM (1999) FLOWERING LOCUS C encodes a novel MADS domain protein that acts as a repressor of flowering. Plant

Cell 11: 949–956.

26. Wood CC, Robertson M, Tanner G, Peacock WJ, Dennis ES, et al. (2006) The Arabidopsis thaliana vernalization response requires a polycomb-like protein complex that also includes VERNALIZATION INSENSITIVE 3. Proc Natl Acad Sci U S A 103: 14631–14636.

27. Yu H, Xu Y, Tan EL, Kumar PP (2002) AGAMOUS-LIKE 24, a dosage-dependent mediator of the flowering signals. Proc Natl Acad Sci U S A 99: 16336–16341.

28. Michaels SD, Ditta G, Gustafson-Brown C, Pelaz S, Yanofsky M, et al. (2003) AGL24 acts as a promoter of flowering in Arabidopsis and is positively regulated by vernalization. Plant J 33: 867–874.

29. Liu C, Chen H, Er HL, Soo HM, Kumar PP, et al. (2008) Direct interaction of AGL24 and SOC1 integrates flowering signals in Arabidopsis. Development 135: 1481–1491.

30. Alexandre CM, Hennig L (2008) FLC or not FLC: the other side of vernalization. J Exp Bot 59: 1127–1135.

31. Amasino R (2010) Seasonal and developmental timing of flowering. Plant J 61: 1001–1013.

32. Mockler TC, Yu X, Shalitin D, Parikh D, Michael TP, et al. (2004) Regulation of flowering time in Arabidopsis by K homology domain proteins. Proc Natl Acad Sci U S A 101: 12759–12764.

33. Veley KM, Michaels SD (2008) Functional redundancy and new roles for genes of the autonomous floral-promotion pathway. Plant Physiol 147: 682–695.

34. Lee JH, Yoo SJ, Park SH, Hwang I, Lee JS, et al. (2007) Role of SVP in the control of flowering time by ambient temperature in Arabidopsis. Genes Dev 21: 397–402.

35. Abe M, Kobayashi Y, Yamamoto S, Daimon Y, Yamaguchi A, et al. (2005) FD, a bZIP protein mediating signals from the floral pathway integrator FT at the shoot apex. Science 309: 1052–1056.

36. Yamaguchi A, Wu MF, Yang L, Wu G, Poethig RS, et al. (2009) The microRNA-regulated SBP-Box transcription factor SPL3 is a direct upstream activator of LEAFY, FRUITFULL, and APETALA1. Dev Cell 17: 268–278.

37. Gu Q, Ferrandiz C, Yanofsky MF, Martienssen R (1998) The FRUITFULL MADS-box gene mediates cell differentiation during Arabidopsis fruit development. Development 125: 1509–1517.

38. Mandel MA, Yanofsky MF (1995) The Arabidopsis AGL8 MADS box

gene is expressed in inflorescence meristems and is negatively regulated by APETALA1. Plant Cell 7: 1763–1771.

39. .Hecht V, Laurie RE, Vander Schoor JK, Ridge S, Knowles CL, et al. (2011) The Pea GIGAS Gene Is a FLOWERING LOCUS T Homolog Necessary for Graft-Transmissible Specification of Flowering but Not for Responsiveness to Photoperiod. Plant Cell 23: 147–161.

40. Ferl RJ, Manak MS, Reyes MF (2002) The 14–3-3s. Genome Biol 3: REVIEWS3010.1–7.

41. International_Brachypodium_Initiative (2010) Genome sequencing and analysis of the model grass Brachypodium distachyon. Nature 463: 763–768.

42. Kim MY, Lee S, Van K, Kim TH, Jeong SC, et al. (2010) Whole-genome sequencing and intensive analysis of the undomesticated soybean (Glycine soja Sieb. and Zucc.) genome. Proc Natl Acad Sci U S A 107: 22032–7.

43. Greb T, Mylne JS, Crevillen P, Geraldo N, An H, et al. (2007) The PHD finger protein VRN5 functions in the epigenetic silencing of Arabidopsis FLC. Curr Biol 17: 73–78.

44. Sung S, Schmitz RJ, Amasino RM (2006) A PHD finger protein involved in both the vernalization and photoperiod pathways in Arabidopsis. Genes Dev 20: 3244–3248.

45. Chen F, Mackey AJ, Vermunt JK, Roos DS (2007) Assessing performance of orthology detection strategies applied to eukaryotic genomes. PLoS One 2: e383.

46. .Summerfield RJ, Roberts EH (1985) Glycine max; Halevy AH, editor. Boca Raton, Florida: CRC Press. pp. 139–148.

47. Levy YY, Mesnage S, Mylne JS, Gendall AR, Dean C (2002) Multiple roles of Arabidopsis VRN1 in vernalization and flowering time control. Science 297: 243–246.

48. Sung S, Amasino RM (2004) Vernalization in Arabidopsis thaliana is mediated by the PHD finger protein VIN3. Nature 427: 159–164.

49. Hartmann U, Hohmann S, Nettesheim K, Wisman E, Saedler H, et al. (2000) Molecular cloning of SVP: a negative regulator of the floral transition in Arabidopsis. Plant J 21: 351–360.

50. Wu G, Park MY, Conway SR, Wang JW, Weigel D, et al. (2009) The sequential action of miR156 and miR172 regulates developmental timing in Arabidopsis. Cell 138: 750–759.

51. Zhang B, Pan X, Stellwag EJ (2008) Identification of soybean microRNAs

and their targets. Planta 229: 161–182.

52. Tian Z, Wang X, Lee R, Li Y, Specht JE, et al. (2010) Artificial selection for determinate growth habit in soybean. Proc Natl Acad Sci U S A 107: 8563–8568.

53. Hu TT, Pattyn P, Bakker EG, Cao J, Cheng JF, et al. (2011) The Arabidopsis lyrata genome sequence and the basis of rapid genome size change. Nat Genet 43: 476–481.

54. Edgar RC (2004) MUSCLE: multiple sequence alignment with high accuracy and high throughput. Nucleic Acids Res 32: 1792–1797.

55. Larkin MA, Blackshields G, Brown NP, Chenna R, McGettigan PA, et al. (2007) Clustal W and Clustal X version 2.0. Bioinformatics 23: 2947–2948.

56. Huson DH, Richter DC, Rausch C, Dezulian T, Franz M, et al. (2007) Dendroscope: An interactive viewer for large phylogenetic trees. BMC Bioinformatics 8: 460.

Chapter 13

GENOME-WIDE ANALYSIS OF THE NADK GENE FAMILY IN PLANTS

Wen-Yan Li, Xiang Wang, Ri Li, Wen-Qiang Li, Kun-Ming Chen

State Key Laboratory of Crop Stress Biology in Arid Areas, College of Life Sciences, Northwest A&F University, Yangling, Shaanxi, China

ABSTRACT

Background

NAD(H) kinase (NADK) is the key enzyme that catalyzes *de novo* synthesis of NADP(H) from NAD(H) for NADP(H)-based metabolic pathways. In plants, NADKs form functional subfamilies. Studies of these families in *Arabidopsis thaliana* indicate that they have undergone considerable evolutionary selection; however, the detailed evolutionary history and functions of the various NADKs in plants are not clearly understood.

Principal Findings

We performed a comparative genomic analysis that identified 74 NADK gene homologs from 24 species representing the eight major plant lineages within the supergroup Plantae: glaucophytes, rhodophytes, chlorophytes, bryophytes, lycophytes, gymnosperms, monocots and eudicots. Phylogenetic and structural analysis classified these NADK genes into four well-conserved subfamilies with considerable variety in the domain organization and gene structure among subfamily members. In addition to the typical NAD_kinase domain, additional domains, such as adenylate kinase, dual-specificity phosphatase, and protein tyrosine phosphatase catalytic domains, were found in subfamily II. Interestingly, NADKs in subfamily III exhibited low sequence similarity (~30%) in the kinase domain within the subfamily and with the other subfamilies. These observations suggest that gene fusion and exon shuffling may have occurred after gene duplication, leading to specific domain organization seen in subfamilies II and III, respectively. Further analysis of the exon/intron structures showed that single intron loss and gain had occurred, yielding the diversified gene structures, during the process of structural evolution of NADK family genes. Finally, both available global microarray data analysis and qRT-RCR experiments revealed that the NADK genes in *Arabidopsis* and *Oryza sativa* show different expression patterns in different developmental stages and under several different abiotic/biotic stresses and hormone treatments, underscoring the functional diversity and functional divergence of the NADK family in plants.

Conclusions

These findings will facilitate further studies of the NADK family and provide valuable information for functional validation of this family in plants.

INTRODUCTION

NAD(H) and NADP(H) are crucial coenzymes and play important and distinguishable roles in all living organisms [1]. NAD(H) is primarily involved in catabolic reactions, whereas NADP(H) participates in anabolic reactions, such as NADP(H)-dependent reductive anabolic pathways, signal transduction, and defense against oxidative stress [1], [2]. Hence, regulation of the intracellular balance of NAD(H) and NADP(H) is critical. NAD(H) kinase (NADK) is the key enzyme in the *de novo* biosynthesis of NADP(H), catalyzing the transfer of a phosphoryl group from ATP to NAD(H), and thus plays an important role in the regulation of intracellular NAD(H)/NADP(H) balance for NADP(H)-based metabolic pathways.

NADK genes have been found in nearly all living organisms, including Archaea, eubacteria and eukaryotes, except for the intracellular parasite *Chlamydia trachomatis* [3]. NADK genes have been cloned from a wide variety of species, including Archaea (*Methanococcus jannaschii* [4]), eubacteria (*Mycobacterium tuberculosis* [5]), *Escherichia coli* [6], yeast (*Saccharomyces cerevisiae* [7], [8]), humans (*Homo sapiens* [9]) and plants (*Arabidopsis thaliana* [10], [11],[12]). Moreover, the number of NADKs in different organisms varies, with most prokaryotic organisms, such as Archaea and eubacteria, having only one NADK, whereas most eukaryotic organisms, such as yeast and plants, have several NADKs. For example, *Euglena gracilis* [13]has two NADKs and *S. cerevisiae* [14], [15] and *Arabidopsis* have three [10], [11], [16].

The enzymatic properties of natural or recombinant NADKs from several organisms, including their substrate specificity and structural properties, especially of the active site, have been well characterized [2], [9], [17], [18], [19].

All characterized NADKs are homomultimers and exhibit species-dependent phosphoryl donor and acceptor specificity. Hence, these enzymes are classified into two groups: ATP-NAD kinases and inorganic polyphosphate poly(P)/ATP-NAD kinases, based on the phosphoryl donor specificity [2]. For example, NADKs from gram-positive bacteria (*M. tuberculosis*) and Archaea (*M. jannaschii*) utilize both ATP and poly(P) as phosphoryl donors and they phosphorylate NAD⁺/NADH, whereas NADKs from gram-negative bacteria (*E. coli*) and eukaryotes (*S. cerevisiae, Arabidopsis* and human) utilize ATP but not poly(P), and phosphorylate NAD⁺ and NAD⁺/NADH, respectively [2]. In addition, analysis of the sequences and crystal structures of NADKs identified three highly conserved and functionally important motifs within the NADK family (a GGDG motif, an NE/D motif and a Gly-rich motif) and a possible phosphate transfer mechanism [20], [21], [22], [23], [24].

The biological significance and physiological functions of NADKs have been characterized from studies of several organisms. For example, mutation of the single NADK gene in *M. tuberculosis* and *Salmonella enterica* is lethal [3], [4]. Similarly, mutation of all three NADK genes in *S. cerevisiae* (*utr1/yef1/pos5*) or two of the three genes (*utr1/pos5*) also causes lethality, whereas the respective NADK gene mutations are not lethal in *Arabidopsis* [11], [16],[25], [26]. In addition, studies of the physiological functions of NADKs in *S. cerevisiae* and humans revealed that NADKs play a major role in protecting living cells against oxidative stress[8], [15], [26], [27] because NADPH is vital in the intracellular anti-oxidative defense system of most organisms [28].

In plants, NADKs are involved in regulating redox balance, biotic and abiotic stress responses and various developmental processes. Notably, NADK was the first protein in plants demonstrated to be regulated by the calcium-sensing protein calmodulin (CaM) [29], and plant NADKs are divided into CaM-independent and CaM-regulated isoforms. CaM-dependent NADK is essential for survival of plants under difficult conditions and for protecting plants against invading pathogens by helping to provide reductants for the NADPH-dependent oxidative burst [29], [30], [31]. For example, CaM-dependent NADK activity, but not CaM-independent NADK activity, increases under cold stress in green bean [32], and decreased in response to high salinity and drought in tomato [33] and wheat [34]. In *Arabidopsis*, three genes encoding NADK (*NADK1/NADK2/NADK3*) have been identified and their physiological functions have been characterized [2], [35]. *AtNADK1* is located in the cytosol and expressed mainly in roots, *AtNADK2* is located in the chloroplasts and expressed mainly in leaves, whereas *AtNADK3* is located in the peroxisome and is strongly expressed in reproductive tissues, such as the stigma, pollen and carpel vasculature [10], [11], [36]. The *AtNADK1*-deficient mutant exhibits sensitivity to γ-irradiation and paraquat-induced oxidative stress [16]. The *AtNADK2*-deletion mutant displays hypersensitivity to environmental stresses that induce oxidative stress, such as UVB-irradiation, drought, heat shock and high salinity [11]. Similarly, the *AtNADK3*-null mutant shows hypersensitivity to oxidative stress, including methyl viologen, high salinity and osmotic shock [10]. Moreover, AtNADK2 plays a vital role in chlorophyll synthesis and protects chloroplasts against oxidative damage [11]. In addition, plants growth and fertility are affected in the NADK-deficient mutants [2], [11], [37].

Although these studies in bacteria, yeast, human and *Arabidopsis* have led to an understanding of the biochemical properties and physiological functions of NADKs, there has been no systematic study of the evolution and functional divergence of the NADK gene family, especially in Plantae. Here, we performed a comprehensive analysis of the NADK gene family in 24 species, representing the eight major plant lineages within the supergroup Plantae. Phylogenetic analysis was performed to delineate the evolutionary history of the NADK family in Plantae, and exon/intron structure analysis was performed to gain insight into the possible mechanisms of the structural diversity of NADK gene family. Finally, the tissue-specificity and inducibility of NADK gene expression in *Arabidopsis* and rice (*Oryza sativa*) were characterized by examining publicly available microarray datasets and by qRT-PCR experiments. The results obtained here will broaden our understanding of the roles of plant NADKs and provide a framework for further functional investigations of these genes in plants.

RESULTS

Identification of NADK family members in plants

To comprehensively investigate and characterize the NADK gene family in plants, 24 species representing the eight major plant lineages within the supergroup Plantae, were selected for analysis (Figure 1). A hidden Markov model (HMM) search was performed with the obtained sequences and 74 NADK homologs were identified (Figure 1, Table S1). Except for the gymnosperm *Picea sitchensis*, for which the full genome sequence was not yet available, two or more NADK genes were identified in each genome of the selected species (Figure 1). Most of the aquatic algae, including glaucophytes, rhodophytes and chlorophytes, contained two NADK genes per genome and only three species in the chlorophytes, *Micromonas pusilla* RCC299, *Chlamydomonas reinhardtii* and *Volvox carteri*, carried three NADK genes per genome (Figure 1). In contrast, three or more NADK genes were found in all land plants, including bryophytes, lycophytes, monocots and eudicots (Figure 1). In addition, only one NADK gene was identified per genome in the cyanophytes (*Acaryochloris marina* MBIC11017 and *Prochlorococcus marinus* MIT 9301) and bacteria (*E. coli* K-12), outgroup (Figure 1).

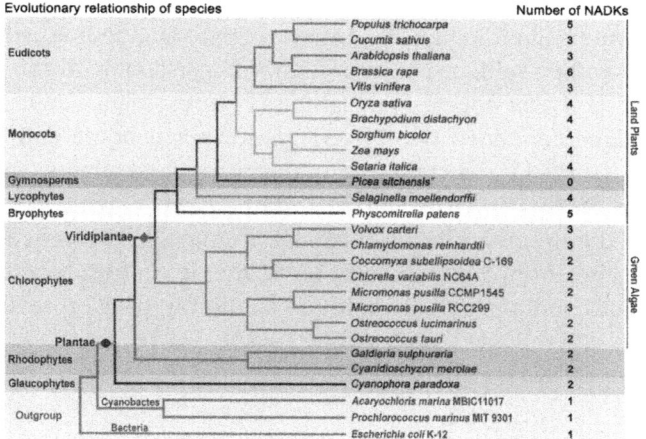

Figure 1: Systematic evolutionary relationships of 24 species among eight lineages within the supergroup Plantae.The numbers of NADK homologs in each species are listed next to the tree. *, the genome sequencing of *Picea sitchensis* is not complete.

doi:10.1371/journal.pone.0101051.g001.

The Pfam and SMART databases were used to analysis the functional domains of the identified NADK candidates. All of the putative NADKs possess a typical NAD_kinase domain (Pfam accession number PF01513) and some also contain other functional domains such as an adenylate kinase domain (ADK; PF00406), a dual-specificity phosphatase, catalytic domain (DSPc; PF00782), and/or a protein tyrosine phosphatase, catalytic domain/ DSPc domain (PTPc/DSPc; SMART accession number SM000012). Only one NADK candidate, CpNADK2, which did not contain a complete NAD_kinase domain, was excluded from the following analysis.

Phylogenetic analysis and classification of the NADK family

To explore the phylogenetic relationships among NADK family members in plants, we first generated a rooted maximum-likelihood phylogenetic tree with the 73 NADKs from the 24 species (Figure 2A), which was inferred from the amino acid sequences of their NAD_kinase domains (Figure S1). Furthermore, phylogenetic trees reconstructed by the neighbor joining, minimum evolution and maximum parsimony methods showed very similar topologies to the maximum-likelihood tree (data not shown). Using *EcNADK1*, *AmNADK1* and *PmNADK1* from bacteria and cyanophytes as the outgroup, all NADK homologs in plants can be classified into four well-conserved subfamilies (I–IV; Figure 2A) with high statistical support, according to the topology and the deep duplication nodes of NADK paralogues in the maximum-likelihood tree. Interestingly, the topological relationship of members within subfamilies was highly consistent with the evolutionary relationships between species in Plantae (Figures 1 and 2A). The majority of aquatic algae contained two NADK genes per genome and grouped into subfamilies II and IV, whereas the majority of land plants, except for the gymnosperm *Picea sitchensis*, carried ≥ three NADK genes per genome and were clustered into subfamilies I, II and III (Figure 2A). Moreover, only one of the NADK genes from green algae (*CrNADK3*) fell into subfamily I and only two (*VcNADK3* and *MpNADK3*) fell into subfamily III (Figure 2A).

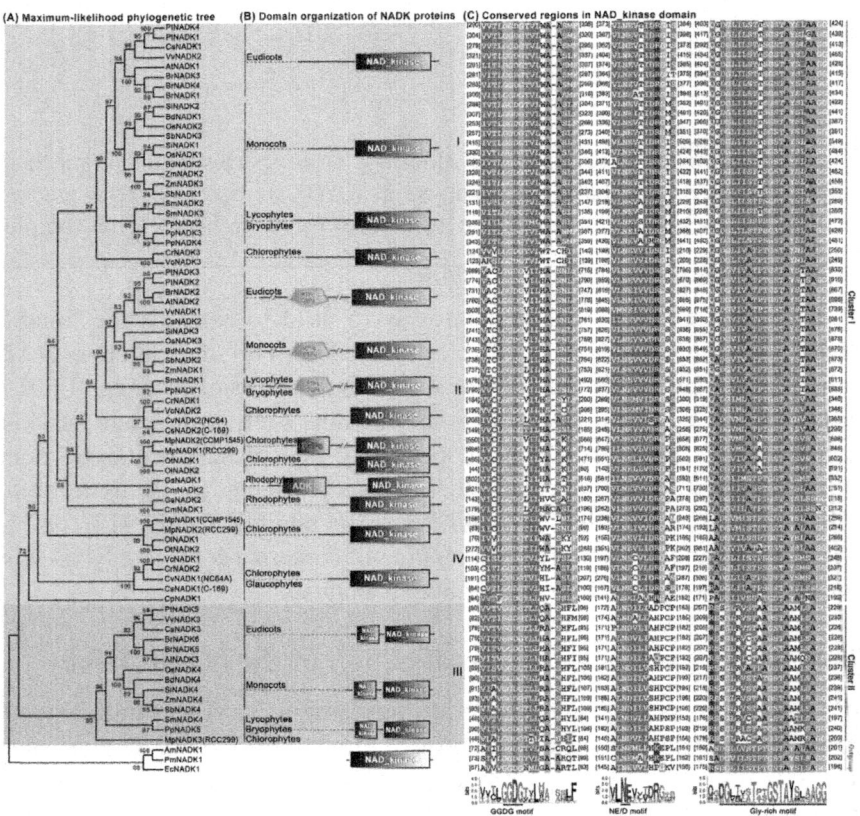

Figure 2: Phylogenetic relationships and domain organization of NADK genes in plants.**(A)** The rooted maximum-likelihood phylogenetic tree of NADK family mem-

bers was inferred from the amino acid sequence alignment of the NAD_kinase domain. Numbers above the nodes represent bootstrap values from 1000 replications. (**B**) Domain organization of the NADKs. (**C**) Amino acid sequence alignment of conserved motifs within the NAD_kinase domain.

doi:10.1371/journal.pone.0101051.g002

Examination of the chromosomal locations of the NADK family genes in the genomes of algae (*Ostreococcus lucimarinus, Ostreococcus tauri, C. reinhardtii* and *V. carteri*), bryophytes (*Physcomitrella patens*), lycophytes (*Selaginella moellendorffii*), monocots (*Brachypodium distachyon,* rice, *Sorghum bicolor* and *Zea mays*) and eudicots (*Arabidopsis, Brassica rapa,Populus trichocarpa* and *Vitis vinifera*) showed that NADK genes are randomly distributed (Figure S3). Moreover, a search for NADK paralogs using the Plant Genome Duplication Database (PGDD; http://chibba.agtec. uga.edu/duplication/) [38] revealed eight paralogous gene pairs in *P. patens*, rice, *Z. mays, B. rapa* and *P. trichocarpa* (Figure S4A), but none in the other species.

To further explore association of positive selection with duplication and divergence of NADK family genes, the rate of non-synonymous substitution (Ka), synonymous substitution (Ks) and the Ka/Ks ratios were calculated for the eight paralogous gene pairs and used to estimate duplication and divergence times. The Ka/Ks ratios varied from 0.18 to 0.39 among the five different species (Figure S4B). The fact that the Ka/Ks ratios were <1 suggests that the NADK family genes have undergone strong negative selection pressure, and the duplication event was estimated to have occurred ~13.8–67.6 million years ago. Divergence of *P. patens*, rice, *Z. mays, P. trichocarpa* and *B. rapa* was estimated to have occurred 67.6, 61.5, 13.8, 20.8 and 23.1–33.1 million years ago, respectively (Figure S4B).

Further analysis of the functional domains showed that domain organization of NADKs in the different subfamilies varied considerably (Figure 2B). All of the identified NADKs contained a typical NAD_kinase domain at the C terminus, but the proteins in subfamily II also carried the additional N-terminal catalytic domains noted earlier (DSPc, PTPc/DSPc or ADK; Figure 2B). Interestingly, the NAD_kinase domain of proteins in subfamily III was not only divided into two parts, but also exhibited low sequence similarity (~30%) between subfamily III members and the other subfamilies (Figures 2C and S1). In addition, based on the NAD_kinase domain sequence similarity, the NADK gene subfamilies could be further divided into two clusters: cluster I (subfamilies I, II and IV) and cluster II (subfamily III).

More detailed analysis of the NAD_kinase domains of the 73 NADKs revealed three highly conserved and functionally important motifs: a GGDG motif, an NE/D motif and a Gly-rich motif (Figures 2C and S1). The GGDG motif is involved in ATP-binding, whereas the NE/D and Gly-rich motifs are involved in NAD(H) binding [2], [20], [21], [23], [24].

Structure analysis of NADK family genes

Intron position is generally very well-conserved in orthologous genes over long evolutionary time intervals, whereas exon/intron structure is slightly less, but sufficiently, conserved in paralogous genes to reveal evolutionary relationships between introns [39], [40], [41]. To investigate the gene structural diversity and possible mechanisms for the structural evolution of NADK homologs in green plants (Viridiplantae, excluding glaucophytes and rhodophytes), we analyzed the exon/intron organization in the coding sequence. Overall, there was considerable diversity in the number of introns (0–11) and the length of introns (50–7452 bp) in the NADK family genes (Figure 3). Interestingly, NADK family members within the same subfamily shared similar gene structure in terms of intron number, exon length, and/or intron phases, with the exception of subfamily IV (Figure 3). For instance, NADK genes in subfamily I had 6–11 introns, 48% (12/25) and 84% (21/25) of which contained 10 and 9–11 introns, respectively. We also investigated intron phases with respect to codons in the NADK genes. The intron phases were remarkably well conserved among subfamily members, whereas the intron arrangement and phases were strikingly distinct between subfamilies (Figure 3B).

To further explore intron loss or gain within the NADK family genes, we next examined the exon/intron organization of paralogs and orthologs in the land plants (Figure 4A). This analysis revealed that single intron loss and gain likely occurred during the structural evolution of NADK family genes in land plants. For example, The paralogous genes *PtNADK1/4* and *SmNADK2/3*showed conserved exon/intron structure in terms of the number of introns and exon length, whereas a single intron appears to have been lost during the evolution of the *BrNADK1/4* and*PpNADK3/4* paralogs (Figures 3B and 4A). By contrast, a single intron gain occurred between the paralogs *PtNADK2* and *3* (Figures 3B and 4A). Among the orthologous NADK genes in land plants, the majority of subfamilies I, II and III members contained 10, 7 and 4 conserved common introns, respectively (Figure 4A), whereas some subfamily I and II NADK orthologs contained fewer introns. For example, *OsNADK1, SiNADK1, BdNADK1/2, ZmNADK2/3*(subfamily I) have only nine conserved common introns. Similar intron loses were seen with*BrNADK1/3/4, SbNADK3* and *SmNADK2/3* (subfamily I), *PpNADK1* (subfamily II), and*SmNADK4* (subfamily III; Figure 4A). Single intron gain was also observed in the NADK genes of land plants. For instance, *OsNADK1*, *SiNADK1*, and *PpNADK2/3/4* in subfamily I and*PtNADK3* in subfamily II appear to have gained an intron, whereas *VvNADK1* in subfamily II appears to have gained two introns (Figure 4A).

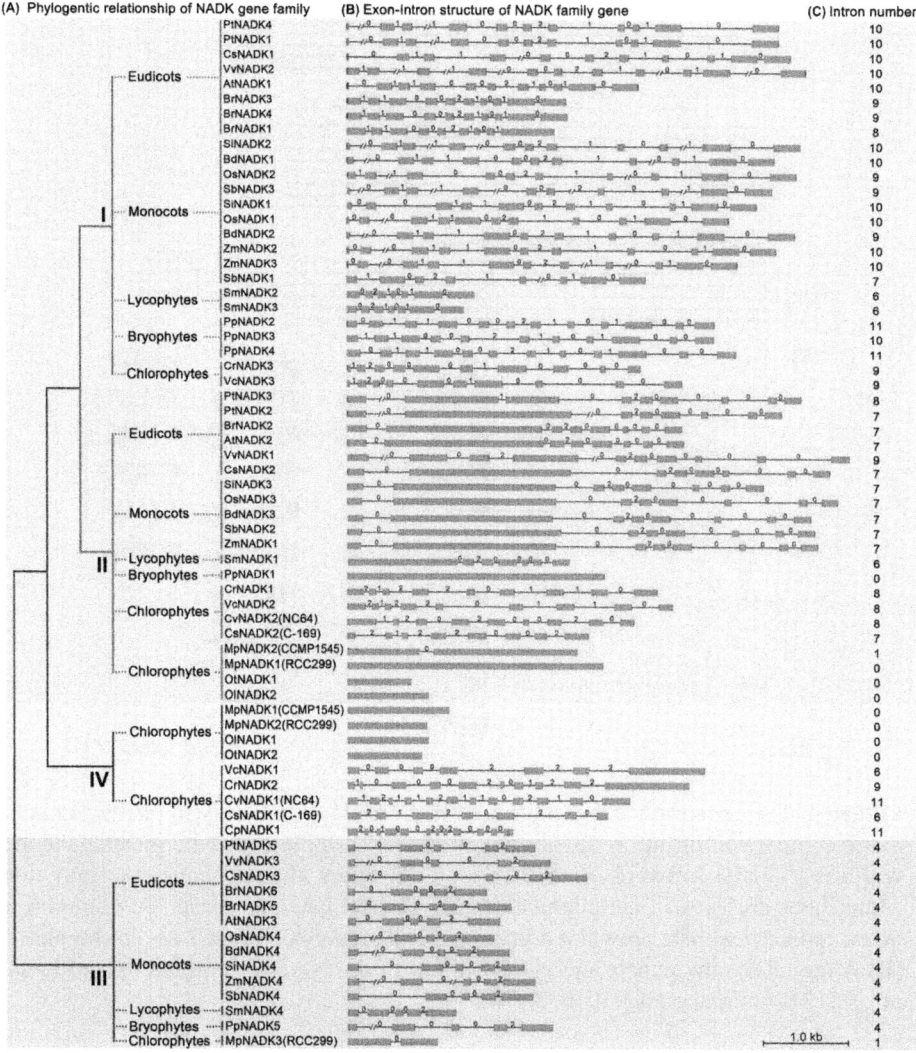

Figure 3: Exon/intron structure of NADK family genes of green plants.Green boxes represent exons; black lines represent introns; numbers 0, 1 and 2 are intron phases. The length of the boxes and lines are scaled relative to the length of the gene, and longer introns are denoted by a double slash.doi:10.1371/journal.pone.0101051.g003.

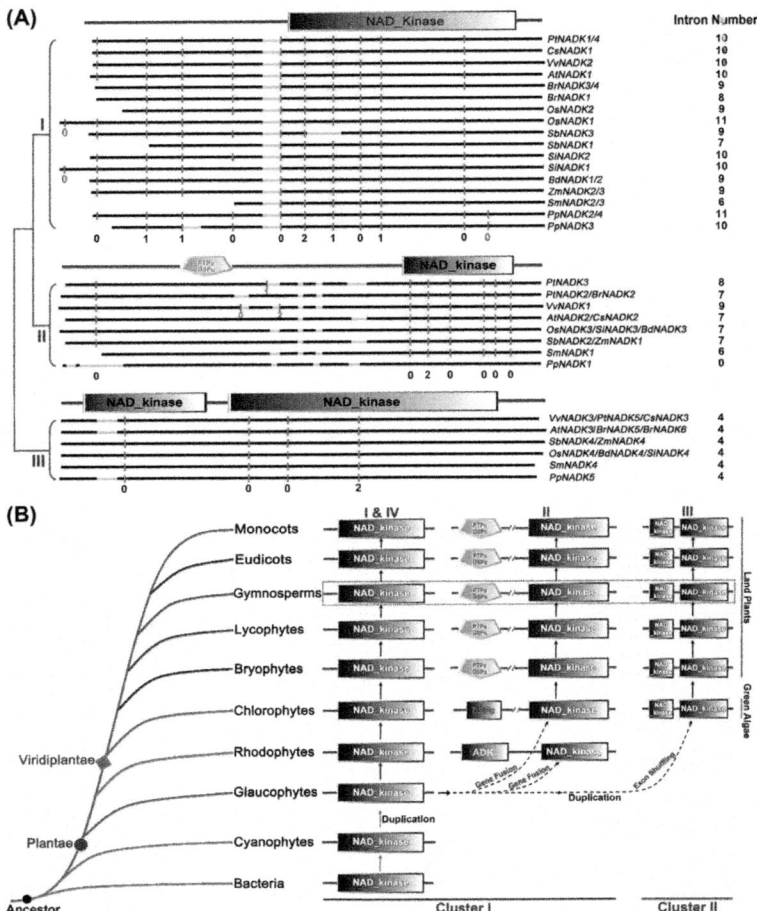

Figure 4: The expansion and evolution of the NADK gene family in plants (**A**) Schematic comparison of intron distribution in NADK orthologs of land plants generated with the CIWOG software. Black horizontal lines are aligned sequences; gray horizontal lines are gaps in the alignment; gray vertical bars are conserved common introns; red vertical bars are gained introns. The numbers 0, 1 and 2 are intron phases. (**B**) A model for the expansion and evolution of the NADK gene family in Plantae. doi:10.1371/journal.pone.0101051.g004.

Tissue-specific expression patterns of NADK genes in *Arabidopsis* and rice

To investigate differences in expression of NADK genes in *Arabidopsis* and rice during plant development, we first analyzed the expression profiles of AtNADK and OsNADK genes available in expressed sequence tag (EST) databases from vegetative and reproductive development stages (Figures 5A and 5B and Tables S3 and S4, respectively). Varying levels of ESTs for each of the AtNADK and OsNADK genes were found the database, indicating that all of the NADK genes in *Arabidopsis* and rice are expressed, and that they are likely differential expressed in different tissues or developmental stages. We then used *Arabidopsis* (ATH1, 22 k array) and rice (Os 51 k array) microarray data in Genevestigator to analyze the expression patterns of the NADK genes in 10 and 9 developmental stages/tissues, respectively (Figure 5C, D and S5). Essentially identical developmental expression profiles for the AtNADK and OsNADK genes were also obtained with data from the *Arabidopsis* and rice eFP browsers in the Bio-Analytic Resource (http://bar.utoronto.ca/welcome. htm) database [42]. *AtNADK3* was not included in the comparison of tissue/ developmental expression patterns because it was not included in the microarray data. The combined expression patterns of *AtNADKs* and *OsNADKs* from the EST and microarray data analyses are summarized in Figure 5E.

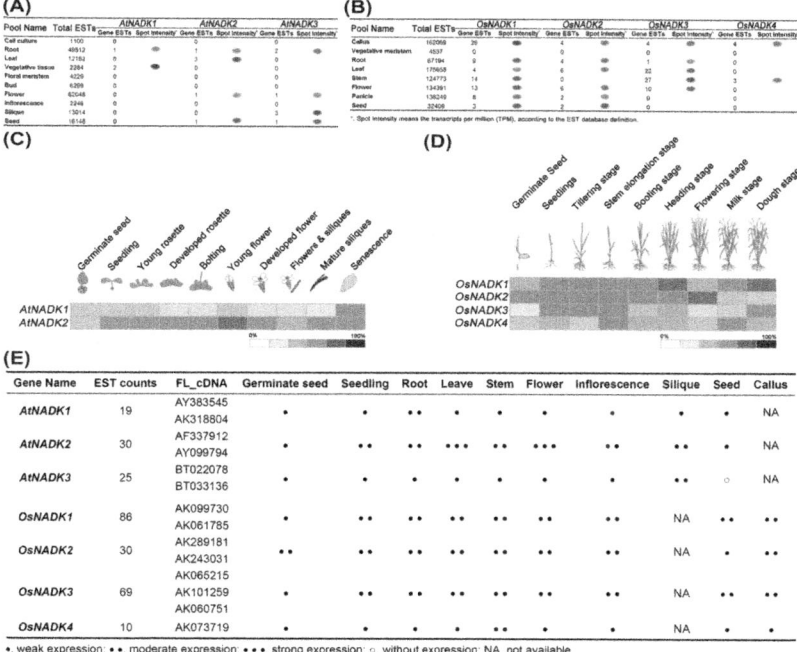

Figure 5: Developmental expression patterns of NADK family genes in *Arabidopsis* and rice. Expression profiles of (**A**) *AtNADKs* and (**B**) *OsNADKs* inferred from public EST data. Expression profiles of (**C**) *AtNADKs* (except *AtNADK3*) and (**D**) *Os-NADKs* in different developmental stages obtained from microarray data reported in Genevestigator. Results are shown as heat maps in white/gray/red (low to high) that re-

flect the percent of expression. (**E**) Summary of the relative expression patterns of At-NADK and OsNADK genes inferred from the combined EST (A, B; Tables S3 and S4) and microarray (C, D) profiles.doi:10.1371/journal.pone.0101051.g005.

Overall, all of the AtNADK and OsNADK genes were expressed during the vegetative and reproductive development stages, and they displayed strong tissue specificity. In *Arabidopsis,AtNADK2* showed higher expression in most tissues and developmental stages than *AtNADK1*(Figures 5C and S5). The highest *AtNADK2* expression is seen in leaves and flowers (Figures 5A, C and S5), consistent with Waller et al. (2010) [36]; whereas *AtNADK1* expression is highest in the dry seed and senescence stages (Figures 5C and S5). *AtNADK3* appears to be mainly expressed in the reproductive tissues such as flowers and siliques (Figure 5A), also consistent with the findings of Waller et al. (2010) [36]. In rice, the *OsNADK1*, *OsNADK2* and *OsNADK3* transcripts tend to accumulate to higher levels than the *OsNADK4* transcript in most tissues/developmental stages (Figures 5B, D and S5), but their individual expression patterns differ. For instance, during the germinate seed stage, *OsNADK1* transcript levels are low compared to *OsNADK2* (Figure 5D), whereas the opposite is observed in callus (Figure 5B). The highest levels of *OsNADK3* transcript are seen in leaf and stem (Figures 5D and S5); whereas relatively low levels of the Os*NADK4* transcript are seen in most tissues/developmental stages except for stem and inflorescence (Figures 5D and S5).

Response profiles of NADK genes under abiotic/biotic stresses and hormone treatments in *Arabidopsis* and rice

To understand the molecular mechanism of *AtNADKs* and *OsNADKs* transcriptional regulation under abiotic/biotic stresses and hormone treatments, we first identified potential *cis*-elements in the promoter regions of each NADK gene using the PlantCARE program (Tables 1 and S2). We found that the promoter regions of AtNADK and OsNADK genes contain response elements for several abiotic/biotic stresses, such as low temperature, heat, drought (MYB binding sites), anaerobic conditions, and pathogens (TC-rich repeats, W box, GCC box, Box S, Box-W1 and EIRE) (Tables 1 and S2). In addition, the AtNADK and OsNADK gene promoters contain *cis*-elements for responding to several hormones, such as auxin, gibberellin, abscisic acid (ABA), ethylene, salicylic acid (SA) and methyl jasmonic acid (MeJA) (Tables 1 and S2). Further analysis showed that all of the *AtNADK* and *OsNADK* promoters contain anaerobic- and ABA-responsive elements; and the majority of them also carry heat stress, drought (MYB binding sites) and MeJA response elements (Tables 1 and S2).

Table 1: Abiotic/biotic stress and hormone response elements in *AtNADK* and*Os-NADK* promoters doi:10.1371/journal.pone.0101051.t001

Cis-elements	Core sequences	Functions of the cis-elements	OsNADK1	OsNADK2	OsNADK3	OsNADK4	AtNADK1	AtNADK2	AtNADK3
Abiotic/biotic stress									
LTR	CCGAAA	low-temperature responsiveness		√		√			
HSE	AAAAAATTTC	heat-stress responsiveness	√		√		√	√	
MBS	TAACTG	MYB binding site, drought inducibility	√		√		√		√
ARE	TGGTTT	anaerobic induction	√		√	√	√	√	√
TC-rich repeats	ATTCTCTAAC	defense and stress responsiveness	√		√				
W box	TTGACC	wound and pathogen responsiveness			√				
Box S	AGCCACC	wound and pathogen responsiveness		√					
GCC box	AGCCGCC	wound and pathogen responsiveness							
EIRE	TTCGACC	Elicitor responsiveness							
Box-W1	TTGACC	fungal elicitor responsiveness			√		√		√
Hormone response elements									
AuxRR-core	GGTCCAT	auxin responsiveness			√				
TGA-element	AACGAC	Auxin responsiveness					√		√
GARE-motif	AAACAGA	Gibberellin responsiveness		√			√		
ABRE	CACGTG	abscisic acid responsiveness	√		√		√		√
ERE	ATTTCAAA	Ethylene responsiveness	√	√	√				
TCA-element	GAGAAGAATA	salicylic acid responsiveness					√		√
CGTCA-motif	CGTCA	MeJA responsiveness	√			√	√	√	√

*the cis-elements were identified with the PlantCARE program (http://bioinformatics.psb.ugent.be/webtools/plantcare/html/) using the sequences 1500 bp upstream from the transcription start site of each NADK gene. The "√" means the NADK gene contains this cis-element in the promoter region.
doi:10.1371/journal.pone.0101051.t001

To further demonstrate that the expression of AtNADK and OsNADK genes is induced by abiotic and biotic stress, we again examined *Arabidopsis* and rice microarray data in the Genevestigator database, as well as by qRT-PCR experiments. As shown in Figures 6, 7 andS6, the expression of AtNADK and OsNADK genes is induced to varying degrees by abiotic stress such as cold, heat, drought (PEG), salt (NaCl) and oxidative (methyl viologen, MV), as well as by biotic stress such as the pathogens *Botrytis cinerea*, *Blumeria graminis*,*Pseudomonas syringae*, *Agrobacterium tumefaciens*, *Mycosphaerella graminicola*,*Magnaporthe oryzae*, and *Magnaporthe grisea*. In *Arabidopsis*, *AtNADK1* is up-regulated in whole plants under heat and MV stresses, whereas it is down-regulated and does not change under PEG and cold treatment, respectively (Figure 6). By contrast, *AtNADK2* is down-regulated or does not change under these conditions (Figure 6). *AtNADK3* is up-regulated under cold stress, while it is down-regulated under other abiotic stresses including heat, PEG, NaCl, and MV (Figure 6). Additionally, both *AtNADK1* and *AtNADK2* can be up-regulated by some biotic stresses, although with different expression profiles under the same stress (Figure S6). In rice, *OsNADK1/2* are up-regulated in shoot and root under cold stress, in root under heat stress and in shoot under MV stress, respectively; while they are slightly down-regulated or does not change under these conditions (Figure 7). *OsNADK3* shows up-regulated under PEG in shoot and down-regulated or does not change under other abiotic stresses such as cold, heat, NaCl and MV (Figure 7). Similarly, *OsNADK4* is up-regulated in root under cold and heat and down-regulated or does not change under other abiotic stresses including PEG, NaCl, and MV (Figure 7). Additionally, *OsNADK1–4* can be induced by biotic stresses. For instance,*OsNADK1* is up-regulated in callus after infection by *A. tumefaciens* and root after infection by*M. oryzae*, whereas they are down-regulated in root and leaf after infection by *M. graminicola*and *M. grisea*, respectively (Figure S6).

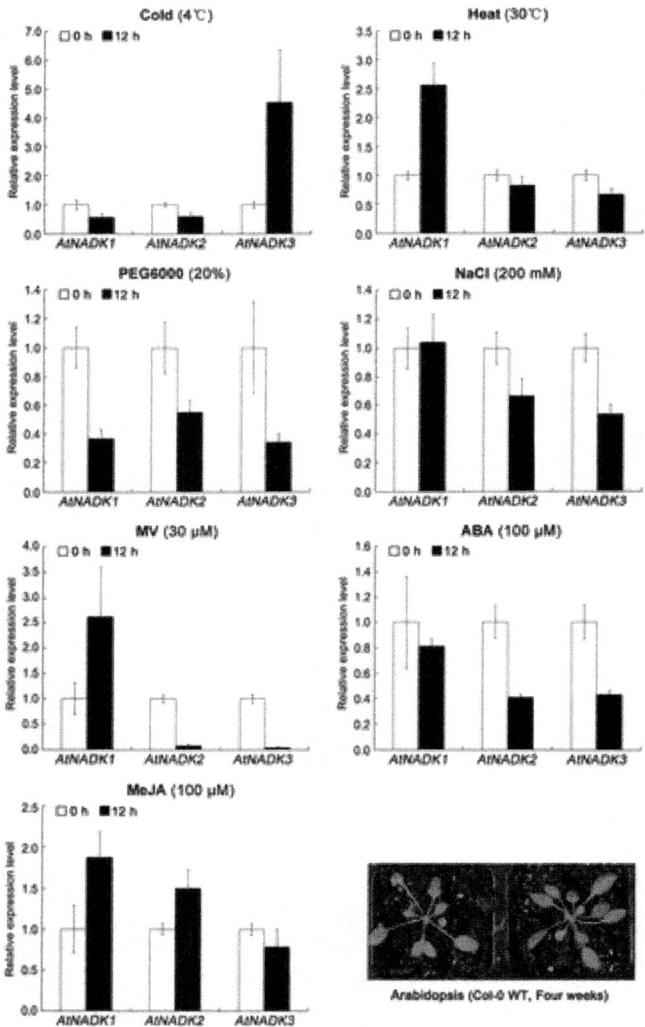

Figure 6: Expression patterns of NADK family genes in *Arabidopsis* under abiotic stress and hormone treatments.Expression levels of *AtNADK1–3* assayed by qRT-PCR under cold (4°C), heat (30°C), drought (20% PEG6000), salt (200 mM NaCl), oxida-

tive (30 μM MV) stresses and MeJA (100 μM), ABA (100 μM) hormone treatments. Data are means ± SD (n=3) and are representative of similar results from three independent experiments.doi:10.1371/journal.pone.0101051.g006.

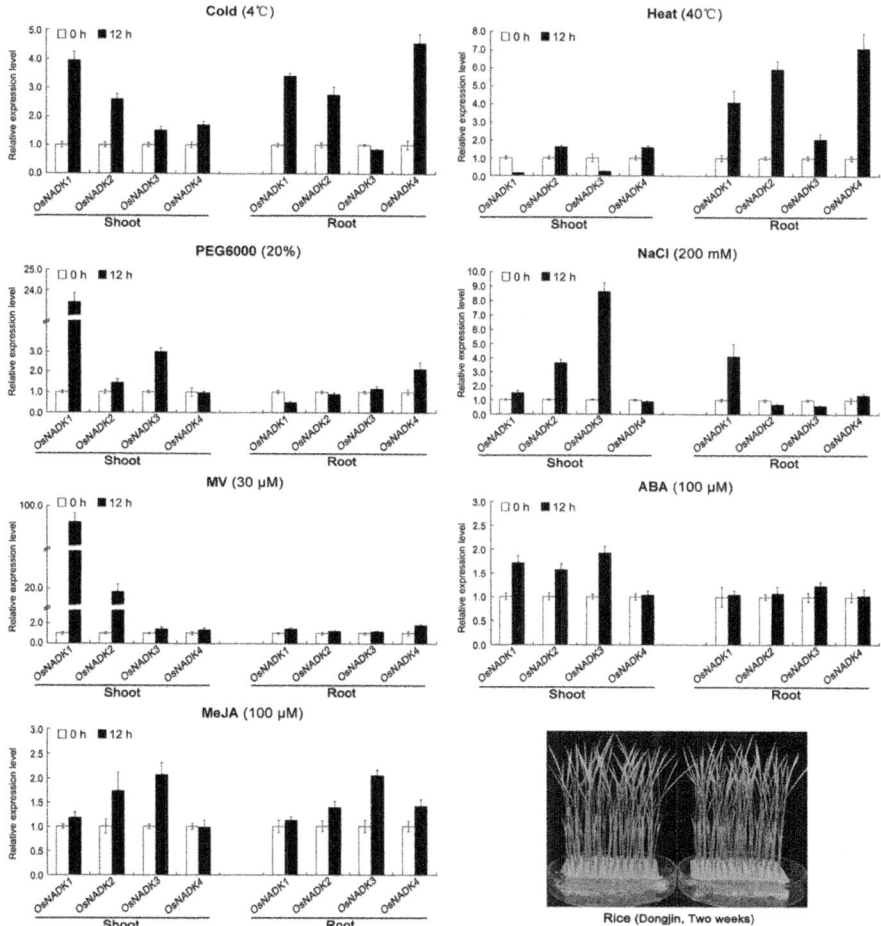

Figure 7: Expression patterns of NADK family genes in rice under abiotic stress and hormone treatments.Expression levels of *OsNADK1–4* assayed by qRT-PCR under cold (4°C), heat (30°C), drought (20% PEG6000), salt (200 mM NaCl), oxidative (30 μM MV) stresses and MeJA (100 μM), ABA (100 μM) hormone treatments. Data are means ± SD (n=3) and are representative of similar results from three independent experiments.doi:10.1371/journal.pone.0101051.g007.

As several hormone response elements were identified in the promoter regions of the AtNADK and OsNADK genes, we also examined their expression profiles under phytohormone treatments in the public *Arabidopsis* and rice microarray data (Figure S7), as well as by qRT-PCR experiments (Figures 6 and 7). *AtNADK1* and *AtNADK2* are differentially expressed in seedlings and cells treated with phytohormones, such as the auxins indole-3-acetic acid (IAA) and naphthaleneacetic acid (NAA), zeatin, SA, ABA, and ABA + SA (Figure S7). Interestingly, *AtNADK1* is up-regulated by ABA, ABA + SA, and ABA + MeJA treatment; whereas *AtNADK2* is down-regulation by ABA and ABA + SA treatment (Figure S7). The expressions of AtNADK genes by qRT-PCR showed *AtNADK1–2* are up-regulated by MeJA treatment and down-regulated or do not change by ABA treatment; *AtNADK3* is down-regulated by ABA treatment and does not change by MeJA treatment (Figure 6). In rice, *OsNADK1–4* are differentially expressed in seedlings and leaves treated with IAA, NAA, zeatin, gibberellin, kinetin, ABA, SA and jasmonic acid (JA) (Figure S7). For instance, *OsNADK3* is up-regulated in roots under *trans*-zeatin treatment, whereas *OsNADK4* is down-regulated in leaves under the same treatment (Figure S7). It is also notable that *OsNADK1–3* can be induced in shoot by ABA treatment and do not change in root, but *OsNADK4* is not obviously changed with the same treatment. All *OsNADK1–4* were up-regulated or not obviously changed by MeJA treatment.

DISCUSSION

Gene fusion and exon shuffling after gene duplication contributed to the expansion and evolution of the NADK family in plants

Gene duplication is a common phenomenon in eukaryotes and contributes to biological diversity during evolution [43], [44], [45]. The NADKs are represented by at least one gene in nearly every living organism. In this study, we found that the 24 representative plant species from the supergroup Plantae examined contain 2–6 NADK homologs, whereas cyanophytes, considered to be the ancestor of Plantae, had only one NADK homolog per genome, suggesting that a single gene duplication leading to the expansion of the NADK gene family, occurred during the divergence of ancestral cyanophytes from Plantae. The scattered distribution of NADK family genes on chromosomes (Figure S3) and the eight paralogous NADK gene pairs found in five land plants (Figure S4) suggest that segmental duplications may have been involved in the expansion of the NADK gene family and caused differences in the number of NADK genes within subfamilies and species of land plants after divergence from aquatic algae.

Gene fusion and exon shuffling after gene duplication are mechanisms that can enhance the functional divergence of duplicated genes by creating additional domains or rearranging the original functional domains [46], [47], [48], [49].

In this study, phylogenetic analysis, together with the domain organizations and gene structures of NADK family genes, showed distinct evolutionary differences among subfamilies (Figure 2). Therefore, a model was constructed to account for the expansion and evolution of the NADK gene family in plants (Figure 4B). In this model, all NADK family members originated from a common ancestor, which contained only the typical NAD_kinase domain and existed in all living organisms from prokaryotic bacteria to eukaryotic angiosperms; cyanophytes were considered to be the ancestor of Plantae, and at least one gene duplication occurred to yield two NADK genes in eukaryotic glaucophytes (Figures 1 and 4B). Moreover, during the evolutionary diversification from glaucophytes to rhodophytes and Viridiplantae, gene fusion events occurred and additional catalytic domains (ADK, DSPc or PTPc/DSPc) were acquired in the N-terminus, leading to the diversified domain organization seen in subfamily II (Figures 2B and 4B). In addition, exon shuffling after gene duplication contributed to the formation of the bifurcated NAD_kinase domain in subfamily III (Figures 2B and 4B).

Single intron loss and gain lead to the diversified gene structures

Gene structural diversity within gene families is another evolutionary mechanism that promotes variability, and intron loss or gain is important in generating structural diversity and complexity[41], [50]. Analyzing the exon/intron structures of NADK genes, we found that the number of introns and intron phases among subfamily members are remarkably conserved, whereas the intron arrangement and intron phases are strikingly distinct between subfamilies (Figures 3 and4). Further analysis of the orthologous and paralogous genes in land plants (Figure 4) suggests that single intron loss as well as intron gain likely occurred and contributed to the diversification of gene structure, and consequent functional diversity and divergence, during the evolution of the NADK family in plants.

NADKs are involved in plant responses to several abiotic and biotic stresses

The three NADK genes of *Arabidopsis* belong to three different subfamilies according to our phylogenetic analysis (Figures 1 and 2). The three *AtNADKs* are known to have anti-oxidative functions [10], [11], [12], [16], [36], but they have distinct mechanisms by which they facilitate plant resistance to oxidative stress. AtNADK1 (subfamily I) is a cytosolic enzyme that indirectly provides cytosolic NADP for plasma membrane NADPH oxidases, which are the key producers of reactive oxygen species under both normal and stress conditions in plants [36], [51], [52],[53]. AtNADK2 (subfamily II) is a chloroplastic enzyme that plays a vital role in chlorophyll synthesis and chloroplast protection against oxidative damage by regulating plastidic NADP-biosynthesis [11], [36]. AtNADK3 (subfamily III) is a peroxisomal enzyme that plays a prominent anti-oxidation role by providing the peroxisomal reductant NADPH [10], [36]. NADK genes in subfamily IV were only present in aquatic algae and grouped with *S. cerevisiae NADK1* (Pos5) in our broader phylogenetic analysis (Figure S2). Pos5 is located in the mitochondria and its deletion causes slow growth, sensitivity to oxidative stress, such as paraquat, hyperoxia, H_2O_2, and Cu^{2+}, and biosynthesis deficiency on iron-sulfur clusters [25], [54], [55]. These observations suggest that plant NADKs in subfamily IV may also localize to mitochondria and protect against oxidative damage by regulating mitochondrial NADP-biosynthesis.

We found that plant NADKs are also involved in responding to several abiotic/biotic stresses, including cold, heat, drought, salt, oxidative and pathogens. *Arabidopsis NADK1* expression is up-regulated by H_2O_2, irradiation and the bacterial pathogen *P. syringae* pv. *tomato*; and the *AtNADK1* deficient mutant is sensitive to γ-irradiation and paraquat-induced oxidative stress[16]. The *AtNADK2* deletion mutant is also hypersensitive to environmental stresses that trigger oxidative stress, such as UVB-irradiation, drought, heat, and high salinity [11]; similarly *AtNADK3* transcription can be induced by MV, high salinity and osmotic shock [10], [36]. In this study, we found that *AtNADK1* and *OsNADK1/2* belongs to subfamily I and can be induced to varying degrees by cold, drought (PEG), salt (NaCl), oxidative (MV) and pathogens (Figures 6,7, S6 and S7). We also found that *AtNADK2*, which belongs to subfamily II, is up-regulated by pathogens such as *P. syringae* pv. *phaseolicola* and *Hyaloperonospora arabidopsidis* (Figure S6); *OsNADK3*, which also belongs to subfamily II, is up-regulated by *Nilaparvata lugens* and down-regulated under anaerobic conditions (Figure S6). *OsNADK4* belongs to subfamily III, and is up-regulated under anaerobic conditions, *N. lugens* and heat in root, and slightly up-regulated by drought or PEG in root (Figures 7 and S6). These observations, together with our analysis of *cis*-elements (Table 1), suggest that NADK genes in subfamilies I, II and III play an important role in plant responses to invading pathogens and abiotic stresses. The functions of NADKs in subfamily IV in abiotic/biotic-stress responses remains to be examined.

NADKs may participate in regulation of hormone signaling

AtNADK3 transcription can be slightly induced by ABA and the mutant is hypersensitive to ABA[10]. In this study, we found that all *AtNADK1–3* showed down-regulated or slightly down-regulated in response to ABA (Figure

6), while *OsNADK1–3* were up-regulated by ABA treatment (Figure 7). It's important to note that response profiles of NADK genes by qRT-PCR are not completely consisted with microarray data or previous studies, because of the different experimental material, processing and analyzing methods used in this study. Moreover, seven AtNADK and OsNADK genes (*AtNADK1–3* and *OsNADK1–4*) contain *cis*-acting ABA responsive elements (Table 1), suggesting that NADKs probably participate in regulation of ABA signaling in plants.

JA and its derivative MeJA are important signaling molecules in plant responses to many abiotic and biotic stresses such as wounding and pathogens [56], [57]. Moreover, JA biosynthesis occurs in peroxisomes [57], [58]. AtNADK3 is also a peroxisomal enzyme and may phosphorylate the NADH derived from β-oxidation to yield NADPH needed for anti-oxidant defense [36]. *AtNADK1/2* and *OsNADK1–4* were up-regulated or slightly up-regulated under MeJA treatment (Figures 6 and 7). Moreover, several of these genes (*AtNADK1/2/3* and*OsNADK1/4*) contain CGTCA-MeJA-responsiveness motifs in their promoter regions (Table 1). These observations suggest that NADKs may also participate in regulation of JA or MeJA signaling in plants.

It should be pointed out that although transcription factors and *cis*-regulatory elements play important roles in regulating gene expression, the relationship is not always direct and the inducibility of expression is often affected by many factors and on multiple levels [59], [60]. For example, ABA-responsive *cis*-regulatory elements were identified in the promoter regions of all seven NADK genes in *Arabidopsis* and rice, but only *OsNADK1–3* were up-regulated or slightly up-regulated by ABA treatment (Figures 7 and S7). Moreover, considering the limitations of the PlantCARE program in predicting *cis*-regulatory elements, further experiments are necessary to verify the relationship between the inducible expression profiles of NADK genes and the *cis*-elements within their promoters.

MATERIALS AND METHODS

Data retrieval and identification of NADK genes

The protein sequences of 24 completely or partially sequenced plant genomes representing the eight major plant lineages were retrieved from public databases. All of the protein sequences were the most current non-redundant sequences from the following sources (Table S1): the glaucophyte *Cyanophora paradoxa* from the *Cyanophora* Genome Project (http://cyanophora.

rutgers.edu/cyanophora/home.php) [61]; the rhodophytes *Cyanidioschyzon merolae* and *Galdieria sulphuraria* from the *Cyanidioschyzon* Genome Project (http://merolae.biol.s.u-tokyo.ac.jp/) [62] and the *Galdieria* Genome Project (http://genomics.msu.edu/galdieria/) [63], respectively; the chlorophytes *O. tauri* (version 2.0)[64], *O. lucimarinus* (version 2.0) [65], *M. pusilla* strain RCC299 (version 3.0) and strain CCMP1545 (version 3.0) [66], *Chlorella variabilis* NC64A (version 1.0) [67], *Coccomyxa subellipsoidea* C-169 (version 2.0) [68], *C. reinhardtii* (version 4.0) [69] and *V. carteri* (version 2.0) and the bryophyte *P. patens* (version 3.0) [70] and lycophyte *S. moellendorffii* (version 1.1)[71] from the Joint Genome Institute (JGI, http://genome.jgi-psf.org/); for the gymnosperm *P. sitchensis* [72] from NCBI (http://www.ncbi.nlm.nih.gov/, partial sequences only because of its genome is only partially sequenced); the monocots *B. distachyon* (version 1.2) [73], *Setaria italica* (version 2.1) [74] and *S. bicolor* (version 2.1) [75] from JGI, rice (version 7.0) and *Z. mays* (version 5.6) from the Institute for Genomic Research Rice Genome Annotation Project (http://rice.plantbiology.msu.edu/index.shtml) [76] and MaizeSequence (http://www.maizesequence.org/index.html) [77], respectively; the eudicots *P. trichocarpa*(version 2.2) [78] and *Cucumis sativus* (version 1.0) [79] from JGI, *Arabidopsis* (version 10.0) from the Arabidopsis Information Resource (http://www.arabidopsis.org/) [80], *B. rapa* (version 1.5) from the *Brassica* Database (http://brassicadb.org/brad/) [81] and *V. vinifera* (version 1.0) from GenoScope (http://www.genoscope.cns.fr/externe/GenomeBrowser/Vitis/) [82]. All of the above protein sequences were integrated into a local protein database for the subsequent identification of NADK homologs.

To identify the NADK genes and their homologs in the Plantae supergroup, HMMER v3.0 [83],[84] was used to perform an HMM search against the local protein database, using the family-specific NAD_kinase domain (PF01513) HMM profile obtained from the Pfam database [85],[86]. The HMM search was performed with the default parameters and an E-value cutoff of $1e^{-5}$. If a candidate gene had multiple alternative splice variants, the longest variant was used to represent the candidate protein. The Pfam and SMART [87], [88] databases were employed to detect conserved domains in the candidate proteins, and the search results were refined manually to eliminate partial NADK domains and other potential false positives.

Sequence alignment and phylogenetic analysis

The NAD_kinase domain sequences of the candidate proteins were aligned using the MUSCLE v3.8 program with the default parameters [89], [90] and the alignments were manually edited using the BioEdit v7.0 program [91]. The rooted maximum-likelihood tree was inferred from the resulting alignments using the Phylip v3.68 package [92] under the γ-corrected Jones–Taylor–Thornton model [93] with default parameters, and the reliability of interior branches was assessed with 1000 bootstrap resamplings. In addition, the MEGA v5.0 program [94] was also used to reconstruct the phylogenetic trees by the neighbor joining, minimal evolution and maximum parsimony methods, and to display the phylogenetic trees.

The rates of non-synonymous substitution (Ka) and synonymous substitution (Ks) were estimated for the orthologous and paralogous gene pairs of NADK family genes using the Codeml program in PAML v4.3 [95] interface tool of PAL2NAL [96], based on the aligned amino acid sequences and the corresponding nucleotide sequences. The duplication and divergence times of each gene pairs were estimated from the Ks of λ substitutions per synonymous site per year as $T=Ks/2\lambda$ ($\lambda=6.5\times10^{-9}$) [97], [98].

Exon/intron structure and conserved motif analysis

The exon/intron structures of individual NADK genes were obtained through the Gene Structure Display Server (http://gsds.cbi.pku.edu.cn) [99] by aligning the coding or cDNA sequences with their corresponding genomic DNA sequences from Phytozome v9.1 (http://www.phytozome.net/) [100]. To illustrate the evolution of introns, gene models were inspected for annotation of introns, and exon/intron boundaries were manually checked. For a subset of genes, predictions pertaining to the types of introns were independently checked using Common Introns Within Orthologous Genes software (http://ciwog.gdcb.iastate.edu/)[101].

The conserved functional motifs within NAD_kinase domains were identified using the MEME v4.9 program (http://meme.sdsc.edu) [102] with the default parameters, and the sequence logos (graphical representations) of these motifs or domains were generated with the WebLogo v3.3 server (http://weblogo.threeplusone.com/create.cgi) [103] based on the results of protein sequence alignments.

Cis-regulatory elements and expression profile analysis

The 1500 bp upstream of the transcription start site of all NADK genes in *Arabidopsis* and rice were obtained from Phytozome v9.1 (http://www.phytozome.net/), and the cis-regulatory elements were identified using the PlantCARE program (http://bioinformatics.psb.ugent.be/webtools/plantcare/html/) [104], [105]. ATH1 22 k and Os 51 k microarray data in the Genevestigator V3 database were used to analyze the tissue-specific and inducible expression profiles of NADK genes in *Arabidopsis* and rice [106], respectively. In addition, EST profiles of each NADK gene in *Arabidopsis* and rice were also analyzed by BLASTN search against the corresponding NCBI (http://www.ncbi.nlm.nih.gov/) EST database, with the coding sequences of the individual NADK gene as query. The BLASTN searches were performed with the following criteria: E-value<$1e^{-10}$ and nucleotide identities >95% over 150 bp.

Plant materials, treatments and quantitative real-time PCR (qRT-PCR) analysis

The *Arabidopsis* (Col-0) seeds were pretreated at 4°C for 3 days and directly sown in Sunshine MVP potting soil, and then grown in growth chambers under 16 h light/8 h dark at 22±1°C; while rice (*Oryza sativa* ssp. *japonica* cv. Dongjin) seeds were pretreated germination at 26±1°C for 4 days and transplanted to 1/2 Hogland nutrient solution, and then grown in growth chambers under 16 h light/8 h dark at 26±1°C. Following two (rice) or four (*Arabidopsis*) weeks of growth, the seedlings were grown under 4°C or 30°C (*Arabidopsis*)/40°C (rice) for 12 h for cold and heat treatments, respectively; submerged in 200 mM NaCl or 20% PEG6000 solutions for 12 h for drought and salt treatments, respectively. For oxidative stress and hormone treatments, solutions of 30 μM MV, 100 μM MeJA and ABA were separately sprayed on seedlings for 12 h. All the MV, MeJA and ABA used for treatments were purchased from Sigma-Aldrich. Samples were collected following treatment and immediately frozen at −80°C.

Total RNA was extracted by using Trizol reagent (Takara, Japan) according to the manufacturer's instructions and treated with RNase-free DNase I (Invitrogen, USA) for 15 min to remove any DNA contamination. RNA concentration and quality were verified by using the NanoDrop 1000 Spectrophotometers (Thermo, USA), and cDNAs were synthesized by using oligo d(T)$_{18}$ reverse primer from 5 μg of total RNA in a total volume of 20 μL by using EasyScript First-Strand cDNA Synthesis SuperMix (TransGen Biotech, China). qRT-PCR reactions were carried out in 96-well (20 μL) format by using the FastStart Essential DNA Green Master (Roche, Switzerland), and were performed in an CFX96 Touch Real-Time PCR Detection System (BIO-RAD, USA). The reactions were repeated three times and the quantitative analysis used the $2^{-\Delta\Delta CT}$ method. All gene-specific primers were designed to avoid the conserved region and span introns or cross an exon-exon junction. The detailed primer sequences are shown in Table S5. The *AtTub6* (AT5G12250) and *OsActin1* (accession ID KC140126) were chosen as the internal control in *Arabidopsis* and rice, respectively.

REFERENCES

1. Ying W (2008) NAD+/NADH and NADP+/NADPH in cellular functions and cell death: regulation and biological consequences. Antioxid Redox Signal 10: 179–206. doi: 10.1089/ars.2007.1672

2. .Kawai S, Murata K (2008) Structure and function of NAD kinase and NADP phosphatase: key enzymes that regulate the intracellular balance of NAD(H) and NADP(H). Biosci Biotechnol Biochem 72: 919–930. doi: 10.1271/bbb.70738

3. .Grose JH, Joss L, Velick SF, Roth JR (2006) Evidence that feedback inhibition of NAD kinase controls responses to oxidative stress. Proc Natl Acad Sci U S A 103: 7601–7606. doi: 10.1073/pnas.0602494103

4. .Kawai S, Fukuda C, Mukai T, Murata K (2005) MJ0917 in archaeon *Methanococcus jannaschii* is a novel NADP phosphatase/NAD kinase. J Biol Chem 280: 39200–39207. doi: 10.1074/jbc.m506426200

5. .Kawai S, Mori S, Mukai T, Suzuki S, Yamada T, et al. (2000) Inorganic Polyphosphate/ATP-NAD kinase of Micrococcus flavus and *Mycobacterium tuberculosis* H37Rv. Biochem Biophys Res Commun 276: 57–63. doi: .1006/bbrc.2000.3433

6. .Kawai S, Mori S, Mukai T, Hashimoto W, Murata K (2001) Molecular characterization of*Escherichia coli* NAD kinase. Eur J Biochem 268: 4359–4365. doi: 10.1046/j.1432-1327.2001.02358.x

7. .Kawai S, Suzuki S, Mori S, Murata K (2001) Molecular cloning and identification of UTR1 of a yeast *Saccharomyces cerevisiae* as a gene encoding an NAD kinase. FEMS Microbiol Lett 200: 181–184. doi: 10.1111/j.1574-6968.2001.tb10712.x

8. Outten CE, Culotta VC (2003) A novel NADH kinase is the mitochondrial source of NADPH in *Saccharomyces cerevisiae*. EMBO J 22: 2015–2024. doi: 10.1093/emboj/cdg211

9. 9.Lerner F, Niere M, Ludwig A, Ziegler M (2001) Structural and functional characterization of human NAD kinase. Biochem Biophys Res Commun 288: 69–74. doi: 10.1006/bbrc.2001.5735

10. Chai MF, Wei PC, Chen QJ, An R, Chen J, et al. (2006) NADK3, a novel cytoplasmic source of NADPH, is required under conditions of oxidative stress and modulates abscisic acid responses in *Arabidopsis*. Plant J 47: 665–674. doi: 10.1111/j.1365-313x.2006.02816.x

11. .Chai MF, Chen QJ, An R, Chen YM, Chen J, et al. (2005) NADK2, n *Arabidopsis*chloroplastic NAD kinase, plays a vital role in both

chlorophyll synthesis and chloroplast protection. Plant Mol Biol 59: 553–564. doi: 10.1007/s11103-005-6802-y

12. .Turner WL, Waller JC, Vanderbeld B, Snedden WA (2004) Cloning and characterization of two NAD kinases from *Arabidopsis*. identification of a calmodulin binding isoform. Plant Physiol 135: 1243–1255. doi: 10.1104/pp.104.040428

13. Stephan C, Renard M, Montrichard F (2000) Evidence for the existence of two soluble NAD(+) kinase isoenzymes in *Euglena gracilis Z*. Int J Biochem Cell Biol. 32: 855–863. doi: 10.1016/s1357-2725(00)00032-7

14. Li YF, Shi F (2006) Partial rescue of pos5 mutants by YEF1 and UTR1 genes in *Saccharomyces cerevisiae*. Acta Biochim Biophys Sin (Shanghai) 38: 293–298. doi: 10.1111/j.1745-7270.2006.00162.x

15. .Shi F, Kawai S, Mori S, Kono E, Murata K (2005) Identification of ATP-NADH kinase isozymes and their contribution to supply of NADP(H) in *Saccharomyces cerevisiae*. FEBS J 272: 3337–3349. doi: 10.1111/j.1742-4658.2005.04749.x

16. Berrin JG, Pierrugues O, Brutesco C, Alonso B, Montillet JL, et al. (2005) Stress induces the expression of AtNADK-1, a gene encoding a NAD(H) kinase in *Arabidopsis thaliana*. Mol Genet Genomics 273: 10–19. doi: 10.1007/s00438-005-1113-1

17. Oganesyan V, Huang C, Adams PD, Jancarik J, Yokota HA, et al. (2005) Structure of a NAD kinase from *Thermotoga maritima* at 2.3 A resolution. Acta Crystallogr Sect F Struct Biol Cryst Commun 61: 640–646. doi: 10.1107/s1744309105019780

18. Mori S, Kawai S, Mikami B, Murata K (2001) Crystallization and preliminary X-ray analysis of NAD kinase from *Mycobacterium tuberculosis* H37Rv. Acta Crystallogr D Biol Crystallogr 57: 1319–1320. doi: 10.1107/s0907444901011362

19. Ando T, Ohashi K, Ochiai A, Mikami B, Kawai S, et al. (2011) Structural Determinants of Discrimination of NAD+ from NADH in Yeast Mitochondrial NADH Kinase Pos5. J Biol Chem 286: 29984–29992. doi: 10.1074/jbc.m111.249011

20. Raffaelli N, Finaurini L, Mazzola F, Pucci L, Sorci L, et al. (2004) Characterization of *Mycobacterium tuberculosis* NAD kinase: functional analysis of the full-length enzyme by site-directed mutagenesis. Biochemistry 43: 7610–7617. doi: 10.1021/bi049650w

21. Garavaglia S, Raffaelli N, Finaurini L, Magni G, Rizzi M (2004) A novel fold revealed by *Mycobacterium tuberculosis* NAD kinase, a key allosteric enzyme in NADP biosynthesis. J Biol Chem 279: 40980–40986. doi:

10.1074/jbc.m406586200

22. Labesse G, Douguet D, Assairi L, Gilles AM (2002) Diacylglyceride kinases, sphingosine kinases and NAD kinases: distant relatives of 6-phosphofructokinases. Trends Biochem Sci 27: 273–275. doi: 10.1016/s0968-0004(02)02093-5

23. Mori S, Yamasaki M, Maruyama Y, Momma K, Kawai S, et al. (2005) NAD-binding mode and the significance of intersubunit contact revealed by the crystal structure of*Mycobacterium tuberculosis* NAD kinase-NAD complex. Biochem Biophys Res Commun 327: 500–508. doi: 10.1016/j.bbrc.2004.11.163

24. Liu J, Lou Y, Yokota H, Adams PD, Kim R, et al. (2005) Crystal structures of an NAD kinase from *Archaeoglobus fulgidus* in complex with ATP, NAD, or NADP. J Mol Biol 354: 289–303. doi: 10.1016/j.jmb.2005.09.026

25. Shianna KV, Marchuk DA, Strand MK (2006) Genomic characterization of POS5, the*Saccharomyces cerevisiae* mitochondrial NADH kinase. Mitochondrion 6: 94–101. doi: 10.1016/j.mito.2006.02.003

26. Bieganowski P, Seidle HF, Wojcik M, Brenner C (2006) Synthetic lethal and biochemical analyses of NAD and NADH kinases in *Saccharomyces cerevisiae*establish separation of cellular functions. J Biol Chem 281: 22439–22445. doi: 10.1074/jbc.m513919200

27. Pollak N, Niere M, Ziegler M (2007) NAD kinase levels control the NADPH concentration in human cells. J Biol Chem 282: 33562–33571. doi: 10.1074/jbc.m704442200

28. Singh R, Mailloux RJ, Puiseux-Dao S, Appanna VD (2007) Oxidative stress evokes a metabolic adaptation that favors increased NADPH synthesis and decreased NADH production in *Pseudomonas fluorescens*. J Bacteriol 189: 6665–6675. doi: 10.1128/jb.00555-07

29. Anderson JM, Charbonneau H, Jones HP, McCann RO, Cormier MJ (1980) Characterization of the plant nicotinamide adenine dinucleotide kinase activator protein and its identification as calmodulin. Biochemistry 19: 3113–3120. doi: 10.1021/bi00554a043

30. Harding SA, Oh SH, Roberts DM (1997) Transgenic tobacco expressing a foreign calmodulin gene shows an enhanced production of active oxygen species. EMBO J 16: 1137–1144. doi: 10.1093/emboj/16.6.1137

31. Zhou CY, Wu GL, Duan ZQ, Wu LL, Gao YS, et al. (2010) H_2O_2-NOX system: an important mechanism for developmental regulation and stress response in plants. Chinese Bulletin of Botany 45: 615–631.

32. Ruiz JM, Sanchez E, Garcia PC, Lopez-Lefebre LR, Rivero RM, et al. (2002) Proline metabolism and NAD kinase activity in green bean plants subjected to cold-shock. Phytochemistry 59: 473–478. doi: 10.1016/s0031-9422(01)00481-2

33. Delumeau O, Paven M-CM-L, Montrichard F, Laval-Martin DL (2000) Effects of short-term NaCl stress on calmodulin transcript levels and calmodulin-dependent NAD kinase activity in two species of tomato. Plant, Cell & Environment 23: 329–336. doi: 10.1046/j.1365-3040.2000.00545.x

34. Zagdanska B (1990) NAD kinase activity in wheat leaves under water deficit. Acta Biochim Pol 1990 37(3): 385–9.

35. Shi F, Li Y, Wang X (2009) Molecular properties, functions, and potential applications of NAD kinases. Acta Biochim Biophys Sin (Shanghai) 41: 352–361. doi: 10.1093/abbs/gmp029

36. Waller JC, Dhanoa PK, Schumann U, Mullen RT, Snedden WA (2010) Subcellular and tissue localization of NAD kinases from *Arabidopsis*: compartmentalization of de novo NADP biosynthesis. Planta 231: 305–317. doi: 10.1007/s00425-009-1047-7

37. Wu LL, Zhou CY, Gao YS, Cong YX, Chen KM, et al. (2011) Cloning and genetic transformation of *OsNADK3* gene in rice. Journal of Nuclear Agricultural Sciences 25: 0863–0870.

38. Lee TH, Tang H, Wang X, Paterson AH (2013) PGDD: a database of gene and genome duplication in plants. Nucleic Acids Res 41: D1152–1158. doi: 10.1093/nar/gks1104

39. 39. Hardison RC (1996) A brief history of hemoglobins: plant, animal, protist, and bacteria. Proc Natl Acad Sci U S A 93: 5675–5679. doi: 10.1073/pnas.93.12.5675

40. olf YI, Sorokin AV, Mirkin BG, Koonin EV (2003) Remarkable interkingdom conservation of intron positions and massive, lineage-specific intron loss and gain in eukaryotic evolution. Curr Biol 13: 1512–1517. doi: 10.1016/s0960-9822(03)00558-x

41. Li W, Liu B, Yu L, Feng D, Wang H, et al. (2009) Phylogenetic analysis, structural evolution and functional divergence of the 12-oxo-phytodienoate acid reductase gene family in plants. BMC Evol Biol 9: 90. doi: 10.1186/1471-2148-9-90

42. Winter D, Vinegar B, Nahal H, Ammar R, Wilson GV, et al. (2007) An "Electronic Fluorescent Pictograph" browser for exploring and analyzing large-scale biological data sets. PLoS One 2: e718. doi: 10.1371/journal.pone.0000718

43. Bowers JE, Chapman BA, Rong J, Paterson AH (2003) Unravelling angiosperm genome evolution by phylogenetic analysis of chromosomal duplication events. Nature 422: 433–438. doi: 10.1038/nature01521

44. Van de Peer Y, Fawcett JA, Proost S, Sterck L, Vandepoele K (2009) The flowering world: a tale of duplications. Trends Plant Sci 14: 680–688. doi: 10.1016/j.tplants.2009.09.001

45. Magadum S, Banerjee U, Murugan P, Gangapur D, Ravikesavan R (2013) Gene duplication as a major force in evolution. J Genet 92: 155–161. doi: 10.1007/s12041-013-0212-8

46. Kolkman JA, Stemmer WP (2001) Directed evolution of proteins by exon shuffling. Nat Biotechnol 19: 423–428. doi: 10.1038/88084

47. Jones CD, Begun DJ (2005) Parallel evolution of chimeric fusion genes. Proc Natl Acad Sci U S A 102: 11373–11378. doi: 10.1073/pnas.0503528102

48. Kaessmann H (2010) Origins, evolution, and phenotypic impact of new genes. Genome Res 20: 1313–1326. doi: 10.1101/gr.101386.109

49. Morgante M, Brunner S, Pea G, Fengler K, Zuccolo A, et al. (2005) Gene duplication and exon shuffling by helitron-like transposons generate intraspecies diversity in maize. Nat Genet 37: 997–1002. doi: 10.1038/ng1615

50. Zhang Z, Kishino H (2004) Genomic background predicts the fate of duplicated genes: evidence from the yeast genome. Genetics 166: 1995–1999. doi: 10.1534/genetics.166.4.1995

51. Wang GF, Li WQ, Li WY, Wu GL, Zhou CY, et al. (2013) Characterization of Rice NADPH Oxidase Genes and Their Expression under Various Environmental Conditions. Int J Mol Sci 14: 9440–9458. doi: 10.3390/ijms14059440

52. Sagi M, Fluhr R (2006) Production of reactive oxygen species by plant NADPH oxidases. Plant Physiol 141: 336–340. doi: 10.1104/pp.106.078089

53. Foreman J, Demidchik V, Bothwell JH, Mylona P, Miedema H, et al. (2003) Reactive oxygen species produced by NADPH oxidase regulate plant cell growth. Nature 422: 442–446. doi: 10.1038/nature01485

54. Strand MK, Stuart GR, Longley MJ, Graziewicz MA, Dominick OC, et al. (2003) POS5 gene of *Saccharomyces cerevisiae* encodes a mitochondrial NADH kinase required for stability of mitochondrial DNA. Eukaryot Cell 2: 809–820. doi: 10.1128/ec.2.4.809-820.2003

55. Pain J, Balamurali MM, Dancis A, Pain D (2010) Mitochondrial NADH

kinase, Pos5p, is required for efficient iron-sulfur cluster biogenesis in *Saccharomyces cerevisiae*. J Biol Chem 285: 39409–39424. doi: 10.1074/jbc.m110.178947

56. Chehab EW, Kaspi R, Savchenko T, Rowe H, Negre-Zakharov F, et al. (2008) Distinct roles of jasmonates and aldehydes in plant-defense responses. PLoS One 3: e1904. doi: 10.1371/journal.pone.0001904

57. Wasternack C, Hause B (2013) Jasmonates: biosynthesis, perception, signal transduction and action in plant stress response, growth and development. An update to the 2007 review in Annals of Botany. Ann Bot 111: 1021–1058. doi: 10.1093/aob/mct067

58. Schaller A, Stintzi A (2009) Enzymes in jasmonate biosynthesis - structure, function, regulation. Phytochemistry 70: 1532–1538. doi: 10.1016/j.phytochem.2009.07.032

59.

60. **59.**Ma Q, Chirn GW, Szustakowski JD, Bakhtiarova A, Kosinski PA, et al. (2008) Uncovering mechanisms of transcriptional regulations by systematic mining of *cis*regulatory elements with gene expression profiles. BioData Min 1: 4. doi: 10.1186/1756-0381-1-4

61. .Jackson RJ, Hellen CU, Pestova TV (2010) The mechanism of eukaryotic translation initiation and principles of its regulation. Nat Rev Mol Cell Biol 11: 113–127. doi: 10.1038/nrm2838

62. .Price DC, Chan CX, Yoon HS, Yang EC, Qiu H, et al. (2012) *Cyanophora paradoxa*genome elucidates origin of photosynthesis in algae and plants. Science 335: 843–847. doi: 10.1126/science.1213561

63. .Matsuzaki M, Misumi O, Shin IT, Maruyama S, Takahara M, et al. (2004) Genome sequence of the ultrasmall unicellular red alga *Cyanidioschyzon merolae* 10D. Nature 428: 653–657. doi: 10.1038/nature02398

64. Barbier G, Oesterhelt C, Larson MD, Halgren RG, Wilkerson C, et al. (2005) Comparative genomics of two closely related unicellular thermo-acidophilic red algae,*Galdieria sulphuraria* and *Cyanidioschyzon merolae*, reveals the molecular basis of the metabolic flexibility of *Galdieria sulphuraria* and significant differences in carbohydrate metabolism of both algae. Plant Physiol 137: 460–474. doi: 10.1104/pp.104.051169

65. .Derelle E, Ferraz C, Rombauts S, Rouze P, Worden AZ, et al. (2006) Genome analysis of the smallest free-living eukaryote *Ostreococcus tauri* unveils many unique features. Proc Natl Acad Sci U S A 103: 11647–11652. doi: 10.1073/pnas.0604795103

66. .Palenik B, Grimwood J, Aerts A, Rouze P, Salamov A, et al. (2007) The tiny eukaryote*Ostreococcus* provides genomic insights into the paradox of plankton speciation. Proc Natl Acad Sci U S A 104: 7705–7710. doi: 10.1073/pnas.0611046104

67. .Worden AZ, Lee JH, Mock T, Rouze P, Simmons MP, et al. (2009) Green evolution and dynamic adaptations revealed by genomes of the marine picoeukaryotes*Micromonas*. Science 324: 268–272. doi: 10.1126/science.1167222

68. Blanc G, Duncan G, Agarkova I, Borodovsky M, Gurnon J, et al. (2010) The *Chlorella variabilis* NC64A genome reveals adaptation to photosymbiosis, coevolution with viruses, and cryptic sex. Plant Cell 22: 2943–2955. doi: 10.1105/tpc.110.076406

69. .Blanc G, Agarkova I, Grimwood J, Kuo A, Brueggeman A, et al. (2012) The genome of the polar eukaryotic microalga *Coccomyxa subellipsoidea* reveals traits of cold adaptation. Genome Biol 13: R39. doi: 10.1186/gb-2012-13-5-r39

70. .Merchant SS, Prochnik SE, Vallon O, Harris EH, Karpowicz SJ, et al. (2007) The*Chlamydomonas* genome reveals the evolution of key animal and plant functions. Science 318: 245–250. doi: 10.1126/science.1143609

71. Rensing SA, Lang D, Zimmer AD, Terry A, Salamov A, et al. (2008) The*Physcomitrella* genome reveals evolutionary insights into the conquest of land by plants. Science 319: 64–69. doi: 10.1126/science.1150646

72. .Banks JA, Nishiyama T, Hasebe M, Bowman JL, Gribskov M, et al. (2011) The*Selaginella* genome identifies genetic changes associated with the evolution of vascular plants. Science 332: 960–963. doi: 10.1126/science.1203810

73. .Ralph SG, Chun HJ, Kolosova N, Cooper D, Oddy C, et al. (2008) A conifer genomics resource of 200,000 spruce (*Picea spp.*) ESTs and 6,464 high-quality, sequence-finished full-length cDNAs for Sitka spruce (*Picea sitchensis*). BMC Genomics 9: 484. doi: 10.1186/1471-2164-9-484

74. International-Brachypodium-Initiative (2010) Genome sequencing and analysis of the model grass *Brachypodium distachyon*. Nature 463: 763–768. doi: 10.3410/f.2203965.1814064

75. Zhang G, Liu X, Quan Z, Cheng S, Xu X, et al. (2012) Genome sequence of foxtail millet (*Setaria italica*) provides insights into grass evolution and biofuel potential. Nat Biotechnol 30: 549–554. doi: 10.1038/nbt.2195

76. Paterson AH, Bowers JE, Bruggmann R, Dubchak I, Grimwood J, et al. (2009) The*Sorghum bicolor* genome and the diversification of grasses.

Nature 457: 551–556. doi: 10.1038/nature07723

77. Goff SA, Ricke D, Lan TH, Presting G, Wang R, et al. (2002) A draft sequence of the rice genome (*Oryza sativa* L. *ssp. japonica*). Science 296: 92–100. doi: 10.1126/science.1068275

78. Schnable PS, Ware D, Fulton RS, Stein JC, Wei F, et al. (2009) The B73 maize genome: complexity, diversity, and dynamics. Science 326: 1112–1115.

79. Tuskan GA, Difazio S, Jansson S, Bohlmann J, Grigoriev I, et al. (2006) The genome of black cottonwood, *Populus trichocarpa* (Torr. & Gray). Science 313: 1596–1604. doi: 10.1126/science.1128691

80. Huang S, Li R, Zhang Z, Li L, Gu X, et al. (2009) The genome of the cucumber,*Cucumis sativus* L. Nat Genet. 41: 1275–1281. doi: 10.1038/ng.475

81. Arabidopsis-Genome-Initiative (2000) Analysis of the genome sequence of the flowering plant *Arabidopsis thaliana*. Nature 408: 796–815. doi: 10.1038/35048692

82. Wang X, Wang H, Wang J, Sun R, Wu J, et al. (2011) The genome of the mesopolyploid crop species *Brassica rapa*. Nat Genet 43: 1035–1039. doi: 10.1038/ng.919

83. Jaillon O, Aury JM, Noel B, Policriti A, Clepet C, et al. (2007) The grapevine genome sequence suggests ancestral hexaploidization in major angiosperm phyla. Nature 449: 463–467. doi: 10.1038/nature06148

84. Zhang Z, Wood WI (2003) A profile hidden Markov model for signal peptides generated by HMMER. Bioinformatics 19: 307–308. doi: 10.1093/bioinformatics/19.2.307

85. Finn RD, Clements J, Eddy SR (2011) HMMER web server: interactive sequence similarity searching. Nucleic Acids Res 39: W29–37. doi: 10.1093/nar/gkr367

86. Finn RD, Mistry J, Schuster-Bockler B, Griffiths-Jones S, Hollich V, et al. (2006) Pfam: clans, web tools and services. Nucleic Acids Res 34: D247–251. doi: 10.1093/nar/gkj149

87. Punta M, Coggill PC, Eberhardt RY, Mistry J, Tate J, et al. (2012) The Pfam protein families database. Nucleic Acids Res 40: D290–301. doi: 10.1093/nar/gkr1065

88. Schultz J, Milpetz F, Bork P, Ponting CP (1998) SMART, a simple modular architecture research tool: identification of signaling domains. Proc Natl Acad Sci U S A 95: 5857–5864. doi: 10.1073/pnas.95.11.5857

89. Letunic I, Doerks T, Bork P (2012) SMART 7: recent updates to the

protein domain annotation resource. Nucleic Acids Res 40: D302–305. doi: 10.1093/nar/gkr931

90. Edgar RC (2004) MUSCLE: multiple sequence alignment with high accuracy and high throughput. Nucleic Acids Res 32: 1792–1797. doi: 10.1093/nar/gkh340

91. Edgar RC (2004) MUSCLE: a multiple sequence alignment method with reduced time and space complexity. BMC Bioinformatics 5: 113.

92. .Hall TA (1999) BioEdit: a user-friendly biological sequence alignment editor and analysis program for Windows 95/98/NT. Nucleic Acids Symposium Series 41: 95–98.

93. Krawetz S, Retief J (1999) Phylogenetic Analysis Using PHYLIP. Bioinformatics Methods and Protocols: Humana Press. 243–258.

94. Jones DT, Taylor WR, Thornton JM (1992) The rapid generation of mutation data matrices from protein sequences. Comput Appl Biosci 8: 275–282. doi: 10.1093/bioinformatics/8.3.275

95. Tamura K, Peterson D, Peterson N, Stecher G, Nei M, et al. (2011) MEGA5: molecular evolutionary genetics analysis using maximum likelihood, evolutionary distance, and maximum parsimony methods. Mol Biol Evol 28: 2731–2739. doi: 10.1093/molbev/msr121

96. Yang Z (2007) PAML 4: phylogenetic analysis by maximum likelihood. Mol Biol Evol 24: 1586–1591. doi: 10.1093/molbev/msm088

97. Suyama M, Torrents D, Bork P (2006) PAL2NAL: robust conversion of protein sequence alignments into the corresponding codon alignments. Nucleic Acids Res 34: W609–612. doi: 10.1093/nar/gkl315

98. Puranik S, Sahu PP, Mandal SN, B VS, Parida SK, et al. (2013) Comprehensive genome-wide survey, genomic constitution and expression profiling of the NAC transcription factor family in foxtail millet (*Setaria italica* L.). PLoS One 8: e64594. doi: 10.1371/journal.pone.0064594

99. Lynch M, Conery JS (2000) The evolutionary fate and consequences of duplicate genes. Science 290: 1151–1155. doi: 10.1126/science.290.5494.1151

100. .Guo AY, Zhu QH, Chen X, Luo JC (2007) GSDS: a gene structure display server. Yi Chuan 29(8): 1023–6. doi: 10.1360/yc-007-1023

101. .Goodstein DM, Shu S, Howson R, Neupane R, Hayes RD, et al. (2012) Phytozome: a comparative platform for green plant genomics. Nucleic Acids Res 40: D1178–1186. doi: 10.1093/nar/gkr944

102. Wilkerson MD, Ru Y, Brendel VP (2009) Common introns within

orthologous genes: software and application to plants. Briefings in Bioinformatics 10: 631–644. doi: 10.1093/bib/bbp051

103. Bailey TL, Boden M, Buske FA, Frith M, Grant CE, et al. (2009) MEME SUITE: tools for motif discovery and searching. Nucleic Acids Res 37: W202–208. doi: 10.1093/nar/gkp335

104. Crooks GE, Hon G, Chandonia JM, Brenner SE (2004) WebLogo: a sequence logo generator. Genome Res 14: 1188–1190. doi: 10.1101/gr.849004

105. Lescot M, Dehais P, Thijs G, Marchal K, Moreau Y, et al. (2002) PlantCARE, a database of plant *cis*-acting regulatory elements and a portal to tools for in silico analysis of promoter sequences. Nucleic Acids Res 30: 325–327. doi: 10.1093/nar/30.1.325

106. Rombauts S, Dehais P, Van Montagu M, Rouze P (1999) PlantCARE, a plant *cis*-acting regulatory element database. Nucleic Acids Res 27: 295–296. doi: 10.1093/nar/27.1.295

107. Hruz T, Laule O, Szabo G, Wessendorp F, Bleuler S, et al. (2008) Genevestigator v3: a reference expression database for the meta-analysis of transcriptomes. Adv Bioinformatics 2008: 420747. doi: 10.1155/2008/420747

CITATION

CHAPTER 1

Ni J, Hu S, Zhang J, Tang W, Lu W, Zhang C (2015) A Preliminary Genetic Analysis of Complement 3 Gene and Schizophrenia. PLoS ONE 10(8): e0136372. doi:10.1371/journal.pone.0136372.

CHAPTER 2

Montaner D, Dopazo J (2010) Multidimensional Gene Set Analysis of Genomic Data. PLoS ONE 5(4): e10348. doi:10.1371/journal.pone.0010348.

CHAPTER 3

Liu W, Fang L, Li M, Li S, Guo S, Luo R, et al. (2012) Comparative Genomics of Mycoplasma: Analysis of Conserved Essential Genes and Diversity of the Pan-Genome. PLoS ONE 7(4): e35698. doi:10.1371/journal.pone.0035698

CHAPTER 4

Kim Y, Park T (2015) Robust Gene-Gene Interaction Analysis in Genome Wide Association Studies. PLoS ONE 10(8): e0135016. doi:10.1371/journal.pone.0135016.

CHAPTER 5

Pingzhao Hu and Andrew D Paterson, Dynamic pathway analysis of genes associated with blood pressure using whole genome sequence data, DOI: 10.1186/1753-6561-8-S1-S106.

CHAPTER 6

Peng Jiang, Hongfang Wang, Wei Li, Chongzhi Zang, Bo Li, Yinling J. Wong, Cliff Meyer, Jun S. Liu, Jon C. Aster and X. Shirley Liu, Network analysis of gene essentiality in functional genomics experiments, DOI: 10.1186/s13059-015-0808-9.

CHAPTER 7

Rivarola M, Foster JT, Chan AP, Williams AL, Rice DW, Liu X, et al. (2011) Castor Bean Organelle Genome Sequencing and Worldwide Genetic Diversity Analysis. PLoS ONE 6(7): e21743. doi:10.1371/journal.pone.0021743.

CHAPTER 8

Tian C, Kosoy R, Lee A, Ransom M, Belmont JW, Gregersen PK, et al. (2008) Analysis of East Asia Genetic Substructure Using Genome-Wide SNP Arrays. PLoS ONE 3(12): e3862. doi:10.1371/journal.pone.0003862.

CHAPTER 9

Loza MJ, McCall CE, Li L, Isaacs WB, Xu J, Chang B-L (2007) Assembly of Inflammation-Related Genes for Pathway-Focused Genetic Analysis. PLoS ONE 2(10): e1035. doi:10.1371/journal.pone.0001035.

CHAPTER 10

Fadhl M. Al-Akwaa (2012). Analysis of Gene Expression Data Using Biclustering Algorithms, Functional Genomics, Dr. Germana Meroni (Ed.), ISBN: 978-953-51-0727-9, InTech, DOI: 10.5772/48150.

CHAPTER 11

Peng Xu, Xiaofeng Zhang, Xumin Wang, Jiongtang Li, Guiming Liu, Youyi Kuang, Jian Xu, Xianhu Zheng, Lufeng Ren, Guoliang Wang, Yan Zhang, Linhe Huo, Zixia Zhao, Dingchen Cao, Cuiyun Lu, Chao Li, Yi Zhou, Zhanjiang Liu, Zhonghua Fan, Guangle Shan, Xingang Li, Shuangxiu Wu, Lipu Song, Guangyuan Hou, Yanliang Jiang, Zsigmond Jeney, Dan Yu, Li Wang, Changjun Shao, Lai Song, Jing Sun, Peifeng Ji, Jian Wang, Qiang Li, Liming Xu, Fanyue Sun, Jianxin Feng, Chenghui Wang, Shaolin Wang, Baosen Wang, Yan Li, Yaping Zhu, Wei Xue, Lan Zhao, Jintu Wang, Ying Gu, Weihua Lv, Kejing Wu, Jingfa Xiao, Jiayan Wu, Zhang Zhang, Jun Yu & Xiaowen Sun Show fewer authors, Genome sequence and genetic diversity of the common carp, Cyprinus carpio, doi:10.1038/ng.3098

CHAPTER 12

Chol-Hee Jung , Chui E. Wong , Mohan B. Singh , Prem L. Bhalla; Comparative Genomic Analysis of Soybean Flowering Genes; DOI: 10.1371/journal.pone.0038250

CHAPTER 13

Li W-Y, Wang X, Li R, Li W-Q, Chen K-M (2014) Genome-Wide Analysis of the NADK Gene Family in Plants. PLoS ONE 9(6): e101051. doi:10.1371/journal.pone.0101051

INDEX

A

Accumulating evidence 7
alternative pathway (AP) 8
ancestry informative markers (AIMs) 136

B

Biological networks 94, 103, 112

C

Castor bean 115, 116, 118, 127
Chinese Americans (CHA) 138
Chloroplast genomic 117
Chloroplast reads 121, 128
Classical clustering 176
Clustering algorithms 181
Common carp 196, 197, 202, 205, 208, 211, 213, 214, 215
Common genetic 90
Complement pathway 1, 2
CRYPTOCHROME2 (CRY2) 221
Cyprinid fish 201
Cyprinid species 196, 208

D

De novo biosynthesis of NADP(H), 252

E

East Asian substructure ancestry informative markers (EASTA-SAIMS) 136
Essential genes 93

F

false discovery rate (FDR) 86
fibroblast growth factor (FGF) 206
Functional genomics 175
Functional implications 15

G

Gene essentiality 93, 94, 95, 96, 97, 98, 99, 106, 107, 284
Gene expression 17
gene-gene (GG) 61, 62
Gene ontology (GO) 16
Gene Ontology (GO) 182
Generalized Multifactor Dimensionality Reduction (GMDR) 62, 64
Genes assigned 83
Gene Set Enrichment Analysis (GSEA) 16
Genetic diversity 42
genome-wide association studies

(GWAS) 62
Genome-wide association studies
    (GWAS) 61
Genomic experimental 101
Genomic techniques 93
Glioblastoma (GBM) 103

H

Hidden Markov model (HMM) 254
Hierarchical clustering 176

I

Immunity contribute 157
inflammation-related genes 153, 163
Inflammatory responses 157, 158, 169

K

Kyoto Encyclopedia of Genes and Ge-
    nomes (KEGG) 205

L

Lateral gene transfer (LGT) 48
linkage disequilibrium (LD) 3, 5

M

methylation filtration (MF) 117
Microbiological applications 46
Multifactor Dimensionality Reduction
    (MDR) 62
Multiple additional 146
Multiple biclusters 179

N

National Institute of Allergy and Immu-
    nologic Diseases (NIAID) 155
Natural selection 42

O

oil-producing plant 115
Order-preserving 181, 192
Orthologue groups (OGs) 222
Ovarian cancer (OV) 104

P

Phylogenetic relationship 48, 55
polymerase chain reaction (PCR) 4
principal-component analysis (PCA)
    204
Principal Components Analyses (PCA)
    135
Protein coding 43, 50

R

Recurrence signature 180

S

Self-replicating 41, 42
Sequence Detection System 130
Simplest definition 175
simulated quantitative trait (SBP) 90
single nucleotide polymorphism (SNP)
    61, 118
single nucleotide polymorphisms (SNPs)
    156, 163
Statistical analysis 175
system lupus erythamatosus (SLE) 156
systolic blood pressure (SBP) 84

T

tagging single nucleotide polymorphism
    (tSNP) 154
teleost-specific genome duplication
    (TSGD) 196
Traditional clustering 177
transcription rates (TR) 30
transcription rate (TR) 18
tumor suppressor genes (TSG) 104

U

U.S. Department of Agriculture (USDA)
    127

V

Various environments 204